Hydrocephalus

Ahmed Ammar

Editor

Hydrocephalus

What do we know?
And what do we still not know?

 Springer

Editor
Ahmed Ammar
Dammam University
King Fahd University Hospital Dammam University
Al Khobar
Saudi Arabia

ISBN 978-3-319-61303-1   ISBN 978-3-319-61304-8   (eBook)
DOI 10.1007/978-3-319-61304-8

Library of Congress Control Number: 2017954303

Disclosure: The editor has nothing to disclose.

Printed on acid-free paper

This Springer imprint is published by Springer Nature
The registered company is Springer International Publishing AG
The registered company address is: Gewerbestrasse 11, 6330 Cham, Switzerland

Professor Mami Yamasaki

A special dedication of this book to a dear friend and a great pediatric neurosurgeon, neuroscientist, and educator, **Prof. Mami Yamasaki.** *She authored Chap. 4, "Understanding Hydrocephalus: Genetic View" in this book, which is her last penned masterpiece.*
Your legacy in neurosurgical field is indelible.

I dedicate this book to all hydrocephalus patients and their parents, and to my mentors
Prof. Kenichiro Sugita
Prof. Shigeaki Kobayashi
Prof. Anthony J Raimondi

Preface

Hydrocephalus is not a ventricles' problem; it is the problem of the whole brain!

Understanding hydrocephalus is one of the main keys to understand the brain and its function. It is scientifically not accurate to extremely simplify the problem and presume that hydrocephalus is a problem of only dilated ventricles treated by simple shunt or ETV or CPC. This wrong assumption is responsible that hydrocephalus remained to be a challenging condition not fully understood. Hydrocephalus is the problem of the whole brain, which involves the genetic profile and disorders, CSF dynamics, ICP, CBF, cerebral metabolism, electrolyte balance, endocrine consideration, and integrity of the cerebral tissues.

The progress in understanding hydrocephalus and its pathophysiology will definitely have a great impact in improving the outcome of this serious disorder. This understanding should take lead in producing new protocols and methods of treatment. Hydrocephalus is among the oldest known human disorders named by Hippocrates in 336 BC while he was in Egypt. 2350 years have passed but many questions regarding hydrocephalus remained unanswered.

The internationally accepted incidence of congenital hydrocephalus is 3 per 1000 live births. This incidence may slightly vary from one geographical area to another. However, the incidence of acquired hydrocephalus is not truly known as hydrocephalus can be caused by vascular abnormalities, tumor, infection, and trauma. VP or VA shunts are the most common methods for the treatment of hydrocephalus. It is interesting to know that nearly 100,000 shunts were implanted every year in North America, Japan, and Europe according to companies of shunt manufacturers. According to Dr. Choux, the incidence of shunt failure is very variable ranging from 0.5 to 90% in cases of premature newborns. Shunt failure or delay in the treatment of hydrocephalus may cause death, severe neurological deficit, and impairment of the cognitive function. **We emphasize that shunt is not the treatment of hydrocephalus. Shunts are a measure to drain CSF and reduce the ICP. It is a treatment for the symptoms and consequences of hydrocephalus.** There is no doubt that we greatly benefit from scientific and technological advancements in the field of medicine. Therefore, it is not acceptable not to have a smart ideal shunt out of the box as a treatment of hydrocephalus. The definite and long-awaited treatment for this condition may come from better understanding the pathophysiology, genetic abnormalities, and CSF and ICP dynamics of hydrocephalus, and research is the only way that will lead us to achieve to this. We cannot emphasize more on the given importance of encouraging hydrocephalus research by all means. The development of new methods of intrauterine interventions such as early diagnosis and early detection of chromosomal and genetic abnormalities may encourage future research for genetic therapy and may carry more hopes.

This book is written to document what we know and what we don't know about hydrocephalus. However, this book also discusses the difficulties to fully understand the pathophysiology, CSF dynamics, and genetic causes of hydrocephalus and may encourage neuroscientists to develop new ideas, perform advanced research, and open new frontiers to defeat this disease to put our children into safety.

I do hope that this book will be the coupling link of a very long chain of studies and efforts to provide good care to our hydrocephalus patients, to document what we have learned, and to open windows for the future.

Ahmed Ammar, M.B.Ch.B., D.M.Sc., F.A.C.S., F.I.C.S., F.A.A.N.S.

Acknowledgments

The editor would like to thank all the contributing authors and co-authors in this book and Mrs. Janice Liwanag-Ventura for all the help and support.

Contents

Part VI Fetal Hydrocephalus

Part I

Introduction to Hydrocephalus

Values-based Medicine: Ethical Issues in Hydrocephalus

Ahmed Ammar

1.1 Values-based Medicine

The patient always is the center of care. The main pillars of good neurosurgical practice are knowledge, skills, and technology. All three basic elements should be integrated to provide a clear vision, strategy, and plan of management for each patient. The vision, strategy, and medical practice should be performed within a rigid frame of values and ethics (Fig. 1.1).

Therefore, it is vital to include medical ethics in every CME training programs. The trainees should not only be encouraged to know about ethics as a knowledge but should consider these ethics in the daily practice [1], such as:

(a) Autonomy: the right of the patient to choose or refuse methods of management of their medical problem. Neurosurgeons and medical doctors should realize that the correct practice of autonomy is to work with the patient as partners in order to decide on the best method of treatment.

(b) Beneficence: to determine that the physician or surgeon considers the patient's utmost benefit as his sole goal in his relationship with the patient. The neurosurgeon should honestly ask himself if the suggested method brings the most benefit for the patient. It should be noted that this is not the same as whether it is in the patient's best interest. These questions are fundamental to ethical practice and should be asked every time a neurosurgeon sees the patient.

(c) Non-maleficence: clearly means "first, do no harm," while neurosurgeons and medical doctors have to do their best not to harm the patient. There may be an element of risks in some procedures necessary to save life. These risks should be explained to the patient, and it should be made sure that every precaution will be practiced to prevent or minimize it.

A. Ammar, M.D., M.B.Ch.B., D.M.Sc.
Department of Neurosurgery, King Fahd University Hospital, Imam Abdulrahman Bin Faisal University, Al Khobar, Saudi Arabia
e-mail: ahmed@ahmedammar.com

© Springer International Publishing AG 2017
A. Ammar (ed.), *Hydrocephalus*, DOI 10.1007/978-3-319-61304-8_1

Fig. 1.1 The concept of values-based medicine, where the patient is the center of care, and every step of management should be performed within the rigid frame of ethics and values

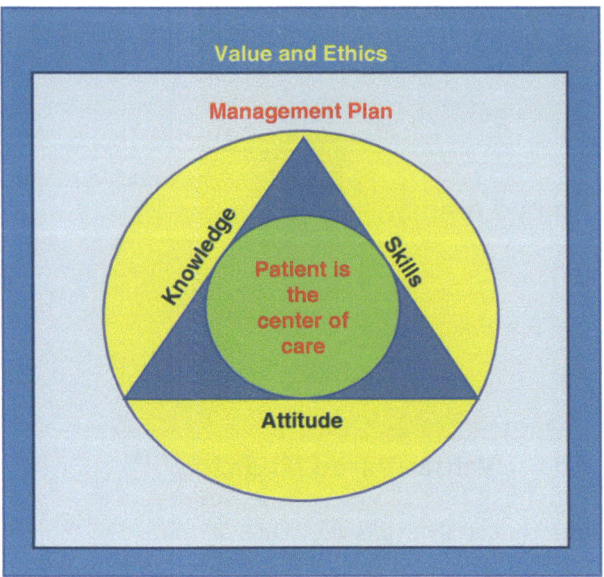

(d) Justice: the application of justice means fairness and equality of medical care for every patient, which includes doctor's time and attention. Every patient has the right to receive the best possible and affordable medical care.

(e) Dignity: it must stream in both directions. The patient and his treating medical team should be treated with respect and dignity. The medical doctors and neurosurgeons should observe the patient's right to confidentiality and privacy.

(f) Truthfulness and honesty: the relationship between a patient and his treating physician or surgeon should be based on honesty and truthfulness. The neurosurgeon should honestly answer every patient's question and should openly admit if any mistake or miscalculation or underestimation of the case is committed.

1.2   Definition of the Topic

There is a general agreement about the tasks and duties of neurosurgeons as stated in several neurosurgical codes, guides for practice, and job descriptions [2–4]. A neurosurgeon should provide the best possible care to his neurosurgical patients. It is vital for every neurosurgeon to update his knowledge and gain new skills. Within the field of neurosurgery, it is known and accepted that following the evidence-based medicine guide is the best way to assist the patient in choosing the best treatment plan. To a certain extent, modern medicine judges and weighs different approaches according to, or in reference to, evidence-based medicine results and gaudiness. The importance of finding and following evidence cannot be overstated, especially to young neurosurgeons or to a neurosurgeon who may find himself working alone in a remote area. Neurosurgeons are also obliged to adhere to medical ethics principles such as:

(a) Autonomy: the patient has the right to refuse or choose their treatment.
(b) Beneficence: a practitioner should act in the best interest of the patient.
(c) Non-maleficence: do no harm.
(d) Justice: the fair distribution of health resources and the decision of who should be treated first.
(e) Dignity: patients and the treating medical and paramedical staff have the right to be treated with dignity.
(f) Truthfulness and honesty between the patient and his treating medical staff.

1.3 Good Doctor or Skillful Doctor?

There is a misconception that a good/great doctor is a skillful doctor. Years of experience and reliable research studies proved that this concept is not always true. In 2002, in a Mayo Clinic study, nearly 200 patients expressed different opinions. They thought that a great doctor should be (in order) confident, sympathetic, humane, forthright, respectful, and thorough. It was repeatedly thought that doctors who love patient care, have compassion for their patients, and put their patients first are the best doctors. It is the duty of different neurosurgical/medical trainers to act as role models for their trainees and teach them to not only learn by hearing but by listening carefully to their patients, respecting their rights and dignity, and truly caring about their well-being. The true measurement of success is not only on successful surgery but also in making patients feel and trust that their neurosurgeon is working for them, prioritizing their care, honest in admitting mistake, and expressing compassion [5–7].

1.4 Continuous Medical Education (CME)

Management of hydrocephalus, either by shunt or endoscopic third ventriculostomy, provoked an ethical and professional discussion on how to make the decision in choosing the shunt and the method to use over the other. There is no evidence that one type of shunt is better than the other types [8]. The same principle applies with neuroendoscopes, as there are numerous good quality and efficient neuroendoscopes in the market that are available.

1.5 Neurosurgeons and New Technology

The use and adaptation of new technology is the backbone of modern neurosurgery. Most neurosurgeons obtain information about new technology from the representatives of industry companies. The use of such new and complex technology requires training, and for the first use in the operating room, many neurosurgeons may feel it is safer for the patient to ask the company representative to be present in the operating room to guide the proper use of the technology. Therefore, it is not unusual to see the representative of these companies in the operating rooms in different hospitals all over the world [9]. This situation creates a very special relationship with the companies beyond any commercial interest. It is an example of good collaboration to provide patients with the most updated and modern medical care. However, this collaboration, direct or indirect participation of a company representative in surgery, creates ethical questions: Who is responsible and should be held accountable if the operation does not go as planned? Was the patient informed that this particular individual would participate in the surgery? Should the patient be required to consent for that? Should the patient see and talk to the representative? According to the principles of medical ethics (autonomy, truthfulness, and honesty), the patient has the right to know who will perform and participate in the surgery, and he/she should consent for that. The neurosurgeon is the one who should be held accountable for any complication on the decision of a particular technique and on choosing the treating team including the company representative in the operation room. However, we believe if the complication is caused by the use of such technology, the manufacturing company and its representative should be held responsible too.

1.6   The Neurosurgeon's Perspective

Recent rapid technological advances demand continuous training and updating of knowledge to be able to provide proper care to patients. Medical ethics principle requires regular and continuous updating of knowledge and skills. Attending workshops, conferences, and seminars is costly. Most hospitals do not have the budget to support this aspect for their medical staff. At the same time, most of these hospitals require that neurosurgeons produce a certain number of credit hours (CH) not only for promotion but also to maintain the license. The tax system in some countries may help by deducting certain percentage of the taxes paid to be allocated for CME. Most of the hospital requires their medical staff, physician, surgeons, and nurses to obtain certain CME hours in order to be promoted or to keep their job. The medical staff members should ask this question: Who should pay for the CME activities? Doctors and nurses consider participation in workshops, seminars, and conferences as essential to gain new knowledge, improve their skills, and at the same time gain the needed CME hours. It's either they fully pay or partially pay for their own CME. However, the majority believes that the process is very costly and that the hospital or universities should pay for it as they are the ones requiring it from them. It may be ethical that CME cost should be covered by the workplaces. The hospital's image on patient care will definitely benefit knowing that the medical care employees are skillful and knowledgeable through continuous learning [10].

Most hospital's budget is under pressure with its constant upgrade to a more modern and expensive equipment. This, in turn, causes a shortfall in budget available for funding CME, meaning that only selective part of the CME can be funded.

Unfortunately, some administrations forget the fact that the true investment is the investment in the manpower, not in surgical tools. Prof. Lars Leksell gives us a great example and lesson on the relationship between the neurosurgeon and new technology. He insisted that users of his stereotactic system or Gamma Knife were fully trained before permitting them to use his systems. He also wrote in his book "The tools used by surgeon must be adapted to the task and where the human brain is concerned they cannot be too refined" [11]. Hospitals, universities, and educational institutes face financial difficulties in organizing courses, conferences, and workshops. Therefore, it may ask financial support from pharmaceutical and equipment industries and companies. The relationship between medical staff, hospital and educational institutes, and commercial companies should be transparent and tightly regulated. It is important and ethically correct to leave the individual neurosurgeons out of that equation. In many countries, the regulation of industry-sponsored CME programs is regulated by an independent body such as ACCME in the USA, which observes that all CME programs must be free from any commercial interest. However, some authors suggest that that regulation could be manipulated easily [1, 12, 13]. There should be a very clear line between commercially oriented sponsored CME programs and educational and scientifically oriented CME programs.

1.7   The Industry's Perspective

Recently, companies have shown interest in sponsoring CME programs. Most of these companies plan these programs as part of their marketing strategies. These companies may prefer to have direct contact with individual neurosurgeons, especially key neurosurgeons in different places. However, in many cases they have to cooperate with the hospital authorities or scientific organizations away from the individual neurosurgeons and they have to obey the regulations guarding such relationships.

Pharmaceuticals and instrument and equipment companies are promoting their products by exposing them to the market and extolling its benefits. One method for achieving this objective is to train neurosurgeons to use their products. These companies usually contract very well-established neurosurgeons to be the faculty of their courses, workshops, and conferences. The presence of such great figures as faculty is attractive enough to encourage many neurosurgeons from different countries to attend such courses or workshops. Some of these great figures endorse their relationship with the companies in their talk. This endorsement is required by many scientific organizations like AANS, CNS, WFNS, and others. The final goal of medical and pharmaceutical companies is to get the medical staff use their products and prescribe their medications. Therefore, these companies have great interest in training medical staff using their equipment and to make them learn about their new drugs and medication. They proposed, sponsor, and sometimes organize workshops, seminars, and conferences excluded from the CME programs itself [9, 14–16]. However, current financial realities may not permit excluding the companies' support for CME programs entirely. The only ethical solution for this dilemma is to have rigid regulation to prevent any conflict of interest, and a possible solution is the formation of an office (education committee, CME committee, etc.) in different hospitals or educational institutes to receive the financial support from different companies. The main task of such offices or committee would be to support all individual neurosurgeons to obtain new skills and knowledge through selected CME programs.

1.8 Learning Curve

Learning curve is a fact and a characteristic of medical and surgical career. Every neurosurgeon must have the experience to perform a certain procedure for the first time or to use special equipment or an industrial or pharmaceutical product for the first time. In these cases, precautions should be taken to guarantee patient's safety. Such precautions include supervision, assistance, and training. New surgical procedures, inventions, and advancement in medical equipment are the corner stones in the development and improving neurosurgical practice. Researches including clinical research are vital for development of equipment; understanding different neurosurgical problems and discovering proper treatments should be encouraged, supported, and governed by ethical codes.

1.9 New Idea and First Use of New Medical Equipment

There are two distinct ethical considerations regarding new products. The first concerns the ethics of approaching a senior staff member or company with an idea. Many times, a young, neurosurgeon faced with the daunting prospect of trying to turn his idea into a reality. In many cases, either the senior or the company wants to have their name attached to the product. If the idea is solely that of one person, it is neither honest nor truthful to attach a second name simply to be able to market the idea. The second consideration is truthfulness with the patient. The patient has the right to be informed that this is the first use of a product and to fully understand the associated risks. Ethical regulations should be made to protect the patient and to assure the safe use of the first product [17–19]. Ethically, neurosurgeons are required to follow this path. However, measures should be taken in order to make this vital "sizable learning curve" safe and beneficial for the patient.

1.10 How to Monitor the Ethical Behavior and Practice of the Trainee. Is It Needed?

Most trainees' evaluation forms in different programs include "attitude or behavior" as an item for evaluation. The difficulty is that how ethics and behavior can be accurately measured or assessed by regular standard forms [5]. Since the earliest times in medicine, the presumption that everyone wearing the white coat must have possessed good values and ethics, honesty, care, and trust has been accepted. Unfortunately, this belief is not always true. The most practical method to assess the behavior and ethics of the trainee is by observing his/her relationship with their patients, colleagues, and paramedical staff. The question arises as to whether patients should be randomly called upon to evaluate the behavior and conduct of trainees. There is a serious ethical debate about whether the behavior of a trainee in the private life should be included or considered in the evaluation. Some researchers believe that ethics and values cannot be put into practice on one side of life and ignored in another. Others say, the behavior of the doctor within the hospital is paramount, and the private life is not and should not be of concern. This debate is an unsolved philosophical debate on good characteristics and values. Philosphocally, there is a clear difference between who holds, believes, and practices good ethics and values and one who practices good values as a duty or out of fear of consequences. Practically, in these days, only the behavior of the trainees inside the hospital should be considered. The doctor's attitude and ethical performance are evaluated according to the existing forms of evaluation as excellent, very good, good, not proper, or lacking ethics. These types of evaluation are not accurate enough; there is a need to develop new forms which should be objective rather being subjective. However, the concept of self-monitor and absolute adherence to good values is very important and is the very core and meaning of Medical Oath.

1.11 Discussion

There is a general consensus that the main task and duty of a neurosurgeon is to provide state-of-the-art neurosurgical care to all patients. Therefore, neurosurgeons should strive to maintain updated in knowledge, skills, and modern techniques in their speciality. The aim of different training programs is to graduate a safe, knowledgeable, and skillful neurosurgeon, thus providing him with a strong base to cultivate on. The structure of these programs is based on clinical engagement, rotations, participation in surgery, courses, academic activities, workshops, and research. Courses and workshops on medical ethics are rarely included in training programs. The trainers may be satisfied with published guidelines for ethics and may believe that medical ethics can be learned from observing more senior staff. The trainees have to know, apply, and practice the principles of evidence-based medicine. In many training programs, there is ample opportunity for the trainee to learn these skills. Equally important, we should remember that from its early history, medicine is value based.

Continuous medical education programs are vital part of modern neurosurgery. Every neurosurgeon should strive to obtain new skills and update his knowledge. The workplace authorities should support their employees to obtain the CME credit hours either financially or by facilitating staff attendance at conferences, workshops, and courses. The industrial companies may play a role in supporting the CME programs. Transparency is the key to prevent conflicts of interest and ethical issues.

Conclusion

Observing the principles of medical ethics should be integrated within every communication and procedure between the patient and the medical staff. The top priority of every neurosurgeon should be to provide the patient with the best medical care. Patients have the right to know and discuss the details of the policy of management. However, the neurosurgeon is ultimately held responsible and accountable for the course of management he chooses for his patient. The neurosurgeon should regularly continue his medical training and push his learning curve safely up. The use of modern technology cannot be neglected and, in fact, it is required. Therefore, a transparent and ethically guarded relationship with the industrial/pharmaceutical companies should be clearly drawn and firmly abided by to avoid any doubts on conflict of interest.

References

1. Institute of Medicine. Conflict of interest in medical research, education, and practice. Washington, DC: National Academies Press; 2009.
2. AANS Board of Directors. AANS Code of Ethics. 13 Apr 2007.
3. AANS. The AANS Professional Conduct Program. Feb 2011.
4. World Federation of Neurological Societies, European Association of Neurological Societies. Good practice: a guide for neurosurgeons. Acta Neurochir. 1999;141:791–9. (Final Version: August 1998).
5. Johnson AG, Johnson PRV. Making sense of medical ethics. Oxford: Oxford University Press; 2007. p. 190–6.
6. West M. What's a good doctor, and how can you make one? BMJ. 1973;325:669–70.
7. Wolpe PR. We are trying to make doctors too good. BMJ. 2002;325:712–4.
8. Drake JM, Kestle JR, Milner R, Cinalli G, Boop F, Piatt J Jr, Haines S, Schiff SJ, Cochrane DD, Steinbok P, MacNeil N. Randomized trial of cerebrospinal fluid shunt valve design in pediatric hydrocephalus. Neurosurgery. 1998;43(2):294–303.
9. McCormick PW. OR friend/courtroom foe: does a product rep in your OR have a duty to your patient? AANS Bulletin. 2005;14(2):15.
10. Accreditation Council for Continuing Medical Education. Standards for Commercial Support: Standards to Ensure Independence in CME Activities. Feb 2012.
11. Leksell L. Sterotaxis and radiosurgery—an operative system. Springfield, IL: Charles C Thomas; 1969. p. 58.
12. Ammar A, Bernstein M. Neurosurgical ethics in practice: value based medicine. New York: Springer; 2014.
13. Minter RM, Angelos P, Coimbra R, Dale P, de Vera ME, Hardacre J, Hawkins W, Kirkwood K, Matthews JB, McLoughlin J, Peralta E, Schmidt M, Zhou W, Schwarze ML. Ethical management of conflict of interest: proposed standards for academic surgical societies. J Am Coll Surg. 2011;213(5):677–82.
14. Morris L, Taitsman JK. The agenda for continuing medical education—limiting industry's influence. N Engl J Med. 2009;361(25):2478–82.
15. Robertson JH. Neurosurgery and industry. J Neurosurg. 2008;109(6):979–88.

16. White AP, Vaccaro AR, Zdeblick T. Counterpoint: physician-industry relationship can be ethically established, and conflicts of interest can be ethically managed. Spine (Phila Pa 1976). 2007;32(11 Supp):S53–7.
17. Ammar A. Influence of different culture on neurosurgical practice. Child's Nervous Systm. 1997;13:91–4.
18. Ammar A. A young man has an idea. What should he do? Neurol Res. 1999;21:8–10.
19. Bernstein M, Bampoe J. Surgical innovation or surgical evolution: an ethical and practical guide to handling noel neurosurgical procedures. J Neurosurg. 2004;100(1):2–7.

Evidence Based Medicine-Hydrocephalus Guideline for Systemic Reviews, Meta-analysis and Evidence Based Medicine

2

Dina El Kayaly, Ignatius Essene, and Ahmed Ammar

2.1 Introduction

Evidence-based clinical medicine can be seen as the conscious incorporation of the best evidence that is currently available into daily clinical practice covering prevention, diagnostics, clinical assessments, treatments, and patient-centered care.

The application of current evidence is a significant challenge for many practitioners since it takes extensive amount of time and effort to keep up with up-to-date improvements in clinical care through reviewing journals, attending conferences, and many other forms of practitioners' professional development.

Moreover, there are challenges associated with convincing successful practitioners to reconsider to coincide with up-to-date evidence and best practices and not to continue practicing the good old way. In developing countries, there are many more significant challenges; some are related to inadequate number of physicians and others related to the limited ability to transfer up-to-date knowledge from developing countries to improve practices in developing countries.

D. El Kayaly, M.D. (✉)
Maastricht School of Management, Cairo, Egypt
e-mail: dina@edgeegypt.com

I. Essene, M.D.
Department of Neurosurgery, Ain Shams University, Cairo, Egypt

A. Ammar, M.D., M.B.Ch.B., D.M.Sc.
Department of Neurosurgery, King Fahd University Hospital, Imam Abdulrahman Bin Faisal University, Al Khobar, Saudi Arabia
e-mail: ahmed@ahmedammar.com

The key tool for closing this gap will be research as it is the proper foundation needed to provide decision support to users at the time they make decisions, which will result in improved quality of care. Every physician/surgeon must know about the various research methodologies to enhance his/her research process leading to better patient care. In this chapter we will give an overview of the research methods used in neurosurgery with special focus on metadata analysis as a media for evidence-based medicine.

Researchers in the field of neurosurgery face some problems; some of them may be unqualified to critically read the medical literature and articles which hinders medical education and ultimately affects patient care. Others might misuse research methods by selecting a wrong sample design or misjudge the sample size or use inappropriate analysis technique. Yet such messed-up articles may get published and cited by other researchers, which aggravates the problem! [5].

2.2 Options for Different Methods to Treat Hydrocephalic Children

There are several methods of treatment for the different types of hydrocephalus considering the age of the child and stage of the disease. Most of the time, confusion arises and questions are asked: what is the best method to treat this particular patient now?

EBV may provide a helpful answer on the matter of choosing the proper treatment method. However, consideration should be given to clinical judgment all the time. Evidence-Based Guidelines Task Force was formed in the USA since 2011 in order to find out the best options for treatment for different pathologies, etiology, and methods of pediatric hydrocephalus. The Task Force followed protocols established by the Joint Guidelines Committee (JGC) of the AANS and the CNS [6].

They presented their valuable studies in the *Journal of Neurosurgery: Pediatric Neurosurgery* supplement in 2014. They used data collected from the US National Library of Medicine PubMed/MEDLINE database and the Cochrane Database of Systematic Reviews that were searched using MeSH headings and keywords relevant to different options of treatment.

Because of the true value of their work, we tabled their results in Table 2.1.

2.3 What Are the Types of Medical Studies?

There are many different types of research studies, where each type has distinct strengths and weaknesses. Medical research can be classified into:

Primary vs. secondary research: Secondary research summarizes available studies using reviews and meta-analyses, while primary research collects data specifically for this study using various types and can be classified into basic or

Table 2.1   Results of the Joint Guidelines Committee (JGC) of the AANS and the CNS

The problem and authors	The option of treatment	The level of evidence and strength
What are the best treatments for posthemorrhagic hydrocephalus in premature infants? (Mazzola et al. [14])	External ventricular drainage (EVD), ventricular subgaleal (VSG) shunt, and lumbar punctures to be used as temporary measures	Level I Clinical judgment is needed
	VSG shunts may reduce the frequency of daily CSF aspiration as compared with VADs	Level II
	Routine and frequent use of lumbar puncture	Level II The daily use of lumbar puncture is not recommended to reduce the need for shunt later on as final treatment of hydrocephalus
	Endoscopic third ventriculostomy (ETV)	Level III There is no enough evidence to recommend ETV in such cases
Do technical adjuvants such as ventricular endoscopic placement, computer-assisted electromagnetic guidance, or ultrasound guidance improve ventricular shunt function and survival? (Flannery et al. [6])	There is insufficient evidence to recommend the use of endoscopic guidance for routine ventricular catheter placement	Level I High degree of certainty
	The routine use of ultrasound-assisted catheter placement is an option	Level III Unclear clinical certainty
	The routine use of computer-assisted electromagnetic (EM) navigation	Level III Unclear clinical certainty
CSF shunt or ETV as options for management of hydrocephalus in children (Limbrick et al. [13])	CSF shunts and ETV demonstrated equivalent outcomes	Level II Moderate clinical certainty
Effect of valve type on cerebrospinal fluid shunt efficacy (Baird et al. [1])	There is insufficient evidence to demonstrate an advantage for one shunt hardware design over another in the treatment of pediatric hydrocephalus	Level I High degree of clinical certainty
	There is insufficient evidence to recommend the use of a programmable valve versus a nonprogrammable valve	Level II Moderate clinical certainty
Preoperative antibiotics for shunt surgery in children with hydrocephalus (Klimo et al. [11])	Routine use of preoperative antibiotics to prevent shunt	Level II Moderate clinical certainty

(continued)

Table 2.1 (continued)

The problem and authors	The option of treatment	The level of evidence and strength
Are antibiotic-impregnated shunts superior to standard shunts at reducing the risk of shunt infection in pediatric patients with hydrocephalus? (Klimo et al. [10])	Antibiotic-impregnated shunt may be associated with a lower risk of shunt infection compared with conventional shunts	Level III Unclear degree of clinical certainty
Management of cerebrospinal fluid shunt infection What is the optimal treatment strategy for CSF shunt infection in pediatric patients with hydrocephalus? (Tamber et al. [16])	The use of antibiotic treatment with partial (externalization) or with complete shunt	Level II Moderate degree of clinical certainty
	Externalization or complete shunt removal	Level III Unclear degree of clinical certainty
	Combination of intrathecal and systemic antibiotics for patients with CSF shunt infection in whom the infected shunt cannot be fully removed or must be removed and immediately replaced, or when the CSF shunt infection is caused by specific organisms. The potential neurotoxicity of intrathecal antibiotic therapy may limit its routine use	Level III Unclear degree of clinical certainty Clinical judgment is required
Effect of ventricular catheter entry point and position Do the entry point and position of the ventricular catheter have an effect on shunt function and survival? (Kemp et al. [9])	There is insufficient evidence to recommend the occipital versus frontal point of entry for the ventricular catheter	Level III Unclear degree of clinical certainty Recommendation: both entry points are options for the treatment of pediatric hydrocephalus
Does ventricle size after treatment have a predictive value in determining the effectiveness of surgical intervention in pediatric hydrocephalus? (Nikas et al. [15])	There is insufficient evidence to recommend a specific change in ventricle size as a measurement of the effective treatment of hydrocephalus	Level III Unclear degree of clinical

exploratory research, clinical or interventional research, and epidemiological or non-interventional research [21, 22].

A researcher needs to know the advantages and disadvantages of each type of research; here is a summary in Table 2.2.

There are important relationships between the various types of medical research studies and the aim of using each type (Fig. 2.1). The researcher should be aware of

Table 2.2 Advantages and disadvantages of various types of medical research

Types of medical research	Advantages	Disadvantages
Secondary research [34]	• Time and cost-effective compared to primary data collection • Extensiveness of data that covers a large spectrum of issues giving a wealth of information if analyzed well • Work as basis to be addressed through primary research	• Secondary researcher needs to understand various parameters and assumptions that primary research had taken while collecting information • Inaccuracy of data depending on someone else's efforts and point of view • Time lag issues where things have changed drastically • May not be specific or related to your research question
Primary research [33]	• Specific issues are addressed. And the researcher has control on the process • Data interpretation is better • Data is more recent reflecting current situations	• Collecting data using primary research is costly • Time consuming because of exhaustive nature of the exercise • In case the research involves taking feedbacks from the targeted audience, there are high chances that feedback given is not correct • Leaving aside cost and time, other resources like human resources and materials too are needed

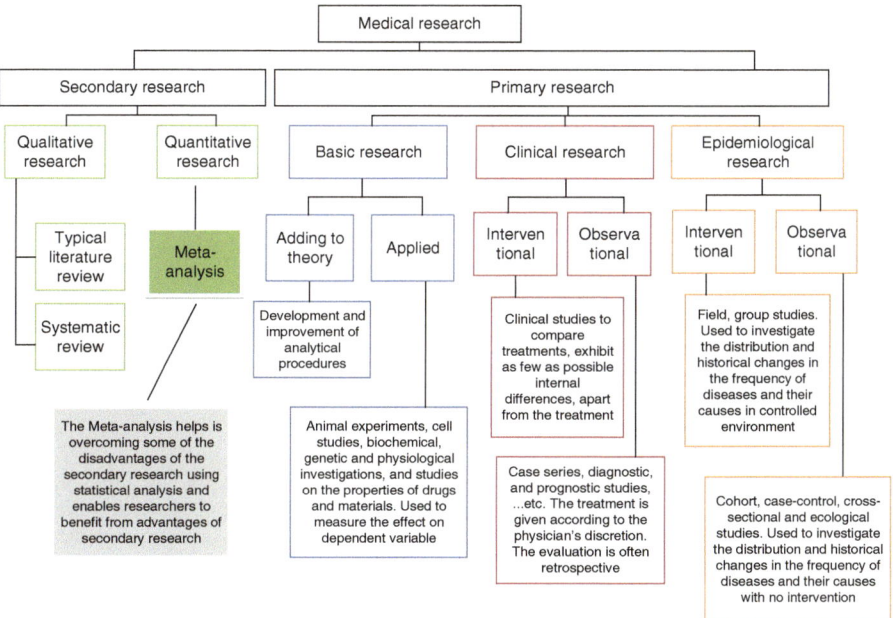

Fig. 2.1 Describes the relationship between the various types of medical research and the aim of using each type

such relationship to choose the type of scientific research which will help his purpose of the study.

Selecting the correct study type is a very important aspect of study design that
can hinder the reliability of the study and its findings. Table 2.3 summarizes some
of the key study types' usage and drawbacks.

Table 2.3 Drawbacks of some of the key study types

	Description	Usage	Drawbacks
Case series and case reports	Collections of reports on the treatment of individual patients or a report on a single patient	Study of a rare exposure or multiple exposures such as exposure to chemicals	No control groups to compare outcomes, have little statistical validity
Case control studies	Patients who already have a specific condition are compared with people who do not have the condition, using patient records and patient recall	Study of rare diseases such as cancer	Less reliable than randomized controlled trials and cohort studies because showing a statistical relationship does not mean that one factor necessarily caused the other Selection and recall bias is high
Cohort studies	Identify a group of patients who are already taking a particular treatment or have an exposure, follow them forward over time, and then compare their outcomes with a similar group that has not been affected by the treatment or exposure being studied	Study of multiple end points, such as mortality from different causes	Observational in nature and not as reliable as randomized controlled studies, since the two groups may differ in ways other than in the variable under study. Need long duration and cost a lot
Randomized controlled clinical trials	Planned experiments that introduce a treatment to study its effect on real patients, use procedures to reduce bias, and allow for comparison between intervention groups and control groups	Study effect of interventions	Planned experiment and can provide sound evidence of cause and effect
Cross-sectional studies	Describe the relationship between diseases and other factors at one point in time in a defined population	Used for comparing diagnostic tests	Lack any information on timing of exposure and outcome relationships and include only prevalent cases Selection bias is moderate, but the recall bias is high

Table 2.3 (continued)

	Description	Usage	Drawbacks
Qualitative research	Uses various methods (focus groups, interviews, observation, etc.) to understand and interpret health-related questions	Describes, explores, and explains the health-related phenomena being studied	It is a complementary technique combined with a less biased technique
Ecological study	Disease rates and exposures are measured in each of a series of populations and their relation is examined	To look for associations between the occurrence of disease and exposure to known or suspected causes in a community	Depends mainly on published statistics and therefore is affected with its quality
Systematic reviews	Focus on a clinical topic	Answer a specific question	Secondary research; studies are reviewed and assessed for quality and the results summarized according to the predetermined criteria of the review question
Meta-analysis	Thoroughly examine a number of valid studies on a topic and mathematically combine the results		Use accepted statistical methodology to report the results as if it were one large study

Source: Compiled by the author depending on multiple resources: Röhrig et al. (2009), Overview of clinical research methods (2011), Gandhi, P, (2011), and Clancy, C. and Cronin, K., (2005)

2.4 What Is Evidence-Based Medicine and Meta-analysis?

In neurosurgery, it is common to find several trials attempting to answer the same question, some of which will fail to show a statistically significant difference between the two interventions, but combining these studies using meta-analysis, then significant difference of treatment may be proven. Meta-analyses are now a symbol of evidence-based medicine.

Evidence-based medicine includes three key components as demonstrated in Fig. 2.2. These key components are (1) research-based evidence, (2) clinician's accumulated experience, knowledge and skills, and (3) patient's value and preferences [23].

Practicing evidence-based medicine is extremely important in today's healthcare environment because it helps clinicians/surgeons to improve quality and improve patient satisfaction while reducing costs. It encourages a dialogue between patients and their clinicians/surgeons; together they can determine an appropriate course of action.

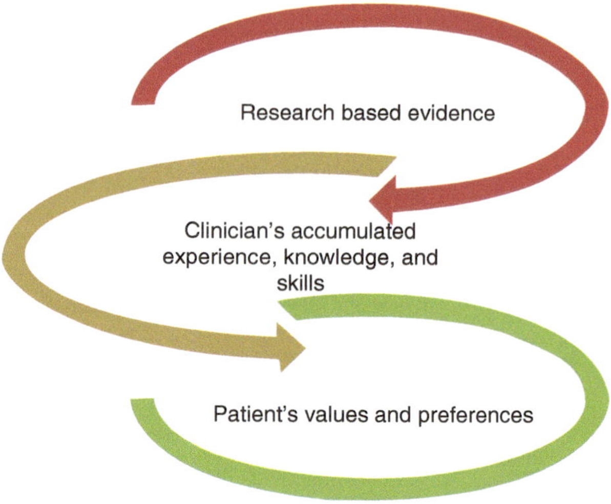

Fig. 2.2 The components
of EBV

In 2004, Lewis and Orland said in their foundation article titled "The Importance and Impact of Evidence-Based Medicine" that using evidence-based medicine helps physicians provide more rational care with better outcomes [12].

2.5   What Are the Benefits of Evidence-Based Medicine Using Meta-analysis?

Evidence-based medicine (EBM) is an approach that aims to optimize decision-making by emphasizing the use of evidence from well-designed and conducted research. Here are some specific benefits [30]:

- *Saves time*: There are enormous published researches which make it difficult for clinicians/surgeons. EBM offers a way to stay current with best practices through the use of standardized, evidence-based protocols [20].
- *Better patient care*: Healthcare workers have much better access to data and knowledge due to technology, and then they can use EBM to provide better patient care based on near real-time data [3].
- *Transparency and accountability*: EBM can help improve transparency of reasoning behind policies and accountability to payers, employers, and patients that are all driving [13].
- *Efficiency*: According to the various case studies published by the Academy of Medical Royal Colleges, UK, in 2013, the use of EBM resulted in more efficient recovery [8, 22].

Yet in 2013, Greenhalgh et al. argued that, although evidence-based medicine has many benefits, it has also some negative unintended consequences [4, 29]:

- Misappropriation of EBM by some commercial interests where pharmaceutical managers use guidelines to control practitioners

- Statistically significant but clinically irrelevant benefits being generated by many trials reporting relative rather than absolute effects
- Evidence being produced that is unsuitable for clinical practice and is in unmanageable volumes

Then the success of the EBM depends on setting up and acting on a realistic bath that reinstates the core values of EBM so that patients benefit from practicing it.

2.6   Steps to Move to an Evidence-Based Medicine Model of Care

Adopting EBM as a way of practice requires clinicians/surgeons to change how they were taught to diagnose and treat patients and introduce a new model of care delivery.

The Centre for Evidence-Based Medicine (CEBM), Oxford, identified the steps to EBM as demonstrated in Fig. 2.3 [29].

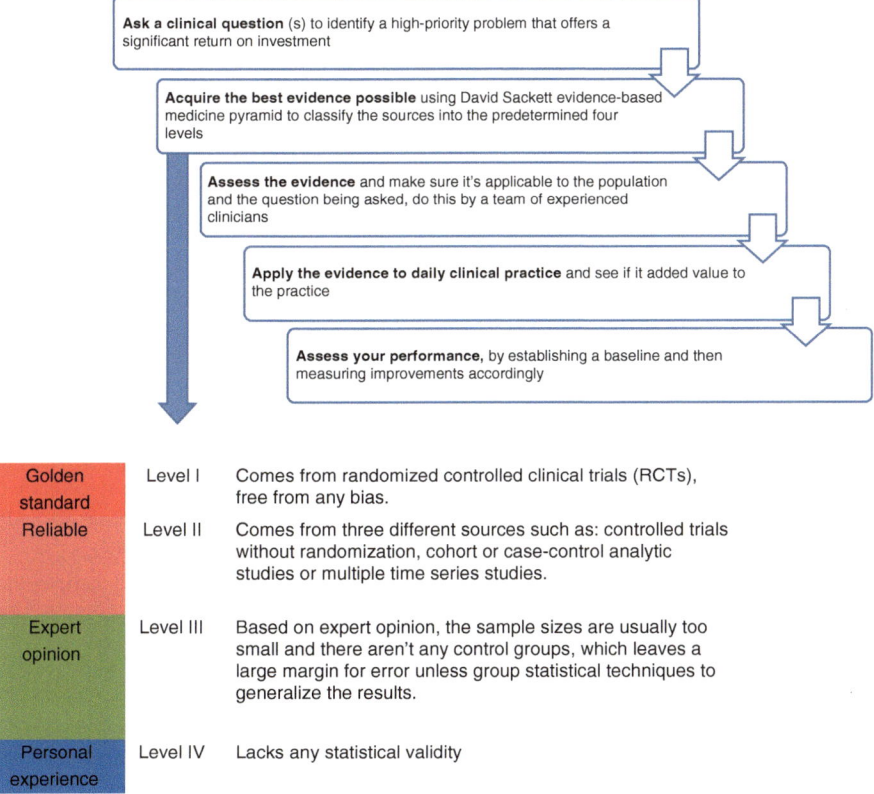

Ask a clinical question (s) to identify a high-priority problem that offers a significant return on investment

Acquire the best evidence possible using David Sackett evidence-based medicine pyramid to classify the sources into the predetermined four levels

Assess the evidence and make sure it's applicable to the population and the question being asked, do this by a team of experienced clinicians

Apply the evidence to daily clinical practice and see if it added value to the practice

Assess your performance, by establishing a baseline and then measuring improvements accordingly

Golden standard	Level I	Comes from randomized controlled clinical trials (RCTs), free from any bias.
Reliable	Level II	Comes from three different sources such as: controlled trials without randomization, cohort or case-control analytic studies or multiple time series studies.
Expert opinion	Level III	Based on expert opinion, the sample sizes are usually too small and there aren't any control groups, which leaves a large margin for error unless group statistical techniques to generalize the results.
Personal experience	Level IV	Lacks any statistical validity

Fig. 2.3 Steps to EBM

The EBM movement will continue its acceleration in the years ahead, and then there is an increasing need for people working in healthcare to learn how to conduct metadata analysis.

2.7   What Is Meta-analysis?

Systematic review methodology is at the core of meta-analysis. The researcher starts by finding all relevant studies (published and unpublished) and assesses the methodological quality of the design and implementation of each study aiming to present an impartial summary of the existing research. Such reviews provide a quantitative statistical estimate of net benefit aggregated over the included studies [31].

Meta-analysis tries to:

- Identify the heterogeneity in effects among multiple studies.
- Increase statistical power to detect an effect.
- Reduce subjectivity of study comparisons.
- Direct future research by identifying the gap [31].

Could we add the data from all trials together?

The answer is definitely no for so many reasons, but the most important reason is that imbalances within trials introduce bias, and this definitely breaks the power of randomization.

2.7.1   Steps to Do Meta-analysis [24, 25, 36]

1. State objectives of the review, and outline eligibility (inclusion/exclusion) criteria for studies, for example, studies with randomized controlled trials.
2. Exhaustively search for studies that seem to meet eligibility criteria.
3. Tabulate characteristics of each study identified and assess its methodological quality.
4. Apply eligibility criteria and justify any exclusions.
5. Classify and code important study characteristics (sample size, length of follow-up, definition of outcome, intervention, etc.).
6. Select and translate results from each study using a common metric.
7. Aggregate findings across studies, generating weighted pooled estimates of effect size.
8. Evaluate the statistical homogeneity of pooled studies.
9. Perform sensitivity analyses to assess the impact of excluding or down-weighting lower-quality, old studies.
10. Prepare a structured report of the review, stating aims, describing materials and methods, and reporting results.

2.7.2   The Requirements of Meta-analysis

1. Calculate effect sizes.
   It is a quantitative measure of the difference between two groups. It is calculated based on the "standardized mean difference" (SMD) between two groups in a trial; in other words, this is the difference between the average score of participants in the intervention group and the average score of participants in the control group.
   One of the most common ways of interpreting effect sizes is based on the work of a man named Cohen, who said that 0.2 and below = small effect size, 0.5 = medium effect size, and 0.8 and above = large effect size [26, 28].
   Another common way is to draw the funnel plot.
2. Check publication bias using a funnel plot.
   Funnel plots display the studies included in the meta-analysis in a plot of effect size against sample size. As smaller studies have more chance variability than larger studies, the expected picture is one of a symmetrical inverted funnel. If the plot is asymmetric, this suggests that the meta-analysis may have missed some trials usually smaller studies showing no effect. Egger's regression test has been widely used to test for publication bias as a more formal test [31].
3. Conduct sensitivity analysis.
   It explores the ways in which the main findings are changed by varying the approach to aggregation by excluding poor-quality articles or older ones, for example. It may also examine results consistency across subgroups such as type of intervention [31].
4. Draw the forest plot.
   It is a graphical representation of a meta-analysis. It has one line going through a box representing each study in the meta-analysis, plotted according to the standardized mean difference. The size of the box is proportional to the precision of the study. A 95% confidence interval is drawn around each of the studies' squares to represent the uncertainty of the estimate of the treatment effect. The aggregate effect size obtained by combining all the studies is usually displayed as a diamond [28].
5. Assess heterogeneity.
   A major concern about meta-analyses is the extent to which they mix studies that are different in kind (heterogeneity). A test which is commonly used is Cochran's Q. If heterogeneity is absent, then size of treatment effect is the same (fixed) across all studies, and the variation seen between studies is due only to the play of chance. If the amount of heterogeneity is large, then use meta-regression to overcome it [31].
6. Meta-regression.
   It is a technique that allows researchers to explore which types of patient-specific factors or study design factors contribute to the heterogeneity. The simplest type of meta-regression uses summary data from each trial, such as the average effect size, average disease severity at baseline, and average length of follow-up. This approach is valuable, but it has only limited ability to identify important factors. Fortunately, using individual patient data will give answers to the important question: what types of patients are most likely to benefit from this treatment? [31]

2.8   Meta-analysis Software

There are many meta-analysis programs; some of them are free programs and others are commercially available programs [27, 38].

Freeware
- **Meta Analysis Calculator** is freeware taking input data and computing meta-analysis statistics developed by Larry C. Lyons of Manassas, VA. Download from http://www.lyonsmorris.com/MetaA/contents.htm.
- **Meta-Analysis Easy to Answer** is freeware developed by David A. Kenny. Moderator variables are not included. A Windows version of the program also is available. Download from http://davidakenny.net/meta.htm.
- **MetaXL** is a tool for meta-analysis in Microsoft Excel. It extends Excel with several functions for input and output of meta-analysis data. It employs the same meta-analysis methods that can be accessed in general statistical packages, but makes two additional methods available: the inverse variance heterogeneity (IVhet) and quality effects (QE) models. In addition, a new way to detect publication bias, the Doi plot, has been implemented. Download from http://www.epigear.com/index_files/metaxl.html.

Commercial
- **Comprehensive Meta-Analysis** is commercial program with a downloadable demonstration program and an academic program version. Download from http://www.meta-analysis.com/pages/comparisons.html.
- **MetaWin** is a commercial program with a downloadable demonstration program and an academic program version. Download demo from http://www.metawinsoft.com/mw2demo.html.

Add-On
- **MIX** (*m*eta-analysis with *i*nteractive *e*xplanations) is Excel-based software. Download from http://www.mix-for-meta-analysis.info/index.html.
- Meta-Analysis website by David B. Wilson includes Excel spreadsheets and SPSS and SAS macros for performing a meta-analysis. Download from http://mason.gmu.edu/~dwilsonb/ma.html.

2.9   Benefits of Meta-analysis

Meta-analysis is powerful but also debatable, because several conditions are critical to a sound meta-analysis, yet Walker et al. (2008) identified a set of benefits [35]:

- Results can be generalized to a larger population.
- The precision and accuracy of estimates can be improved as more data is used. This, in turn, may increase the statistical power to detect an effect.

- Inconsistency of results across studies can be quantified and analyzed.
- Hypothesis testing can be applied on summary estimates.
- Moderators can be included to explain variation between studies.
- The presence of publication bias can be investigated.

2.10   Strengths and Weaknesses of Meta-analysis

Although the use of meta-analysis became very popular recently, yet a logical evaluation of the technique is still relevant. Finckh and Tramèr (2008) published an article discussing the strengths and weaknesses of meta-analysis as summarized in Table 2.4 [26].

2.11   A Step-by-Step Procedure to Conduct Meta-analysis

For the purpose of drafting the step-by-step procedure conducting meta-analysis, we used MetaXL [38] which is an add-in for meta-analysis in Microsoft Excel for Windows. It supports all major meta-analysis methods, plus, uniquely, the inverse variance heterogeneity and quality effects models, producing output in table and graphical format.

We created a factitious data set that we used to demonstrate the procedure.

The Requirements of Meta-analysis
- Data entry.
  We created a factitious data set using one of the example files of MetaXL, reflecting a post- and pretest, and so the key indicator is the mean difference represented in column 2 in Table 2.5, and we run the analysis accordingly.
- Check publication bias using a funnel plot.
  Funnel plots measures the magnitude of the treatment effect of individual studies (odds ratio) is plotted against the sample size or precision of the studies (standard error) may be used to detect publication biases. A symmetrical inverted funnel as shown implies that the studies found are likely to be inclusive (Fig. 2.4).

Table 2.4 Strengths and weaknesses of meta-analysis [32]

Strengths of meta-analysis	Weaknesses of meta-analysis
• Discipline the process of summing up research findings	• Requires perseverance and hard work
• Represents findings in an objective manner than reviews	• Ignores qualitative distinctions between studies
• Can detect relationships across studies that are hidden	• Comparing "apples to oranges" is the main criticism
• Protects against over-interpreting differences across studies	• Most meta-analyses include faulty studies
• Can handle a large numbers of studies	• Selection bias is a clear threat
	• Heavy reliance on published studies, which may increase the effect as it is very hard to publish studies that show no significant results

Table 2.5 Data set

Study	MD	LCI 95%	HCI 95%	Weight
AAAA	−0.400	−1.066	0.266	1590
BBBB	−0.150	−0.953	0.653	2.969
CCCC	−2.200	−9.749	5.349	2.024
DDDD	−0.860	−1.677	−0.043	4.479
HHHH	−1.200	−1.851	−0.549	5.320
LLLL	−0.390	−0.709	−0.071	8.081
MMMM	−0.150	−1.066	0.766	5.614
NNNN	−0.200	−0.680	0.280	7.252
WWWW	−0.620	−1.758	0.518	5.055
QQQQ	−2.250	−5.409	0.909	3.117
RRRR	0.010	−0.733	−0.733	8.098
YYYY	0.070	−0.617	0.757	3.319
OOOO	−0.500	−1.571	0.571	3.869
PPPP	−0.580	−2.153	0.993	3.413
SSSSS	−0.600	−1.270	0.070	3.386
FFFF	−0.360	−1.008	0.288	1.622
JJJJJ	−0.600	−0.824	−0.376	22.419
KKKK	−0.170	−0.597	0.257	8.373
Pooled	−0.525	−0.764	−0.285	100.00
Statistics				
I-squared	0.542	0.000	50.239	
Cochran's Q	17.093			
Chi2	0.448			
Q-index	46.575			

Source: Developed by the researcher using free downloadable add-in to Excel (MetaXL)

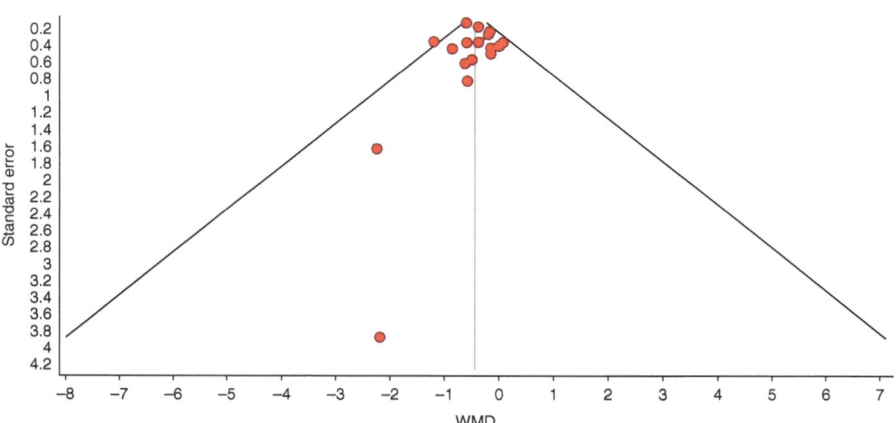

Fig. 2.4 Funnel plot

But when an asymmetrical plot is found, then it suggests that small, negative, or neutral studies have been omitted. The vertical line represents the pooled estimate of the treatment effect of all the included studies [37].

The below figure is the Doi plot (Fig. 2.5) which is another way to detect publication bias, and it indicates a minor asymmetric leading us to confirm the insights we reached using the funnel plot.

- Conduct sensitivity analysis (Fig. 2.6).
  It explores the ways in which the main findings are changed by varying the approach to aggregation by excluding poor-quality articles or older ones, for

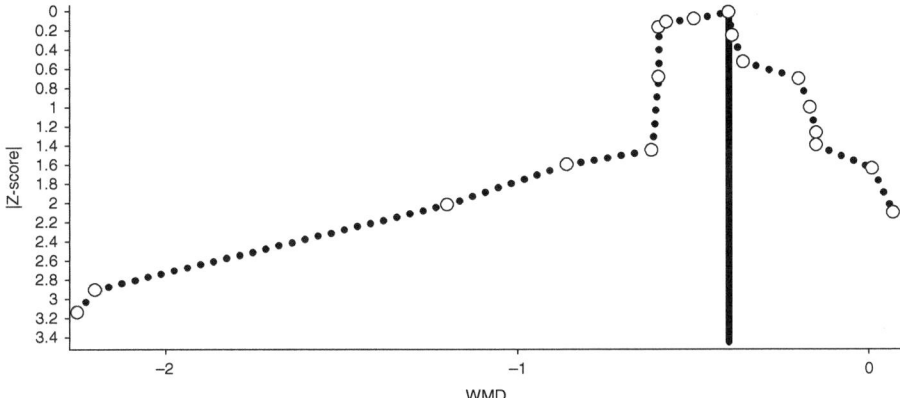

Fig. 2.5 Doi plot

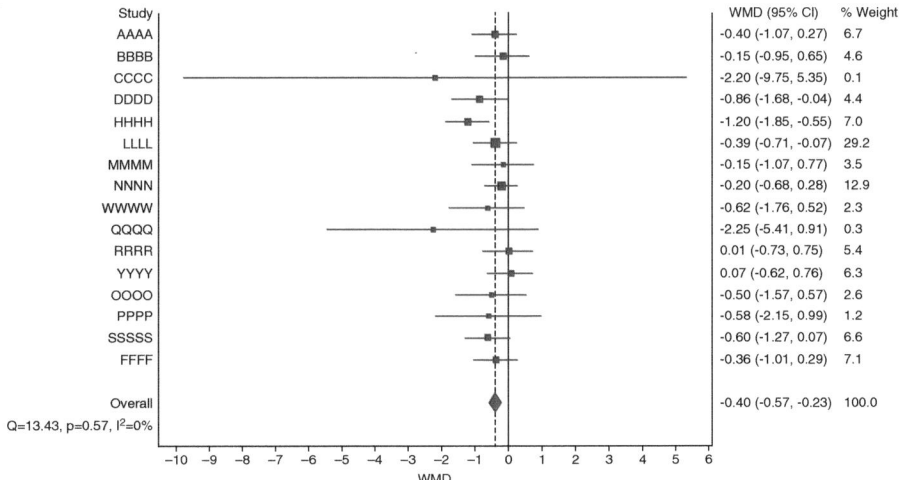

Fig. 2.6 Sensitivity analysis

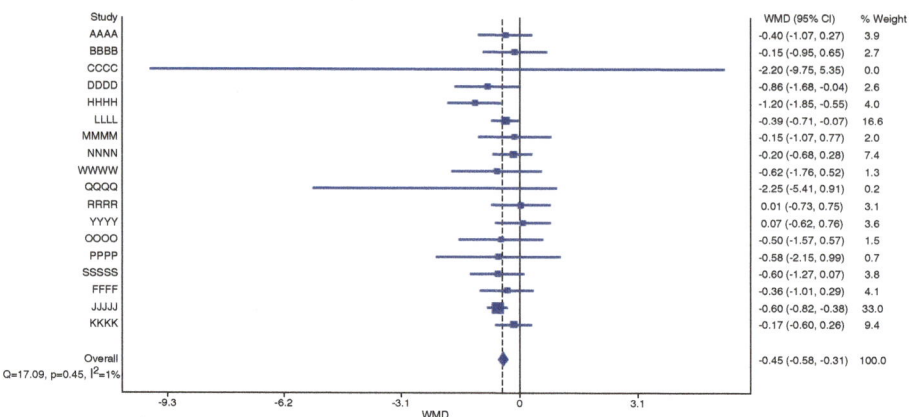

Fig. 2.7 Forest plot

example. We excluded the last two cases and recalculated and found almost similar results indicating consistency. The following forest plot and indicators (lower left corner) prove our argument.

- Draw the forest plot (Fig. 2.7).
  It is a graphical representation of a meta-analysis. It has one line going through a box representing each study in the meta-analysis, plotted according to the standardized mean difference. The size of the box is proportional to the precision of the study. A 95% confidence interval is drawn around each of the studies' squares to represent the uncertainty of the estimate of the treatment effect. The aggregate effect size obtained by combining all the studies is usually displayed as a diamond. We can see more studies to the left of the line of no effect, suggesting that there is a difference between the two treatments. However, the diamond at the bottom of the forest plot represents the pooled data from all of the studies and demonstrates limited confidence interval. Therefore, the conclusion of the meta-analysis is that there is difference between the two treatments [31].

- Assess heterogeneity.
  A major concern about meta-analyses is the extent to which they mix studies that are different in kind (heterogeneity). A test which is commonly used is Cochran's Q. If heterogeneity is absent (since $p > 0.05$), then we conclude that the size of treatment effect is the same across all studies and the variation seen between studies is due only to the play of chance [17].

  Another test is I-squared, and the rule of thumb to interpret it is 25% corresponding to low heterogeneity, 50% to moderate, and 75% to high. So, in our case it moderates heterogeneity and can be because of small size studies [31].

  Based on the two mentioned tests, there is no need for meta-regression.

Conclusion

There is no option but to continue striving and searching for the best treatment for the different types of hydrocephalus. Most of the available evidences are classified as Level II or Level III. Therefore, there is utmost need to continue searching to understand the pathophysiology of the hydrocephalus and to find new treatment not only to save life of such children but also to preserve or restore their normal neurological and cognitive function and allow them to have a good future. The search should extend to prevent or at least to reduce the complications and consequences of the different methods of treatment.

Meta-analysis is a quantitative, formal, epidemiological study design used to systematically assess previous research studies to derive conclusions about that body of research; it is a cornerstone in EBM by making findings from individual studies more applicable to clinical practice. Outcomes from a meta-analysis may include a more precise estimate of the effect of treatment or risk factor for disease, than any individual study. Rigorously conducted meta-analyses are useful tools in evidence-based medicine. The need to integrate findings from many studies ensures that meta-analytic research is desirable and the large body of research now generated makes the conduct of this research feasible.

References

1. Baird LC, Mazzola CA, KI A, Klimo P Jr, Flannery AM, Pediatric Hydrocephalus Systematic Review and Evidence-Based Guidelines Task Force. Pediatric hydrocephalus: systematic literature review and evidence-based guidelines. Part 5: effect of valve type on cerebrospinal fluid shunt efficacy. J Neurosurg Pediatr. 2014;14(Suppl 1):35–43.
2. Berry JG, Toomey SL, Zaslavsky AM, Jha AK, Nakamura MM, Klein DJ, et al. Pediatric readmission prevalence and variability across hospitals. JAMA. 2013;309:372–80. (Erratum in JAMA 309:986, 2013).
3. Clancy C, Cronin K. Evidence-based decision making: global evidence, local decisions. Health Aff. 2005;24(1):151–62.
4. Croft P, Malmivaara A, van Tulder V. The pros and cons of evidence-based medicine. Spine (Phila Pa 1976). 2011;36(17):E1121–5. doi:10.1097/BRS.0b013e318223ae4c.
5. Esene I, El-Shehaby A, Baeesa S. Essentials of research methods in neurosurgery and allied sciences for research, appraisal and application of scientific information to patient care (Part I). Neurosciences. 2016;21(2):97–107.
6. Flannery AM, Duhaime AC, Tamber MS, Kemp J, Pediatric Hydrocephalus Systematic Review and Evidence-Based Guidelines Task Force. Pediatric hydrocephalus: systematic literature review and evidence-based guidelines. Part 3: endoscopic computer-assisted electromagnetic navigation and ultrasonography as technical adjuvants for shunt placement. J Neurosurg Pediatr. 2014;14(Suppl 1):24–9.
7. Gandhi P. Clinical research methodology. Indian J Pharm Educ Res. 2011;45(2):199–209.
8. Gaspar L. The role of whole brain radiation therapy in the management of newly diagnosed brain metastases: a systematic review and evidence-based clinical practice guideline. J Neuro-Oncol. 2010;96(1):17–32.
9. Kemp J, Flannery AM, Tamber MS, Duhaime AC, Pediatric Hydrocephalus Systematic Review and Evidence-Based Guidelines Task Force. Pediatric hydrocephalus: systematic literature review and evidence-based guidelines. Part 9: effect of ventricular catheter entry point and position. J Neurosurg Pediatr. 2014;14(Suppl 1):72–6.
10. Klimo P Jr, Thompson CJ, Baird LC, Flannery AM, Pediatric Hydrocephalus Systematic Review and Evidence-Based Guidelines Task Force. Pediatric hydrocephalus: systematic literature review and evidence-based guidelines. Part 7: antibiotic-impregnated shunt systems versus conventional shunts in children: a systematic review and meta-analysis. J Neurosurg Pediatr. 2014;14(Suppl 1):53–9.
11. Klimo P Jr, Van Poppel M, Thompson CJ, Baird LC, Duhaime AC, Flannery AM, Pediatric Hydrocephalus Systematic Review and Evidence-Based Guidelines Task Force. Pediatric

hydrocephalus: systematic literature review and evidence-based guidelines. Part 6: preoperative antibiotics for shunt surgery in children with hydrocephalus: a systematic review and meta-analysis. J Neurosurg Pediatr. 2014;14(Suppl 1):44–52.

12. Lewis S, Orland B. The importance and impact of evidence-based medicine. J Manag Care Pharm. 2004;10(5):S3–5.

13. Limbrick DD, Baird LC, Klimo P Jr, Riva-Cambrin J, Flannery AM. Pediatric hydrocephalus: systematic literature review and evidence-based guidelines. Part 4: cerebrospinal fluid shunt or endoscopic third ventriculostomy for the treatment of hydrocephalus in children. J Neurosurg Pediatr. 2014;14(Suppl 1):30–4.

14. Mazzola CA, Choudhri AF, Auguste KI, Limbrick DD Jr, Rogido M, Mitchell L, Flannery AM, Pediatric Hydrocephalus Systematic Review and Evidence-Based Guidelines Task Force. Pediatric hydrocephalus: systematic literature review and evidence-based guidelines. Part 2: management of posthemorrhagic hydrocephalus in premature infants. J Neurosurg Pediatr. 2014;14(Suppl 1):8–23.

15. Nikas DC, Post AF, Choudhri AF, Mazzola CA, Mitchell L, Flannery AM, Pediatric Hydrocephalus Systematic Review and Evidence-Based Guidelines Task Force. Pediatric hydrocephalus: systematic literature review and evidence-based guidelines. Part 10: change in ventricle size as a measurement of effective treatment of hydrocephalus. Neurosurg Pediatr. 2014;14(Suppl 1):77–81.

16. Tamber MS, Klimo P Jr, Mazzola CA, Flannery AM, Pediatric Hydrocephalus Systematic Review and Evidence-Based Guidelines Task Force. Pediatric hydrocephalus: systematic literature review and evidence-based guidelines. Part 8: management of cerebrospinal fluid shunt infection. J Neurosurg Pediatr. 2014;14(Suppl 1):60–71.

17. Sedgwick P. Meta-analyses: heterogeneity and sub-group analysis. BMJ. 2013;346:f4040.

18. Röhrig B, et al. Types of study in medical research. Dtsch Arztebl Int. 2009;106(15):262–8.

19. Williams MA, McAllister JP, Walker ML, Kranz DA, Bergsneider M, Del Bigio MR, et al. Priorities for hydrocephalus research: report from a National Institutes of Health-sponsored workshop. J Neurosurg. 2007;107(5 Suppl):345–57.

20. Youngblut J, Brooten D. Evidence-based nursing practice: why is it important? AACN Clin Issues. 2001;12(4):468–76. Flannery AM, Mitchell L. Pediatric hydrocephalus: systematic literature review and evidence-based guidelines. Part 1: introduction and methodology. J Neurosurg Pediatr. 2014;14(Suppl 1):3–7.

Online Sources

21. Overview of clinical research methods, compiled by Crowe Associates Ltd for James Lind Alliance December 2011: https://www.nottingham.ac.uk/research/groups/cebd/documents/patientscarers/overview-of-clinical-research-methods.pdf. Accessed 5 Oct 2016.

22. Sense About Science, Academy of Medical Royal Colleges: www.senseaboutscience.org. Accessed 5 Oct 2016.

23. Himmelfarb Health Science Library: www.libguides.gwumc.edu. Accessed 5 Oct 2016.

24. DeCoster, J.: Meta-analysis notes. http://www.stat-help.com/notes.html (2004). Accessed 5 Oct 2016.

25. Haidich, A.: Meta-analysis in medical research. https://www.ncbi.nlm.nih.gov/pmc/articles/PMC3049418/ (2010). Accessed 5 Oct 2016.

26. Meta-analysis Research Methodology: www.academia.edu. Accessed 5 Oct 2016.

27. Computer Programs for Meta-Analysis. www.commfaculty.fullerton.edu. Accessed 5 Oct 2016.

28. For Practitioners/What is a good evidence. www.cebi.ox.ac.uk.

29. Greenhalgh T, et al.: Evidence based medicine: a movement in crisis? www.nbci.nlm.nih.gov (2013).

30. 5 Reasons the Practice of Evidence-Based Medicine Is a Hot Topic. www.healthcatalyst.com. Accessed 30 Oct 2016.
31. What is meta-analysis? www.whatisseries.co.uk. Accessed 5 Oct 2016.
32. Centre for Evidence-Based Medicine (CEBM) Oxford, UK, The steps to EBM include: http://www.cebm.net Accessed 5 Oct 2016.
33. Primary research advantages and disadvantages. http://www.ianswer4u.com/2012/02/. Accessed 5 Oct 2016.
34. Secondary research advantages and disadvantages. http://www.ianswer4u.com/2012/05/. Accessed 5 Oct 2016.
35. Meta-analysis: its strengths and limitations. Cleveland Clinic Journal of Medicine. http://www.ccjm.org/. Accessed 5 Oct 2016.
36. Primer: strengths and weaknesses of meta-analysis. https://www.researchgate.net/publication/. Accessed 5 Oct 2016.
37. Statistics V.: Introduction to clinical trials and systematic reviews Abdul Ghaaliq Lalkhen and Anthony McCluskey. http://ceaccp.oxfordjournals.org/. Accessed 11 Oct 2016.
38. MetaXL—EpiGear: www.epigear.com. Accessed 11 Oct 2016.

Part II

Etiology and Pathophysiology of Hydrocephalus

Pathophysiology of Hydrocephalus

3

Deepak Gupta, Raghav Singla, and Chinmay Dash

3.1 Introduction

Hydrocephalus has been defined as an active ventricular distension due to an inadequate passage of cerebrospinal fluid (CSF) from its point of production within the cerebral ventricles to its point of absorption into the systemic circulation [1]. The classical hypothesis of CSF hydrodynamics suggests a unidirectional flow of CSF from its production point in the ventricles to the cortical subarachnoid space and further absorption into the venous sinuses.

In his landmark work of choroid plexectomy, Dandy had concluded that it was the only absolute proof that cerebrospinal fluid was formed from the choroid plexus. At the same time, he hypothesized that the ependyma lining the ventricles was not concerned in the production of cerebrospinal fluid [2].

CSF physiology is based on three key premises:

1. Active secretion of CSF
2. Passive absorption of CSF
3. Unidirectional flow of CSF from the place of formation to the place of absorption (Fig. 3.1)

3.2 Formation of Cerebrospinal Fluid

Results from isolated choroid plexus preparations would indicate that 80% or more of CSF production is derived from this source alone [3]. The passage of an ultrafiltrate of the plasms through the fenestrated endothelium of choroid plexus

D. Gupta, M.D. (✉) • R. Singla, M.D. • C. Dash, M.D.
Department of Neurosurgery, Neurosciences Centre and Association, JPN Apex Trauma Centre AIIMS, Delhi, India
e-mail: drdeepakgupta@gmail.com

© Springer International Publishing AG 2017
A. Ammar (ed.), *Hydrocephalus*, DOI 10.1007/978-3-319-61304-8_3

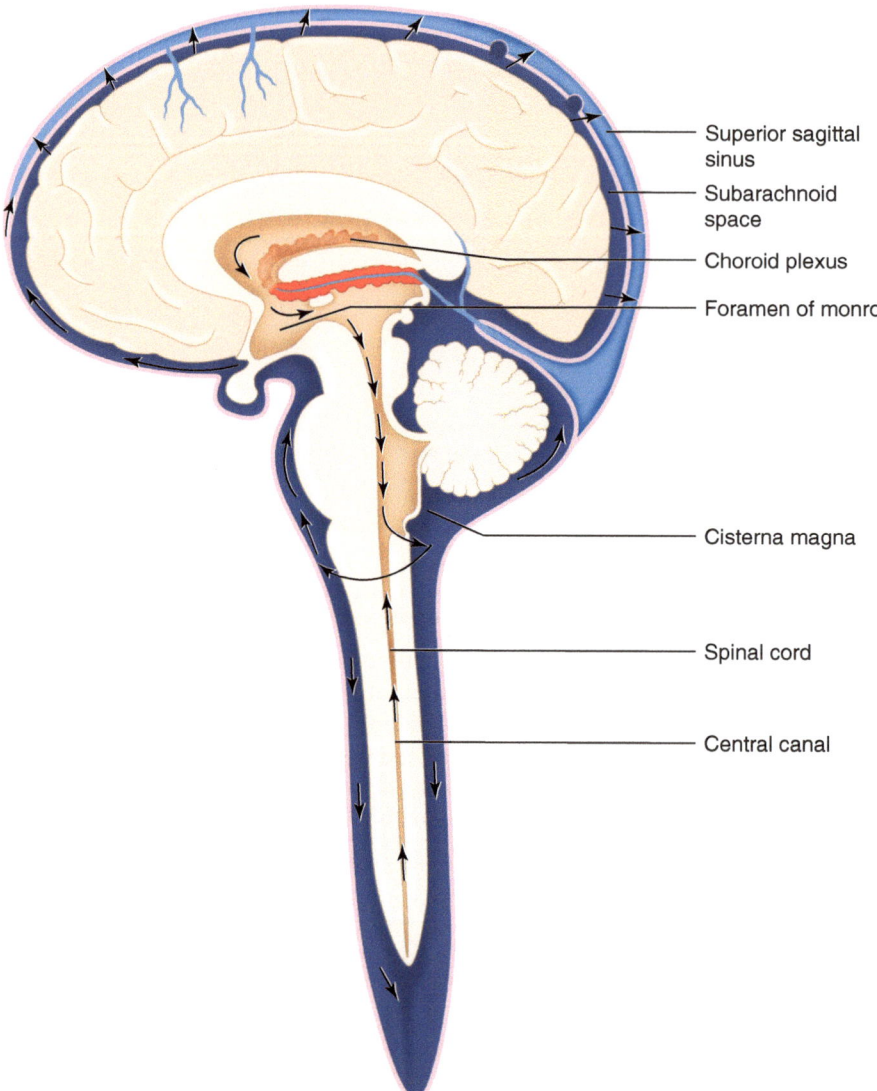

Fig. 3.1 Classical hypothesis. Adapted from "Orešković D, Klarica M. Development of hydro-cephalus and classical hypothesis of cerebrospinal fluid hydrodynamics: facts and illusions"

constitutes the first step in the formation of CSF. This is facilitated by hydrostatic pressure. The ultrafiltrate then passes through the choroidal epithelium, which is an active metabolic process that transforms the ultrafiltrate into a secretion product. Perfusion of a portion of the ventricular system devoid of choroid plexus has demonstrated that 30–60% of CSF is produced from nonchoroidal sources [4], which may explain the failure of choroid plexectomy in the clinical setting to control progressive hydrocephalus. Parenchyma may be the main source of nonchoroidal CSF formation.

3.2.1 Absorption of CSF

The rate of CSF absorption is dependent on pressure and is relatively linear over a fairly wide physiologic range that has been well established. Absorption of CSF occurs via the arachnoid villi, brain, and blood vessels [5, 6]. It is believed that CSF is passively absorbed from the cranial subarachnoid space to the cranial venous blood by means of a hydrostatic gradient [7].

3.2.2 Unidirectional Flow

The cerebrospinal fluid circulates in a to-and-fro movement with a net caudal flow through the ventricles to the subarachnoid space [8]. The CSF flows unidirectionally from the lateral brain ventricles through the foramina of Monro, then through the third ventricle and the aqueduct of Sylvius into the fourth ventricle, and finally through the foramina of Luschka and Magendie into the subarachnoid space. The pumping action of the choroid plexus is believed to play a major role and that pulsation of the CSF is generated mainly by the filling and draining of the choroid plexuses [9, 10].

Therefore, the CSF physiology conceived this way has been presented as the classical hypothesis of CSF hydrodynamics, i.e., CSF is actively produced (secreted) mainly from the choroid plexuses (Fig. 3.2) inside the brain ventricles, and then it circulates from the ventricles toward the subarachnoid space and is absorbed passively by the arachnoid villi into the venous sinuses (Fig. 3.3). This means that in physiological conditions, the same CSF volume, which is actively formed by choroid plexus, must be passively absorbed into the cortical subarachnoid space.

An imbalance in this CSF production and absorption, leading to net accumulation of fluid and an enlargement of the brain ventricles, leads to hydrocephalus. The balance between production and absorption of CSF is critically important. The resulting pressure of the fluid against the brain tissue causing signs and symptoms of raised pressure is what constitutes hydrocephalus.

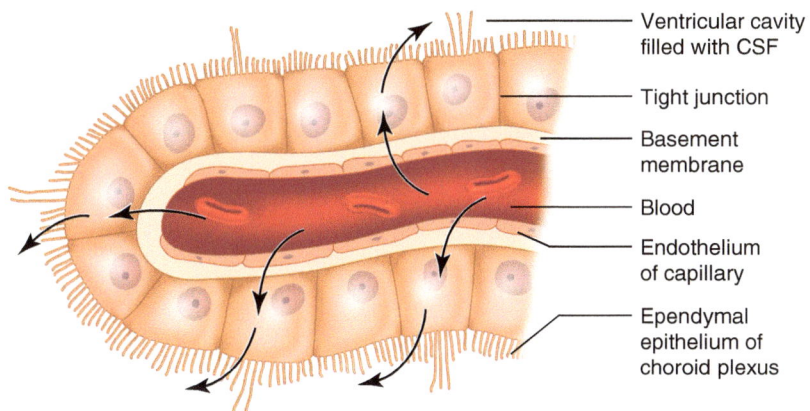

Ventricular cavity filled with CSF

Tight junction

Basement membrane

Blood

Endothelium of capillary

Ependymal epithelium of choroid plexus

Fig. 3.2 Microscopic structure of the choroid plexus showing the path taken by fluids in the formation of cerebrospinal fluid

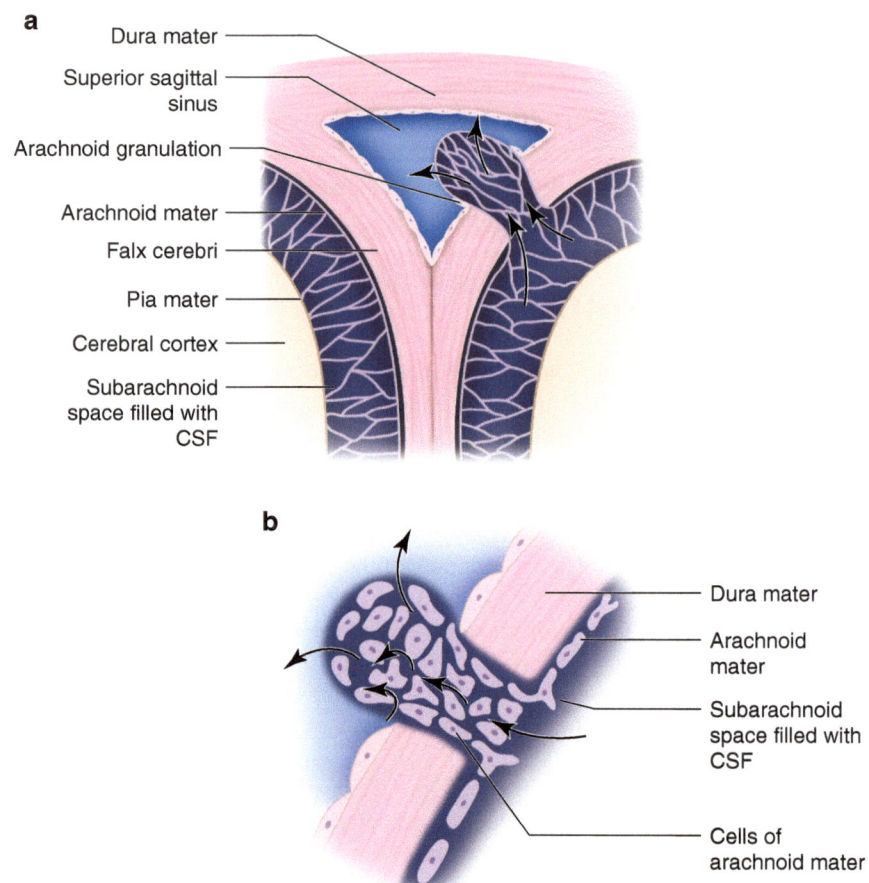

Fig. 3.3 (a) Coronal section of the superior sagittal sinus showing an arachnoid granulation. (b) Magnified view of an arachnoid granulation showing the path taken by the cerebrospinal fluid into the venous system (Reproduced from Snell's clinical neuroanatomy but drawn)

The classical theory has been challenged in recent literature. According to a new hypothesis, CSF is not formed mainly by the choroid plexuses. Rather it appears and disappears throughout the entire CSF system as a result of the hydrostatic and osmotic forces between the CSF, interstitial fluid (ISF), and blood capillaries. Osmotic and hydrostatic forces are crucial to the regulation of ISF–CSF volume [11, 12]. It suggests that there is a constant exchange of fluids and substances between CSF and tissue that is finely regulated by the various (patho)physiological conditions that predominate within these compartments. The much larger surface area of cerebral capillaries as compared to choroid plexus suggests that the "formation" and "absorption" of CSF mainly take place at the cerebral capillaries [13, 14]. Thus, a rapid turnover of water, which constitutes 99% of ISF–CSF volume, continuously takes place between the plasma and ISF–CSF at all places [11].

As it occurs in the rest of the body, the movement of water inside the CNS is influenced by osmotic and hydrostatic forces and depends on physiological or pathophysiological processes (trauma, ischemia, inflammation, hydrocephalus, etc.), which can cause changes in fluid osmolarity and hydrostatic pressure. According to this new hypothesis, a disruption of the balance between ISF and CSF is essential to cause hydrocephalus. It suggests that additional pathophysiological changes are of crucial importance in "obstructive hydrocephalus" and a blockade of the CSF pathways alone, without other active pathological processes, would not cause clinically manifested hydrocephalus (Fig. 3.4).

The classical CSF hypothesis can explain most etiology of hydrocephalus. This primarily involves the active CSF production or overproduction by "CSF pumps"

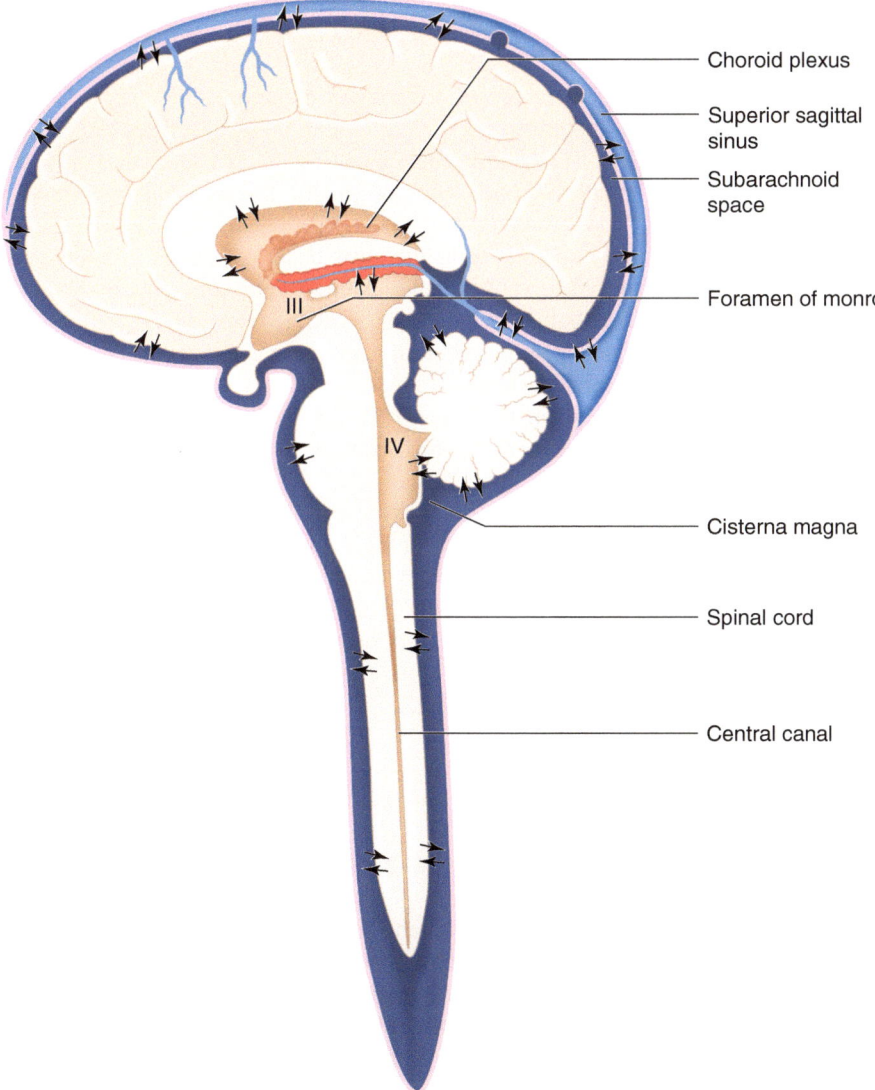

Fig. 3.4 Orešković and Klarica hypothesis

Table 3.1 Proposed classification of hydrocephalus by Rexate et al.

Site of obstruction	Pathology	Hydrocephalus type
None	Choroid plexus papilloma	Communicating
Foramen of Monro	Tumor	Obstructive
Aqueduct of Sylvius	Tumors, congenital	Obstructive
Fourth ventricular outlet	Meningitis, Chiari	Obstructive
Basal cisterns	Meningitis, post-SAH	Obstructive
Arachnoid granulations	Hemorrhage, infections	Obstructive
Venous outflow	Skull base anomaly, heart disease	Obstructive

(choroid plexuses), impaired circulation, inhibited absorption, and increased hydrostatic CSF pressure. However, it is far from perfect and has been questioned by the Orešković and Klarica hypothesis.

Based on the classical hypothesis of CSF hydrodynamics, hydrocephalus may develop as a result of an obstruction of the circulating pathways, a reduction in the ability to absorb the CSF, or an overproduction of CSF. Dandy identified the site of production of CSF to be choroid plexus and originally described the two varieties: "communicating" and "noncommunicating" [15]. Since then, this classification has been the accepted system and still forms the basis of classification in many situations. On the basis of ventriculography, Dandy defined communicating hydrocephalus as that form of hydrocephalus in which the injected dye could be recovered from the spinal subarachnoid space and noncommunicating variety as one in which the dye did not reach the spinal subarachnoid space.

The classification is based on the major CSF pathway, with noncommunicating hydrocephalus being characterized by a block between the ventricles and the lumbar subarachnoid space.

Ransohoff et al. renamed noncommunicating hydrocephalus as intraventricular obstructive hydrocephalus and thought communicating hydrocephalus to be caused by an obstruction at the level of the basal cisterns or arachnoid villi and called it "extraventricular obstructive hydrocephalus" [16].

A recent classification by Rekate et al. proposes only cases of overproduction of CSF by choroid plexus papilloma as communicating hydrocephalus and considers all other forms as obstructive (Table 3.1) [1].

3.3 Experimental Models of Hydrocephalus

3.3.1 Animal Models of Congenital Hydrocephalus

H-Tx strain is the congenital rat model which is most frequently used in experimental research, which develops obstructive hydrocephalus from aqueductal stenosis in the perinatal period [17, 18]. Non-Mendelian mechanisms are attributed to the development of aqueductal stenosis in the inbred strains of Wistar-Lewis (LEW/ Jms) rats, as early as day 17 in a 21-day gestational period [19]. Various factors make these rat models an excellent option for studies of neonatal and juvenile hydrocephalus, especially that these include a natural occurrence of ventriculomegaly, in which the brain is of adequate size to enable customized shunting, the amenability of rats to behavioral testing, the relatively low cost, and the availability of ample data for correlation. Various mouse models have thrown light on the cause of ventriculomegaly [17, 18, 20]. The most widely used models include the SUMS/ NP [21], hy3 [22], transforming growth factor (TGF)-β1 overexpression [23–25], hyh with point mutation in α-SNAP and ependymal denudation that precedes aqueductal stenosis [26–28], fibroblast growth factor (FGF)-2 [4], L1-cell adhesion molecule deficient [29], aquaporin deficient [30], hpy [31], members of the conserved forkhead/winged helix transcription factor gene (previously *Mf1*) [32, 33], heparin-binding epidermal growth factor, and collagen deficiency [34]. Sweger and colleagues [35] have developed a double transgenic mouse model that allows expression of the G1-coupled Ro1 receptor, which are exclusive to astrocytes. By controlling Ro1 expression with a tetracycline-on promoter in drinking water, these mice develop enlarged ventricles, partial ependymal denudation, morphologic changes in the subcommissural organ, and obliteration of the cerebral aqueduct at designated times. This represents a powerful experimental model because the pathogenesis of hydrocephalus can be studied with neonatal, juvenile, and adult onset. The main disadvantage of these mouse models is that their size limits the use of CSF shunts and invasive physiologic sensors.

3.3.2 Animal Models of Acquired Hydrocephalus

Kaolin has been used commonly in animal experiments for inducing hydrocephalus. An inert silicate in nature, it has been used in a wide variety of infant and adult animals (mouse, rat, rabbit, hamster, cat, dog) via injections into the CSF to induce hydrocephalus. Injections of kaolin into the cisterna magna or fourth ventricle or both usually produce obstructive (noncommunicating or intraventricular) hydrocephalus. It has been more difficult to produce communicating extraventricular hydrocephalus with acquired approaches. Only moderate amounts of ventriculomegaly have been seen after kaolin injections into the cortical subarachnoid injections of kaolin or silicone [36] in adult rats [37, 38] and dogs [39, 40]. Silastic has been infused into the basal cisterns of monkeys to produce mild forms of communicating hydrocephalus in a similar manner [41]. Li and colleagues [42] were able to induce a communicating form of hydrocephalus by injecting kaolin into the basal cisterns of adult rats. Viral [43] and bacterial [44, 45] inoculations (nonmechanical methods) have also been used to induce communicating hydrocephalus. Davis et al. [43] found that following intracerebral viral inoculation, a transient inflammation of the meninges, ependyma, and choroid plexus was found followed by hydrocephalus. Growth factors such as TGF-β and FGF-1 and FGF-2 [25], as well as neurotoxins [46], have all been successful to varying degrees after intrathecal or intracisternal injections in animal models.

3.4 Mathematical Models of Hydrocephalus

Several mathematical models have revealed possible mechanisms associated with hydrocephalus [47–52]. Pang et al. [51] gave the following postulates to explain the low-pressure hydrocephalic state in 12 shunted patients:

1. Low-pressure hydrocephalic state related to alteration of the viscoelastic modulus of the brain, secondary to expulsion of extracellular water from the brain parenchyma and to structural changes in brain tissues due to prolonged overstretching.
2. Certain patients are susceptible to developing low-pressure hydrocephalic state because of an innate low brain elasticity due to bioatrophic changes.
3. Low-pressure hydrocephalic state symptoms are due not to pressure changes but to brain tissue distortion and cortical ischemia secondary to severe ventricular distortion and elevated radial compressive stresses within the brain.
4. Treatment must be directed toward allowing the entry of water into the brain parenchyma and the restoration of baseline brain viscoelasticity.

The mathematical model describes the brain as a quasi-linear viscoelastic tissue, i.e., in response to ventriculomegaly, there is an initial elastic response followed by a different long-term viscoelastic response. The flaw that prevents this model and all others from being predictive is that the material properties of the brain are not known precisely enough to be useful.

Magnetic resonance elastography [53, 54] measures the elastic properties of brain tissue in vivo, based on the propagation of waves through deformed tissue detected with magnetic resonance imaging (MRI). Tissues with different material properties conduct these waves differently. Although this method holds great potential for noninvasive tissue property measurement, the true material properties of the brain (healthy or hydrocephalic) have yet to be measured directly, so the accuracy of this method has not been verified. In addition, the device used to perform the mechanical perturbation of the tissue requires a flat surface [54], so more developed gyral hemispheres are problematic.

3.5   Pathophysiologic Mechanisms

3.5.1   Gliosis and Inflammation

Gliosis is a consistent finding in hydrocephalus, and inflammation and glial scar formation could play a major role in creating the chronic problems that plague hydrocephalic patients. It has been suggested by many investigators that scar formation is a permanent fixture in hydrocephalic brains [55, 56].

Previous studies in both congenital and acquired models of hydrocephalus have shown that glial fibrillary acidic protein (GFAP) RNA and protein levels increase with the progression of hydrocephalus. Additionally, Mangano and colleague [55] illustrated that microglial cell proliferation and activation increased in regions distant from the cortical "lesion," suggesting that neuroinflammation is related to damage throughout the cortical pathways. Experimental models of hydrocephalus have demonstrated that shunting can reduce the amount of GFAP and RNA present in the cerebral cortex, but these levels begin to rise over time [57], suggesting that reactive astrocytosis is highly sensitive to suboptimal CSF drainage. Clinically, increased levels of GFAP have been found in the CSF of patients with normal-pressure hydrocephalus and those who developed secondary hydrocephalus due to subarachnoid hemorrhage. It is likely that gliosis may dramatically change the mechanical properties of the brain so that it becomes more rigid (less compliant) and resistant to increases in CSF pressure and flow.

Most recently, an interesting and potentially powerful relationship has been suggested between astrocytes and aquaporin (AQP) channels, which can have major impact on CSF absorption. Two AQPs in the central nervous system, AQP1 and AQP4, might play a role in hydrocephalus and are thus potential drug targets. AQP1 is expressed in the ventricular-facing membrane of choroid plexus epithelial cells, where it facilitates the secretion of cerebrospinal fluid (CSF). AQP4 is expressed in astrocyte foot processes and ependymal cells lining ventricles, where it appears to facilitate the transport of excess water out of the brain. Altered expression of these AQPs in experimental animal models of hydrocephalus and limited human specimens suggests their involvement in the pathophysiology of hydrocephalus, as do data in knockout mice demonstrating a protective effect of AQP1 deletion and a deleterious effect of AQP4 deletion in hydrocephalus [58]. Expression of AQP4 (but not AQP1 or AQP9) increases in the cerebral cortex and hippocampus of rats with mechanically induced hydrocephalus, suggesting that this water channel plays a compensatory role in CSF absorption during ventriculomegaly. One structural response of endothelial cells to chronic hypoperfusion is an abnormal basal lamina [59], so it is not surprising that aquaporin changes have been reported in experimental hydrocephalus.

3.5.2 Intracranial Pulsatility

Di Rocco et al. [60] have shown that ventricular dilation can be caused by abnormal intracranial pulsatility in the absence of a physical obstruction to CSF flow. Intraventricular pulsatility, detected as CSF pulsations in the aqueduct on cine MRI, is often dramatically elevated in hydrocephalus [61] and can be a useful diagnostic criterion in normal-pressure hydrocephalus. Although authors have posited theories on the cause of these elevated pulsations [61, 63, 64], no study has provided a clear link between abnormal pulsations and the underlying pathophysiology of hydrocephalus. Furthermore, the dissipation of cerebral arterial pulsatility, resulting in minimal (homeostatic) capillary and venous pulse pressure, is believed to be critical for normal cerebrovascular function [63]. The major arteries entering the cranium are located within the CSF-filled subarachnoid spaces, allowing the efficient coupling and transfer of pulsation. Pulsations can be transferred out of the cranium either directly via CSF flow through the foramen magnum into the compliant spinal subarachnoid space or indirectly via coupling to the sagittal sinus within the convexity. An important component of the pathophysiology of hydrocephalus is a change in intracranial compliance, which may lead to a redistribution of the pulsation dissipation mechanism [62, 63]. Increased capillary pulsatility may have several pathophysiologic consequences. Structural responses may lead to the loss of parenchymal microvessels, and in fact, decreased capillary density has been shown in experimental hydrocephalus [65, 66]. In addition, the integrity of the blood–brain barrier is compromised in hydrocephalus [67, 68]. Both these effects might explain the marked loss of cerebral perfusion that has been well documented in clinical [69–71] and experimental hydrocephalus [56, 72]. Importantly, it was recently shown that excessive pulsatile stress can impair hemodynamics through changes in endothelial cell homeostasis [73, 74] mediated by nitric oxide [75]. A marked increase in nitric oxide synthase immunoreactivity has also been reported in a kaolin hydrocephalus rat model during the first few weeks of ventriculomegaly [76]. These examples all provide a compelling case for the importance of pulsation dissipation in maintaining normal capillary function.

3.5.3   Cerebrospinal Fluid Absorption: Lymphatic, Arachnoid, and Microvascular

The traditional view of CSF absorption exclusively via arachnoid granulations into the superior sagittal sinus or the veins adjacent to the spinal roots has been challenged by a series of experiments [77–79]. Initially, the results of these experiments were challenged because they were performed on animals or human cadavers and consisted of latex Microfil infusions to reveal potential CSF pathways from the subarachnoid space. However, the latest in vivo studies clearly demonstrate that cranial lymphatics, most notably the nasal pathways surrounding the olfactory nerves, are capable of transporting large volumes of CSF. Most importantly, these pathways are impaired in adult rats with communicating hydrocephalus.

3.6   Commonly Encountered Clinical Conditions

3.6.1   Infantile Post-hemorrhagic Hydrocephalus

The incidence of IVH increases inversely with decreasing birth weight or EGA. In a study of infants born in the mid-1990s weighing less than 1500 g, 22% had IVH, and one-fourth of those infants had progressive ventricular dilation. However, one-third of those who survived required CSF diversion [80]. The germinal matrix is a transient structure in the subependyma of the ventricular walls that generates neural cell progenitors for the overlying cortex, primarily from 8 to 28 weeks of gestation. Involution of the germinal matrix is complete by 32 weeks of gestation. The proliferative germinal matrix is a metabolically active area with friable, immature vessels prone to hemorrhage.

Multiple factors contribute to the germinal matrix's propensity to hemorrhage, and the pathophysiology is incompletely understood. Preterm infants have impaired cerebral autoregulation due to their immaturity. There is a transfer of systemic fluctuations in blood volume, flow, and pressure that commonly occur in preterm infants to the friable germinal matrix, and there is no buffer from standard autoregulation. Low levels of fibronectin and other extracellular matrix also likely contributes to the propensity to hemorrhage [81]. After IVH occurs, the absorption of cerebrospinal fluid (CSF) is impaired, likely caused by a combination of decreased absorption across the arachnoid villi into veins and from ependyma into parenchyma [82].

3.6.2 Post-infectious Hydrocephalus

TORCH infections (toxoplasma, others (varicella, HIV, syphilis), rubella, cytomega-lovirus, and herpes simplex) are responsible for antenatal brain infections and may present with a wide range of consequences—from asymptomatic to having a fatal outcome. The parasites invade and destroy the ependyma of lateral ventricles, and the debris fall into the ventricle and cause obstruction of CSF flow pathway [83]. The other widely accepted hypothesis is that the leptomeningeal inflammation is the main cause of hydrocephalus in such patients. In bacterial meningitis, neutrophil migration into the subarachnoid space follows bacterial invasion. The resultant puru-lent exudate tends to collect in the Rolandic and Sylvanian sulci over the cerebral hemispheres and in the basal cisterns, where the subarachnoid space is deepest and where, presumably, cerebrospinal fluid flow is most sluggish. The exudate interferes with absorption of cerebrospinal fluid by the arachnoid villi and may also cause obstructive hydrocephalus by obstructing the foramina of Luschka and Magendie. Typically, the obstruction occurs toward the end of the second week of the illness, when neutrophils begin to degenerate and fibroblasts proliferate in the exudate [84]. In tubercular meningitis, hydrocephalus occurs because of the inflammatory exudate occupying the subarachnoid spaces or the ventricular pathways. In the earlier stages of the disease, the thick gelatinous exudates block the subarachnoid spaces in the base of the brain (notably the interpeduncular and ambient cisterns) leading to hydrocephalus. The exudates lead to a dense scarring of the subarachnoid spaces in the later stages of the disease thereby leading to hydrocephalus. Hydrocephalus can also result from the exudates blocking the arachnoid granulations which prevent the absorption of cerebrospinal fluid (CSF) [85]. The inflammation of the choroid plexus and ependyma also leads to an overproduction of CSF in the acute phase of the ill-ness. This also contributes to the hydrocephalus and raised intracranial pressure. Fourth ventricular outlets may be blocked by the exudates or leptomeningeal scar tissue or when there is obstruction of the aqueduct either due to a strangulation of the brain stem by exudates or by a subependymal tuberculoma thereby resulting in hydrocephalus [85]. The etiology of hydrocephalus in NCC is multifactorial. This includes the presence of intraventricular cysts, racemose cysts in the basal subarach-noid cisterns, obstruction at the outlet of the fourth ventricle due to basal cysticer-cotic meningitis, and fibrosis or ependymitis even in the absence of cysts [86].

3.6.3 Post-traumatic Hydrocephalus (PTH)

Ventriculomegaly is common after severe TBI [87]. Post-traumatic ventricular dila-tion may have a wide range of etiological factors: starting from neuronal loss due to head trauma and possible secondary ischemic insults to obstruction of CSF circula-tion resulting in hydrocephalus. It is important to differentiate between post-traumatic hydrocephalus and gliosis. However, the diagnosis is not always easy [88].

PTH may develop acutely in the presence of subarachnoid hemorrhage. Aseptic meningitis may lead to occlusion of basal cisterns. Similarly, meningitis following head injury can lead to communicating hydrocephalus. Hydrocephalus may be seen acutely in the presence of intracerebral hematoma causing obstruction to CSF flow by mass effect. Typically, PTH develops in the post-acute phase of head injury, weeks or months later. Development of PTH is presumed to be due to subarachnoid hemorrhage. Subarachnoid spaces in patients after head injury and hydrocephalus are obliterated with fibrosis; ependymal destruction and presence of subependymal gliosis together with loss of white matter especially around the ventricles are other prominent findings [89]. Experimental studies have shown that there is an increase in the outflow resistance and CSF pressure in rats following subarachnoid infusion of plasma or whole blood, and this is due to the decrease in transport of fluid across the epithelium of arachnoid villi. However, direct evidence of obstruction to CSF pathways in man is sparse. In patients dying soon after subarachnoid hemorrhage, fibrin is rarely seen in the drainage channel within arachnoid granulations. The degree to which fibrosis and obliteration of the subarachnoid granulations/subarachnoid space are contributory in the late onset of PTH is unknown [90].

Studies suggested that hydrocephalus following decompressive craniectomy may be related to a reduction in pulsatile CSF flow [91]. The effect of the skull and dura on CSF hydrodynamics has been explored experimentally: the resistance to CSF outflow after craniectomy decreases twofold and brain compliance (expressed using the pressure–volume index, PVI) increases [92]. Waziri et al. reported that DC may play a dampening role in the dicrotic CSF pulse wave in patients undergoing DC because of transmission of a pressure pulse out through the cranial defect. Arachnoid granulation function is dependent on the pressure difference between the subarachnoid space and draining venous supply. The disruption of pulsatile intracranial pressure (ICP) dynamics secondary to craniectomy results in altered CSF outflow and absorption, leading to hydrocephalus [93] (Fig. 3.5).

Some studies have suggested that altered venous outflow as a result of craniectomy close to midline is associated with post-traumatic hydrocephalus [94]. Age, severity of injury, CSF infection, intraventricular hemorrhage, the presence of subarachnoid blood, and development of subdural hygromas have also been reported as predisposing factors for post-traumatic hydrocephalus [88, 89, 95–97]. However, the risk factors for hydrocephalus following decompressive craniectomy remain controversial, and no consensus has been reached at present [89, 98].

3.6.4 Congenital Hydrocephalus

Congenital and neonatal hydrocephalus can be caused by a wide variety of developmental abnormalities or insults; the primary culprits are neural tube defects, infection, intraventricular hemorrhage, trauma, and tumors. The pathophysiology of congenital hydrocephalus almost always includes two separate mechanisms: primary genetic abnormalities that may affect outcome individually and secondary injury mechanisms that occur mainly as a result of expanding ventricles and/or altered CSF physiology [99]. Forty percent of hydrocephalus cases are estimated to have possible genetic etiology [100]. In humans, X-linked hydrocephalus (HSAS1, OMIM) comprises approximately 5–15% of the congenital cases with a genetic cause [101, 102]. L1CAM (L1 protein) at Xq28 has been implicated in X-linked human congenital hydrocephalus. Recently, an X-linked adult-onset NPH [103] and a form of familial NPH that is transmitted in autosomal dominant fashion [104]

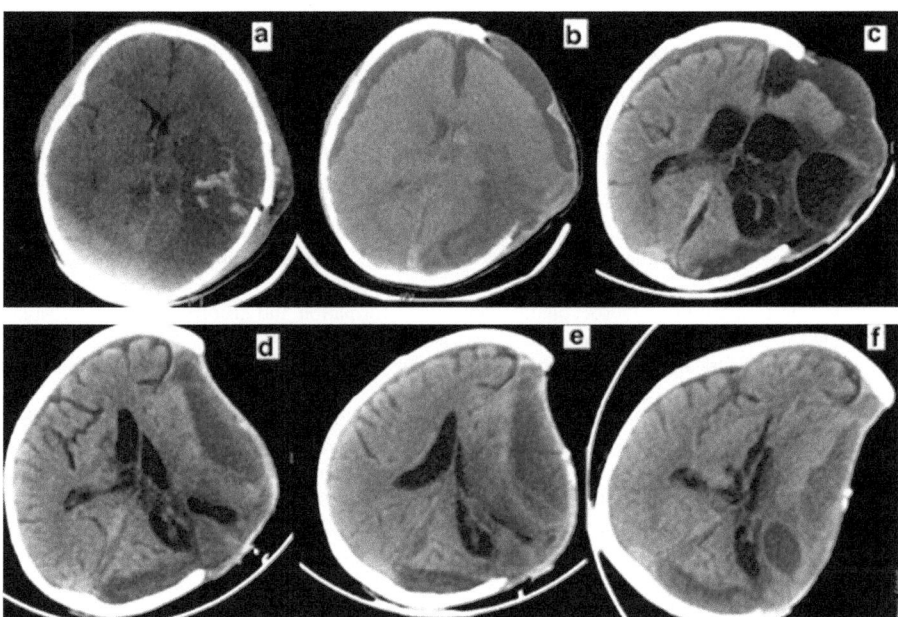

Fig. 3.5 (**a-f**) show serial scans of a 4-year-old male child who had a left parietotemporal contusion following a road traffic accident. He was subject to left FTP decompressive craniectomy. Later, went on to develop subdural hygroma and hydrocephalus that improved following theco-peritoneal shunt

have been reported. However, the gene locus for the same has not been identified. Human hydrocephalus genetic research has lags far behind animal models. Only 1 hydrocephalus gene has been identified in humans out of 43 mutant genes of hydrocephalus identified [99]. The pathological studies of hydrocephalus clearly suggest that impaired and abnormal brain development in the early development stage caused by altered neural cell formation and perturbed regulation of cellular proliferation and apoptosis lead to hydrocephalus [105–107].

On a cellular level, a number of mechanisms have been implicated in pathophysiology of congenital hydrocephalus. These include disruption of the cell membrane proteins such a L1 protein, malfunction of ependymal cells, malfunction of mesenchymal cells, abnormalities in growth factor signaling, and disruption of the extracellular matrix [27, 108–114]. Any of these disturbances could lead to a sequential alteration of CSF dynamics leading to congenital hydrocephalus.

References

1. Rekate HL. The definition and classification of hydrocephalus: a personal recommendation to stimulate debate. Cerebrospinal Fluid Res. 2008;5:2.
2. Dandy WE. Experimental hydrocephalus. Ann Surg. 1919;70(2):129–42.
3. Brown PD, Davies SL, Speake T, Millar ID. Molecular mechanisms of cerebrospinal fluid production. Neuroscience. 2004;129(4):957–70.
4. Pollay M, Curl F. Secretion of cerebrospinal fluid by the ventricular ependyma of the rabbit. Am J Phys. 1967;213(4):1031–8.

5. Alksne JF, Lovings ET. The role of the arachnoid villus in the removal of red blood cells from the subarachnoid space. An electron microscope study in the dog. J Neurosurg. 1972;36(2):192–200.
6. Weed LH. Studies on cerebro-spinal fluid. No. III: the pathways of escape from the subarachnoid spaces with particular reference to the Arachnoid Villi. J Med Res. 1914;31(1):51–91.
7. Brodbelt A, Stoodley M. CSF pathways: a review. Br J Neurosurg. 2007;21(5):510–20.
8. Johanson CE, Duncan JA, Klinge PM, Brinker T, Stopa EG, Silverberg GD. Multiplicity of cerebrospinal fluid functions: new challenges in health and disease. Cerebrospinal Fluid Res. 2008;5:10.
9. Bering EA. Circulation of the cerebrospinal fluid. Demonstration of the choroid plexuses as the generator of the force for flow of fluid and ventricular enlargement. J Neurosurg. 1962;19:405–13.
10. Bering EA, Sato O. Hydrocephalus: changes in formation and absorption of cerebrospinal fluid within the cerebral ventricles. J Neurosurg. 1963;20:1050–63.
11. Bulat M, Klarica M. Recent insights into a new hydrodynamics of the cerebrospinal fluid. Brain Res Rev. 2011;65(2):99–112.
12. Orešković D, Klarica M. Development of hydrocephalus and classical hypothesis of cerebrospinal fluid hydrodynamics: facts and illusions. Prog Neurobiol. 2011;94(3):238–58.
13. Crone C. The permeability of capillaries in various organs as determined by use of the "indicator diffusion" method. Acta Physiol Scand. 1963;58:292–305.
14. Raichle ME. Neurogenic control of blood-brain barrier permeability. Acta Neuropathol Suppl. 1983;8:75–9.
15. Pudenz RH. The surgical treatment of hydrocephalus—an historical review. Surg Neurol. 1981;15(1):15–26.
16. Ransohoff J, Shulman K, Fishman RA. Hydrocephalus: a review of etiology and treatment. J Pediatr. 1960;56:399–411.
17. Johanson C, Del Bigio M, Kinsman S, Miyan J, Pattisapu J, Robinson M, et al. New models for analysing hydrocephalus and disorders of CSF volume transmission. Br J Neurosurg. 2001;15(3):281–3.
18. Kohn DF, Chinookoswong N, Chou SM. A new model of congenital hydrocephalus in the rat. Acta Neuropathol (Berl). 1981;54(3):211–8.
19. Yamada H, Oi SZ, Tamaki N, Matsumoto S, Sudo K. Prenatal aqueductal stenosis as a cause of congenital hydrocephalus in the inbred rat LEW/Jms. Childs Nerv Syst. 1991;7(4):218–22.
20. Jones HC, Yehia B, Chen G-F, Carter BJ. Genetic analysis of inherited hydrocephalus in a rat model. Exp Neurol. 2004;190(1):79–90.
21. Jones HC, Dack S, Ellis C. Morphological aspects of the development of hydrocephalus in a mouse mutant (SUMS/NP). Acta Neuropathol (Berl). 1987;72(3):268–76.
22. Raimondi AJ, Bailey OT, McLone DG, Lawson RF, Echeverry A. The pathophysiology and morphology of murine hydrocephalus in Hy-3 and Ch mutants. Surg Neurol. 1973;1(1):50–5.
23. Aliev G, Miller JP, Leifer DW, Obrenovich ME, Shenk JC, Smith MA, et al. Ultrastructural analysis of a murine model of congenital hydrocephalus produced by overexpression of transforming growth factor-beta1 in the central nervous system. J Submicrosc Cytol Pathol. 2006;38(2–3):85–91.
24. Wyss-Coray T, Feng L, Masliah E, Ruppe MD, Lee HS, Toggas SM, et al. Increased central nervous system production of extracellular matrix components and development of hydrocephalus in transgenic mice overexpressing transforming growth factor-beta 1. Am J Pathol. 1995;147(1):53–67.
25. Tada T, Kanaji M, Kobayashi S. Induction of communicating hydrocephalus in mice by intrathecal injection of human recombinant transforming growth factor-beta 1. J Neuroimmunol. 1994;50(2):153–8.
26. Wagner C, Batiz LF, Rodríguez S, Jiménez AJ, Páez P, Tomé M, et al. Cellular mechanisms involved in the stenosis and obliteration of the cerebral aqueduct of hyh mutant mice developing congenital hydrocephalus. J Neuropathol Exp Neurol. 2003;62(10):1019–40.
27. Jiménez AJ, Tomé M, Páez P, Wagner C, Rodríguez S, Fernández-Llebrez P, et al. A programmed ependymal denudation precedes congenital hydrocephalus in the hyh mutant mouse. J Neuropathol Exp Neurol. 2001;60(11):1105–19.

28. Bátiz LF, Páez P, Jiménez AJ, Rodríguez S, Wagner C, Pérez-Fígares JM, et al. Heterogeneous expression of hydrocephalic phenotype in the hyh mice carrying a point mutation in alpha-SNAP. Neurobiol Dis. 2006;23(1):152–68.
29. Kamiguchi H, Hlavin ML, Lemmon V. Role of L1 in neural development: what the knockouts tell us. Mol Cell Neurosci. 1998;12(1–2):48–55.
30. Bloch O, Auguste KI, Manley GT, Verkman AS. Accelerated progression of kaolin-induced hydrocephalus in aquaporin-4-deficient mice. J Cereb Blood Flow Metab. 2006;26(12): 1527–37.
31. Bryan JH. The immotile cilia syndrome. Mice versus man. Virchows Arch A Pathol Anat Histopathol. 1983;399(3):265–75.
32. Topczewska JM, Topczewski J, Solnica-Krezel L, Hogan BL. Sequence and expression of zebrafish foxc1a and foxc1b, encoding conserved forkhead/winged helix transcription factors. Mech Dev. 2001;100(2):343–7.
33. Blackshear PJ, Graves JP, Stumpo DJ, Cobos I, Rubenstein JLR, Zeldin DC. Graded phenotypic response to partial and complete deficiency of a brain-specific transcript variant of the winged helix transcription factor RFX4. Development (Cambridge, England). 2003;130(19):4539–52.
34. Utriainen A, Sormunen R, Kettunen M, Carvalhaes LS, Sajanti E, Eklund L, et al. Structurally altered basement membranes and hydrocephalus in a type XVIII collagen deficient mouse line. Hum Mol Genet. 2004;13(18):2089–99.
35. Sweger EJ, Casper KB, Scearce-Levie K, Conklin BR, McCarthy KD. Development of hydrocephalus in mice expressing the G(i)-coupled GPCR Ro1 RASSL receptor in astrocytes. J Neurosci. 2007;27(9):2309–17.
36. Del Bigio MR, Bruni JE. Cerebral water content in silicone oil-induced hydrocephalic rabbits. Pediatr Neurosci. 1987;13(2):72–7.
37. Cosan TE, Gucuyener D, Dundar E, Arslantas A, Vural M, Uzuner K, et al. Cerebral blood flow alterations in progressive communicating hydrocephalus: transcranial Doppler ultrasonography assessment in an experimental model. J Neurosurg. 2001;94(2):265–9.
38. Cosan TE, Guner AI, Akcar N, Uzuner K, Tel E. Progressive ventricular enlargement in the absence of high ventricular pressure in an experimental neonatal rat model. Childs Nerv Syst. 2002;18(1–2):10–4.
39. Deo-Narine V, Gomez DG, Vullo T, Manzo RP, Zimmerman RD, Deck MD, et al. Direct in vivo observation of transventricular absorption in the hydrocephalic dog using magnetic resonance imaging. Investig Radiol. 1994;29(3):287–93.
40. James AE, Burns B, Flor WF, Strecker EP, Merz T, Bush M, et al. Pathophysiology of chronic communicating hydrocephalus in dogs (Canis familiaris). Experimental studies. J Neurol Sci. 1975;24(2):151–78.
41. Diggs J, Price AC, Burt AM, Flor WJ, McKanna JA, Novak GR, et al. Early changes in experimental hydrocephalus. Investig Radiol. 1986;21(2):118–21.
42. Li J, McAllister JP, Shen Y, Wagshul ME, Miller JM, Egnor MR, et al. Communicating hydrocephalus in adult rats with kaolin obstruction of the basal cisterns or the cortical subarachnoid space. Exp Neurol. 2008;211(2):351–61.
43. Davis LE. Communicating hydrocephalus in newborn hamsters and cats following vaccinia virus infection. J Neurosurg. 1981;54(6):767–72.
44. Wiesmann M, Koedel U, Brückmann H, Pfister HW. Experimental bacterial meningitis in rats: demonstration of hydrocephalus and meningeal enhancement by magnetic resonance imaging. Neurol Res. 2002;24(3):307–10.
45. Brandt CT, Simonsen H, Liptrot M, Søgaard LV, Lundgren JD, Ostergaard C, et al. In vivo study of experimental pneumococcal meningitis using magnetic resonance imaging. BMC Med Imaging. 2008;8:1.
46. Fiori MG, Sharer LR, Lowndes HE. Communicating hydrocephalus in rodents treated with beta,beta'-iminodipropionitrile (IDPN). Acta Neuropathol (Berl). 1985;65(3–4):209–16.
47. Tenti G, Drake JM, Sivaloganathan S. Brain biomechanics: mathematical modeling of hydrocephalus. Neurol Res. 2000;22(1):19–24.
48. Dutta-Roy T, Wittek A, Miller K. Biomechanical modelling of normal pressure hydrocephalus. J Biomech. 2008;41(10):2263–71.

49. Miller K, Chinzei K. Mechanical properties of brain tissue in tension. J Biomech. 2002;35(4):483–90.
50. Davis GB, Kohandel M, Sivaloganathan S, Tenti G. The constitutive properties of the brain paraenchyma Part 2. Fractional derivative approach. Med Eng Phys. 2006;28(5):455–9.
51. Pang D, Altschuler E. Low-pressure hydrocephalic state and viscoelastic alterations in the brain. Neurosurgery. 1994;35(4):643–55; discussion 655–6.
52. Peña A, Bolton MD, Whitehouse H, Pickard JD. Effects of brain ventricular shape on periventricular biomechanics: a finite-element analysis. Neurosurgery. 1999;45(1):107–16; discussion 116–8.
53. Muthupillai R, Lomas DJ, Rossman PJ, Greenleaf JF, Manduca A, Ehman RL. Magnetic resonance elastography by direct visualization of propagating acoustic strain waves. Science. 1995;269(5232):1854–7.
54. Yin M, Rouvière O, Glaser KJ, Ehman RL. Diffraction-biased shear wave fields generated with longitudinal magnetic resonance elastography drivers. Magn Reson Imaging. 2008;26(6):770–80.
55. Mangano FT, McAllister JP, Jones HC, Johnson MJ, Kriebel RM. The microglial response to progressive hydrocephalus in a model of inherited aqueductal stenosis. Neurol Res. 1998;20(8):697–704.
56. da Silva MC, Michowicz S, Drake JM, Chumas PD, Tuor UI. Reduced local cerebral blood flow in periventricular white matter in experimental neonatal hydrocephalus-restoration with CSF shunting. J Cereb Blood Flow Metab. 1995;15(6):1057–65.
57. Miller JM, McAllister JP. Reduction of astrogliosis and microgliosis by cerebrospinal fluid shunting in experimental hydrocephalus. Cerebrospinal Fluid Res. 2007;4:5.
58. Verkman AS, Tradtrantip L, Smith AJ, Yao X. Aquaporin water channels and hydrocephalus. Pediatr Neurosurg. 2016. doi:10.1159/000452168.
59. Mao X, Enno TL, Del Bigio MR. Aquaporin 4 changes in rat brain with severe hydrocephalus. Eur J Neurosci. 2006;23(11):2929–36.
60. Di Rocco C, Di Trapani G, Pettorossi VE, Caldarelli M. On the pathology of experimental hydrocephalus induced by artificial increase in endoventricular CSF pulse pressure. Childs Brain. 1979;5(2):81–95.
61. Bradley WG, Scalzo D, Queralt J, Nitz WN, Atkinson DJ, Wong P. Normal-pressure hydrocephalus: evaluation with cerebrospinal fluid flow measurements at MR imaging. Radiology. 1996;198(2):523–9.
62. Greitz D. Radiological assessment of hydrocephalus: new theories and implications for therapy. Neurosurg Rev. 2004;27(3):145–65; discussion 166–7.
63. Egnor M, Zheng L, Rosiello A, Gutman F, Davis R. A model of pulsations in communicating hydrocephalus. Pediatr Neurosurg. 2002;36(6):281–303.
64. Wagshul ME, Chen JJ, Egnor MR, McCormack EJ, Roche PE. Amplitude and phase of cerebrospinal fluid pulsations: experimental studies and review of the literature. J Neurosurg. 2006;104(5):810–9.
65. Del Bigio MR, Bruni JE. Changes in periventricular vasculature of rabbit brain following induction of hydrocephalus and after shunting. J Neurosurg. 1988;69(1):115–20.
66. Jones HC, Bucknall RM, Harris NG. The cerebral cortex in congenital hydrocephalus in the H-Tx rat: a quantitative light microscopy study. Acta Neuropathol (Berl). 1991;82(3):217–24.
67. Castejon OJ. Transmission electron microscope study of human hydrocephalic cerebral cortex. J Submicrosc Cytol Pathol. 1994;26(1):29–39.
68. Nakagawa Y, Cervós-Navarro J, Artigas J. Tracer study on a paracellular route in experimental hydrocephalus. Acta Neuropathol (Berl). 1985;65(3–4):247–54.
69. Nakamura T, Xi G, Hua Y, Schallert T, Hoff JT, Keep RF. Intracerebral hemorrhage in mice: model characterization and application for genetically modified mice. J Cereb Blood Flow Metab. 2004;24(5):487–94.
70. Brooks DJ, Beaney RP, Powell M, Leenders KL, Crockard HA, Thomas DG, et al. Studies on cerebral oxygen metabolism, blood flow, and blood volume, in patients with hydrocephalus before and after surgical decompression, using positron emission tomography. Brain J Neurol. 1986;109(Pt 4):613–28.
71. Chang C-C, Kuwana N, Ito S, Yokoyama T, Kanno H, Yamamoto I. Cerebral haemodynamics in patients with hydrocephalus after subarachnoid haemorrhage due to ruptured aneurysm. Eur J Nucl Med Mol Imaging. 2003;30(1):123–6.

72. Dombrowski SM, Schenk S, Leichliter A, Leibson Z, Fukamachi K, Luciano MG. Chronic hydrocephalus-induced changes in cerebral blood flow: mediation through cardiac effects. J Cereb Blood Flow Metab. 2006;26(10):1298–310.
73. Ziegler T, Bouzourène K, Harrison VJ, Brunner HR, Hayoz D. Influence of oscillatory and unidirectional flow environments on the expression of endothelin and nitric oxide synthase in cultured endothelial cells. Arterioscler Thromb Vasc Biol. 1998;18(5):686–92.
74. Rubanyi GM, Freay AD, Kauser K, Johns A, Harder DR. Mechanoreception by the endothelium: mediators and mechanisms of pressure- and flow-induced vascular responses. Blood Vessels. 1990;27(2–5):246–57.
75. Bilfinger TV, Stefano GB. Human aortocoronary grafts and nitric oxide release: relationship to pulsatile pressure. Ann Thorac Surg. 2000;69(2):480–5.
76. Klinge PM, Samii A, Mühlendyck A, Visnyei K, Meyer G-J, Walter GF, et al. Cerebral hypoperfusion and delayed hippocampal response after induction of adult kaolin hydrocephalus. Stroke J Cereb Circ. 2003;34(1):193–9.
77. Boulton M, Flessner M, Armstrong D, Hay J, Johnston M. Lymphatic drainage of the CNS: effects of lymphatic diversion/ligation on CSF protein transport to plasma. Am J Phys. 1997;272(5 Pt 2):R1613–9.
78. Koh L, Zakharov A, Johnston M. Integration of the subarachnoid space and lymphatics: is it time to embrace a new concept of cerebrospinal fluid absorption? Cerebrospinal Fluid Res. 2005;2:6.
79. Nagra G, Koh L, Zakharov A, Armstrong D, Johnston M. Quantification of cerebrospinal fluid transport across the cribriform plate into lymphatics in rats. Am J Physiol Regul Integr Comp Physiol. 2006;291(5):R1383–9.
80. Murphy BP, Inder TE, Rooks V, Taylor GA, Anderson NJ, Mogridge N, et al. Posthaemorrhagic ventricular dilatation in the premature infant: natural history and predictors of outcome. Arch Dis Child Fetal Neonatal Ed. 2002;87(1):F37–41.
81. Xu H, Hu F, Sado Y, Ninomiya Y, Borza D-B, Ungvari Z, et al. Maturational changes in laminin, fibronectin, collagen IV, and perlecan in germinal matrix, cortex, and white matter and effect of betamethasone. J Neurosci Res. 2008;86(7):1482–500.
82. Cherian S, Whitelaw A, Thoresen M, Love S. The pathogenesis of neonatal post-hemorrhagic hydrocephalus. Brain Pathol Zurich Switz. 2004;14(3):305–11.
83. Olariu TR, Remington JS, McLeod R, Alam A, Montoya JG. Severe congenital toxoplasmosis in the United States: clinical and serologic findings in untreated infants. Pediatr Infect Dis J. 2011;30(12):1056–61.
84. Edmond K, Clark A, Korczak VS, Sanderson C, Griffiths UK, Rudan I. Global and regional risk of disabling sequelae from bacterial meningitis: a systematic review and meta-analysis. Lancet Infect Dis. 2010;10(5):317–28.
85. Dastur DK, Manghani DK, Udani PM. Pathology and pathogenetic mechanisms in neurotuberculosis. Radiol Clin N Am. 1995;33(4):733–52.
86. Sinha S, Sharma BS. Intraventricular neurocysticercosis: a review of current status and management issues. Br J Neurosurg. 2012;26(3):305–9.
87. Poca MA, Sahuquillo J, Mataró M, Benejam B, Arikan F, Báguena M. Ventricular enlargement after moderate or severe head injury: a frequent and neglected problem. J Neurotrauma. 2005;22(11):1303–10.
88. Licata C, Cristofori L, Gambin R, Vivenza C, Turazzi S. Post-traumatic hydrocephalus. J Neurosurg Sci. 2001;45(3):141–9.
89. Honeybul S, Ho KM. Incidence and risk factors for post-traumatic hydrocephalus following decompressive craniectomy for intractable intracranial hypertension and evacuation of mass lesions. J Neurotrauma. 2012;29(10):1872–8.
90. Modi NJ, Agrawal M, Sinha VD. Post-traumatic subarachnoid hemorrhage: a review. Neurol India. 2016;64(Suppl):S8–13.
91. Kaen A, Jimenez-Roldan L, Alday R, Gomez PA, Lagares A, Alén JF, et al. Interhemispheric hygroma after decompressive craniectomy: does it predict posttraumatic hydrocephalus? J Neurosurg. 2010;113(6):1287–93.
92. Czosnyka M, Whitehouse H, Smielewski P, Simac S, Pickard JD. Testing of cerebrospinal compensatory reserve in shunted and non-shunted patients: a guide to interpretation based on an observational study. J Neurol Neurosurg Psychiatry. 1996;60(5):549–58.

93. Waziri A, Fusco D, Mayer SA, McKhann GM, Connolly ES. Postoperative hydrocephalus in patients undergoing decompressive hemicraniectomy for ischemic or hemorrhagic stroke. Neurosurgery. 2007;61(3):489–93; discussion 493–4.
94. De Bonis P, Pompucci A, Mangiola A, Rigante L, Anile C. Post-traumatic hydrocephalus after decompressive craniectomy: an underestimated risk factor. J Neurotrauma. 2010;27(11): 1965–70.
95. Tian H-L, Xu T, Hu J, Cui Y, Chen H, Zhou L-F. Risk factors related to hydrocephalus after traumatic subarachnoid hemorrhage. Surg Neurol. 2008;69(3):241–6; discussion 246.
96. Yuan Q, Wu X, Yu J, Sun Y, Li Z, Du Z, et al. Subdural hygroma following decompressive craniectomy or non-decompressive craniectomy in patients with traumatic brain injury: clinical features and risk factors. Brain Inj. 2015;29(7–8):971–80.
97. Ki HJ, Lee H-J, Lee H-J, Yi J-S, Yang J-H, Lee I-W. The risk factors for hydrocephalus and subdural hygroma after decompressive craniectomy in head injured patients. J Korean Neurosurg Soc. 2015;58(3):254–61.
98. Takeuchi S, Nawashiro H, Wada K, Takasato Y, Masaoka H, Hayakawa T, et al. Ventriculomegaly after decompressive craniectomy with hematoma evacuation for large hemispheric hypertensive intracerebral hemorrhage. Clin Neurol Neurosurg. 2013;115(3):317–22.
99. Zhang J, Williams MA, Rigamonti D. Genetics of human hydrocephalus. J Neurol. 2006;253(10):1255–66.
100. Haverkamp F, Wölfle J, Aretz M, Krämer A, Höhmann B, Fahnenstich H, et al. Congenital hydrocephalus internus and aqueduct stenosis: aetiology and implications for genetic counselling. Eur J Pediatr. 1999;158(6):474–8.
101. Halliday J, Chow CW, Wallace D, Danks DM. X linked hydrocephalus: a survey of a 20 year period in Victoria, Australia. J Med Genet. 1986;23(1):23–31.
102. Kuzniecky RI, Watters GV, Watters L, Meagher-Villemure K. X-linked hydrocephalus. Can J Neurol Sci. 1986;13(4):344–6.
103. Katsuragi S, Teraoka K, Ikegami K, Amano K, Yamashita K, Ishizuka K, et al. Late onset X-linked hydrocephalus with normal cerebrospinal fluid pressure. Psychiatry Clin Neurosci. 2000;54(4):487–92.
104. Portenoy RK, Berger A, Gross E. Familial occurrence of idiopathic normal-pressure hydrocephalus. Arch Neurol. 1984;41(3):335–7.
105. Bruni JE, Del Bigio MR, Cardoso ER, Persaud TV. Hereditary hydrocephalus in laboratory animals and humans. Exp Pathol. 1988;35(4):239–46.
106. Ulfig N, Bohl J, Neudörfer F, Rezaie P. Brain macrophages and microglia in human fetal hydrocephalus. Brain Dev. 2004;26(5):307–15.
107. Mori F, Tanji K, Yoshida Y, Wakabayashi K. Thalamic retrograde degeneration in the congenitally hydrocephalic rat is attributable to apoptotic cell death. Neuropathology. 2002;22(3):186–93.
108. Okamoto N, Del Maestro R, Valero R, Monros E, Poo P, Kanemura Y, et al. Hydrocephalus and Hirschsprung's disease with a mutation of L1CAM. J Hum Genet. 2004;49(6):334–7.
109. Takano T, Rutka JT, Becker LE. Overexpression of nestin and vimentin in ependymal cells in hydrocephalus. Acta Neuropathol (Berl). 1996;92(1):90–7.
110. Fernández-Llebrez P, Grondona JM, Pérez J, López-Aranda MF, Estivill-Torrús G, Llebrez-Zayas PF, et al. Msx1-deficient mice fail to form prosomere 1 derivatives, subcommissural organ, and posterior commissure and develop hydrocephalus. J Neuropathol Exp Neurol. 2004;63(6):574–86.
111. Ohmiya M, Fukumitsu H, Nitta A, Nomoto H, Furukawa Y, Furukawa S. Administration of FGF-2 to embryonic mouse brain induces hydrocephalic brain morphology and aberrant differentiation of neurons in the postnatal cerebral cortex. J Neurosci Res. 2001;65(3):228–35.
112. Galbreath E, Kim SJ, Park K, Brenner M, Messing A. Overexpression of TGF-beta 1 in the central nervous system of transgenic mice results in hydrocephalus. J Neuropathol Exp Neurol. 1995;54(3):339–49.
113. Zechel J, Gohil H, Lust WD, Cohen A. Alterations in matrix metalloproteinase-9 levels & tissue inhibitor of matrix metalloproteinases-1 expression in a transforming growth factor-beta transgenic model of hydrocephalus. J Neurosci Res. 2002;69(5):662–8.
114. Crews L, Wyss-Coray T, Masliah E. Insights into the pathogenesis of hydrocephalus from transgenic and experimental animal models. Brain Pathol Zurich Switz. 2004;14(3):312–6.

Understanding Hydrocephalus: Genetic View

4

Mami Yamasaki

4.1 Introduction

We live in a time when molecular biology is being widely and effectively applied in every field of medicine. Technical advances have led to the development of new molecular mechanism-based target drugs for several diseases. In the field of pediatric neurosurgery, the causative genes for some diseases have been identified, and this information has enabled genetic diagnosis and improved disease classifications.

Diseases treated by pediatric neurosurgeons, including the various forms of hydrocephalus, can be classified according to the available molecular genetics knowledge base and its clinical applications, as follows. In the first group, genetic testing has been clinically established and is already used in disease diagnosis, disease classification, carrier detection, and prenatal diagnosis. In the second group, the molecular cause has not been firmly established, and clinical tools are not yet available, but some causative genes have been identified, and/or there are recent molecular findings regarding the pathophysiology and classification. In the third group, the molecular basis of the disease is still uncertain, and the clinical significance of current research findings has not been established.

Of hydrocephalus and its related diseases, X-linked hydrocephalus is categorized in the first group, holoprosencephaly and porencephaly in the second, and the Dandy-Walker syndrome and myelomeningocele in the third. In this paper, I will mention about X-linked hydrocephalus.

M. Yamasaki, M.D., Ph.D.
Department of Pediatric Neurosurgery, AIJINKAI Healthcare Corporation, Takatsuki General Hospital, 1-3-13 Kosobecho, Takatsuki City, Osaka 569-1192, Japan
e-mail: myamasaki@ajk.takatsuki-hp.or.jp

© Springer International Publishing AG 2017
A. Ammar (ed.), *Hydrocephalus*, DOI 10.1007/978-3-319-61304-8_4

4.2   X-Linked Hydrocephalus

X-linked hydrocephalus (XLH) was first described by Bickers and Adams in 1949, as HSAS, the acronym for *H*ydrocephalus due to *S*tenosis of the *A*queduct of *S*ylvius (MIM 307000) [1]. Since the first family of XLH carriers with a gene mutation in neural cell adhesion molecule L1CAM (L1) was reported in 1992 [2], there have been many advances in the genetics of XLH. L1 is a member of the immunoglobulin (Ig) superfamily of cell adhesion molecules and is expressed predominantly in developing neurons. Mutations in the L1 gene were found to be responsible for many cases of XLH, mental retardation, adducted thumbs, shuffling gait and aphasia (MASA) syndrome, certain forms of X-linked spastic paraplegia (SPG1), and X-linked agenesis of the corpus callosum (ACC). Therefore, these syndromes have been reclassified and grouped together as L1 syndrome [3].

The *L1* gene is located on the X chromosome in humans and is composed of 28 exons. The open reading frame has 3825 base pairs (bp) and encodes a protein of 1257 amino acids. According to an updated (May 2014) database of *L1* gene mutations (website maintained by Yvonne Vos from the Department of Genetics, University Medical Center Groningen, Groningen), 211 mutations have been found in 254 unrelated families with L1 syndrome [4–6]. In Japan, Kanemura et al. conducted a nationwide investigation of L1 gene mutations and identified *L1* mutations in 90 unrelated families [7].

The sites and types of the *L1* gene mutations in families with XLH are almost always different, regardless of race. Genotype and phenotype correlations have been reported [8, 9]. Yamasaki et al. revealed a striking correlation between the mutation class and the severity of ventricular dilatation. Class I mutations affect only the cytoplasmic domain (CD) of L1. Class II mutations consist of missense point mutations and deletions in the extracellular domain (ED) that result in a predicted protein that should remain associated with the plasma membrane. Class III mutations include nonsense or frameshift mutations that produce a premature stop codon in the L1ED. The mutant molecules in this class do not remain associated with the cell membrane and therefore lose all the normal functions of L1. Mutations in the noncoding region are divided into splice-site mutations and others. Splice-site mutations result in the same L1 protein structure as Class III mutations and cause loss of function [10].

Class II mutations can be divided into two subclasses based on molecular modeling studies. One subgroup includes mutations affecting the key residues in L1ED that are responsible for maintaining the conformation of the domains, and the other subgroup includes mutations that affect residues on the protein's surface [11]. Patients whose ventricles showed severe dilatation had Class III mutations, exon 1–26 splice-site mutations, or Class II L1ED key residue mutations. All of these mutations cause the loss of L1ED function and are therefore referred to as L1-LF mutations.

All of the patients with L1-LF mutations had severe ventricular dilatation (Fig. 4.1a) and required a VP shunt in the early days of life; however, most of these patients show severe developmental delays, including a lack of independent locomotion and an undetectably low IQ. To look at the characteristics of radiological

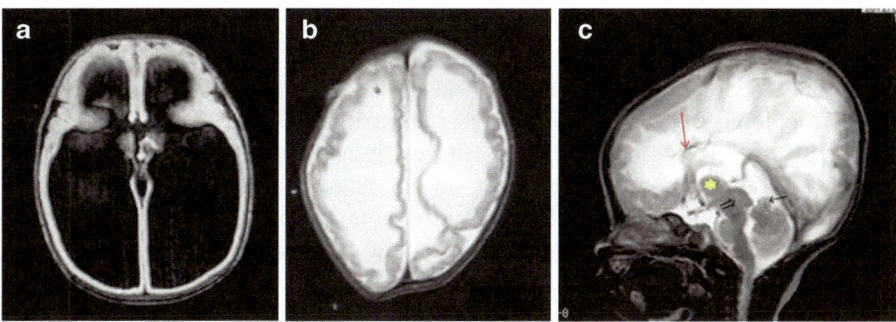

Fig. 4.1 X-linked hydrocephalus with *L1* gene mutation. (**a**) MRI (T1) shows severe ventricular dilatation. (**b**) MRI (T2) after VP shunt shows rippled ventricular wall. (**c**) MRI (T2) shows an enlarged massa intermedia (*star*), an anterior vermis hypoplasia (*leftward arrow*), and a large quadrigeminal plate (*rightward double arrow*), and callosal dysplasia was observed (*downward arrow*)

findings, they showed a rippled ventricular wall after shunting (Fig. 4.1b), which is not seen in other kinds of hydrocephalus and is an absolute characteristic of the severe hydrocephalus that accompanies L1-LF mutations. This unique neuroradiological finding can be used as one of the clinical diagnostic criteria of XLH with an L1-LF mutation.

Notably, most of the patients with L1-LF mutations also had an enlarged massa intermedia, enlarged quadrigeminal plate, and hypoplasia of the cerebellar vermis (Fig. 4.1c). These results are consistent with reported pathological findings of XLH, which include an enlarged masa intermedia or fused thalami (Fig. 4.1c). Hypoplasia of the cerebellar vermis has also been reported in XLH patients [12] and observed in L1-knockout mice [13].

There are two possible explanations for the ventricular dilatation caused by *L1* mutations. First, a decrease in elasticity in the white matter might increase the vulnerability of the ventricular system to alteration by cerebrospinal fluid (CSF) pressure. Second, maldevelopment of the midline structure might cause narrowing of the CSF pathway. Both mechanisms may be at work, and both are compatible with the loss of L1 functions, such as L1-mediated cell adhesion and cell migration, being responsible for the ventricular dilatation.

These characteristic radiological findings could be explained by a decrease in white matter elasticity, resulting from the loss of axons and axon adhesion, which could also explain the rippled appearance of the ventricular wall. In addition, maldevelopment of the midline structure could be induced by the disturbance of L1-mediated cell migration, which could cause the enlargement of the massa intermedia and quadrigeminal plate. In support of this mechanism, Kamiguchi et al. reported in vitro findings that neuronal migration from the subependymal zone in the songbird brain depends on a heterophilic interaction between neuronal L1 and a radial glial cell receptor [14]. The enlarged massa intermedia would cause narrowing of the third ventricle, and the enlarged quadrigeminal plate would cause narrowing of the aqueduct.

4.2.1   Application for Clinical Use of Molecular Biology in XLH

L1 genetic testing has been clinically established and is already used in prenatal and postnatal diagnosis of the disease and in carrier detection [15]. Renier reported that XLH shows the poorest outcome of any prenatal noncommunicating hydrocephalus [16]. For a mother carrying an L1 mutation, 50% of the male fetuses could have severe hydrocephalus; therefore, prenatal molecular genetic diagnosis with genetic counseling is extremely beneficial for a family with XLH. The prenatal molecular genetic diagnosis of L1 syndrome was first reported by Jouet and Kenwrick in 1995 [17]. Later, prenatal L1 gene analysis was systematically organized by Yamasaki et al. [15].

4.3   Subjects and Methods

4.3.1   Patients and Protocol for Prenatal Gene Testing

We conducted a nationwide L1 gene analysis and identified L1 mutations in 60 patients and 41 carriers in 56 families [18]. Genetic counselors informed women carrying L1 gene mutations of the guidelines for prenatal genetic analysis, and after a full explanation of the procedure, informed written consent was obtained from carriers who were pregnant. In 2004, new clinical guidelines for genetic testing were established by ten Japanese genetic-medicine-related societies (Guideline 2004). This guideline recommends that carriers be tested voluntarily with their informed consent, so carrier testing in childhood is prohibited. Therefore, since the release of Guideline 2004, only male fetuses have been tested (Fig. 4.2).

This study was approved by the Institutional Review Board of Osaka National Hospital and was carried out in accordance with the principles of the Declaration of Helsinki as well as the ethics guidelines for human genome/gene analysis research by the Ministry of Education, Culture, Science, and Technology; the Ministry of Health, Labor, and Welfare; and the Ministry of Economy, Trade, and Industry of Japan.

4.3.2   DNA Sampling and Karyotype Testing

Fetal tissues or cells were isolated by chorionic villus sampling (CVS) between 10 and 12 weeks gestation and amniocentesis (AC) from 15 to 16 weeks gestation. Fetal cells were cultured in vitro, and their karyotypes examined by standard techniques. DNA was then extracted from cultured fetal cells (Fig. 4.2).

Fig. 4.2 Protocol for prenatal L1 mutation analysis (Osaka National Hospital) CVS; chorionic villus sampling

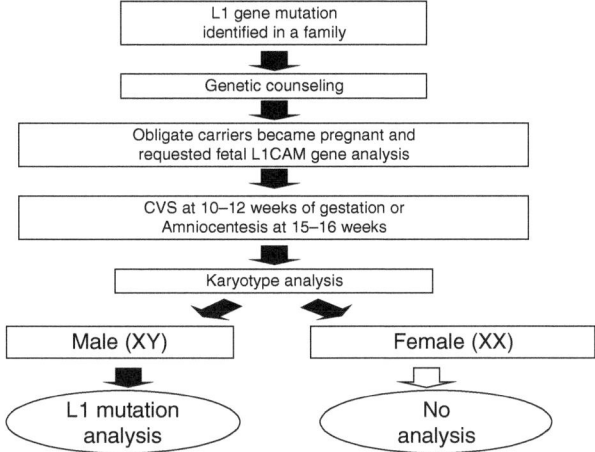

4.3.3 Genetic Analysis

Each mutation site of the L1 gene was amplified by polymerase chain reaction (PCR). Full details were described previously [8]. Briefly, 100 ng of genomic DNA was used for PCR amplification. Each PCR product was separated by electrophoresis on a 2% agarose gel and purified using the PowerPrep™ Express Gel Extraction System (Marligen Biosciences, Inc. Rockville, MD). The purified products were examined with direct sequencing using the BigDye™ Terminator v1.1 Cycle Sequencing Kit (Applied Biosystems), followed by analysis on a DNA sequencer (ABI PRISM® 3100 Genetic Analyzer, Applied Biosystems). DNA sequencing was carried out in both directions for each template.

4.4 Results

In the cases reported here, nine obligate carriers requested prenatal testing 14 times in total (Table 4.1). The fetuses were tested for sex, L1 mutations, or both using CVS or AC. Of the 14 fetuses, four were determined to be male, and ten were female. Prenatal genetic testing was performed on seven of the fetuses, three female (prior to 2004), and four male. Of the four male fetuses tested, only one (HC53) had an L1 gene mutation (c.1146C>A, p.Tyr382X) (Table 4.1). The mother terminated the pregnancy. Of the three female fetuses tested, one did not carry the mutation, and two (HC16, HC13) were heterozygous carriers for mutations (c.665delA, c.694+5G>A). Prenatal genetic testing was not performed on the other seven female fetuses, as directed by Guideline 2004. In 13 of the 14 cases, the mothers continued

Table 4.1 Case summary of prenatal L1 mutation analysis

	Origin	DNA	L1 mutation	Sex	L1 mutation	Outcome
1	HC17 Osaka City General Hospital	10 weeks CVS	c.1963A>G p.Lys655Glu	F	Not detected	37 weeks, 2820 g, healthy
2		9 weeks CVS	c.924C>T p.Gly308Gly	F	Not done	
3		11 weeks CVS		F	Not done	
4	HC22 Tokyo University	12 weeks CVS	c.2872+1G>A	M	Not detected	40 weeks, 3326 g, healthy
5		12 weeks CVS		F	Not done	40 weeks, 3165 g
6	HC16 Nagaoka Red Cross Hospital	12 weeks CVS	c.665delA p.Lys222fs	F	c.665delA heterozygous	40 weeks, 3215 g, healthy
7	HC13 Nagoya City University	11 weeks CVS	c.694+5G>A	F	c.694+5G>G/A, heterozygous	Healthy
8	HC28 Hokkaido University	16 weeks AC	c.1373T>A p.Val458Asp	F	Not done	
9	HC36 Nagoya City University	10 weeks CVS	c.2278C>T p.Arg760X	F	Not done	
10	HC53 Hirosaki University	10 weeks CVS	c.1146C>A p.Tyr382X	M	c.1146C>A p.Tyr382X	Termination of pregnancy
11		10 weeks CVS		F	Not done	
12	HC45 CRIFM	16 weeks AC	c.817-819del p.Thr273del	M	Not detected	39 weeks, 2652 g, healthy
13	HC27 Nagoya City University	11 weeks CVS	c.1829-1G>C	F	Not done	
14		11 weeks CVS		M	Not detected	37 weeks, 3260 g, healthy

CVS chorionic villus sampling, *AC* amniocentesv

their pregnancies and delivered healthy babies without an XLH phenotype. The diagnosis was made with perfect accuracy. In 12 cases out of 14, DNA was obtained by CVS between 10 and 12 weeks gestation, and in two cases it was obtained by AC between 15 and 16 weeks gestation. No maternal or fetal complications occurred during either CVS or AC. The results of genetic analysis were available within 3 weeks after CVS or AC.

4.5 Illustrative Cases

4.5.1 Family HC27

A first-born boy (III-1) showed severe hydrocephalus with adducted thumbs. The mother's family history was unremarkable. LI gene analysis of the patient (III-1), his mother (II-1), her sister (II-2), and their maternal cousin (II-3) was performed. DNA examination revealed that the patient's L1 gene had a G1829 to A transition located 1 bp downstream from the 5′end of intron 14. This transition was at a splice site, which would cause aberrant splicing in the sixth Ig-like domain of L1. The mother was heterozygous for the same mutation. The mother's sister and their cousin (II-2, II-3) did not carry the mutation. The mother (II-1) was pregnant and requested prenatal genetic testing under genetic counseling. At 11 weeks gestation, CVS and karyotype testing showed that the fetus was female. In accordance with Guideline 2004, genetic analysis was not performed. The mother continued her pregnancy and delivered a healthy girl. The mother (II-1) became pregnant again and requested prenatal genetic testing. At 11 weeks gestation, DNA was obtained by CVS, and karyotype testing showed that the fetus was male. DNA analysis determined that the fetus did not have an L1 gene mutation. The mother continued her pregnancy and delivered a healthy boy (Fig. 4.3a–c are for case summary of HC27).

4.5.2 Family HC45

The mother (I-1) was in her third pregnancy. The mother's family history was unremarkable. Her first pregnancy had aborted spontaneously at 8 weeks gestation. The second baby, a female, died in utero due to a twisted umbilical cord at 35 weeks. The third baby was male (II-1), and ventricular dilatation was detected at 20 weeks gestation. An ultrasound echogram revealed bilaterally adducted thumbs. The fetus was diagnosed with severe hydrocephalus with bilateral adducted thumbs, without family history. The mother chose to terminate her pregnancy. LI genetic analysis of fetal tissues (II-1) and a sample from the mother (I-1) were performed. We found a deletion of three nucleotides, CCA (817–819), in exon 8 of the fetal L1 gene, which would have deleted threonine 273, in the Ig3 domain of L1. The mother was heterozygous for the same mutation. The mother became pregnant again and requested prenatal genetic analysis under genetic counseling. DNA was obtained by AC at 16 weeks gestation. The fetus was male and did not have a mutation of the L1 gene. The mother continued her pregnancy and delivered a healthy boy. The third baby (II-1) was not included in this series, because the fetus was diagnosed with X-linked hydrocephalus by morphological findings from fetal ultrasonic images, not through prenatal genetic testing (Fig. 4.4a–c shows case summary of HC45).

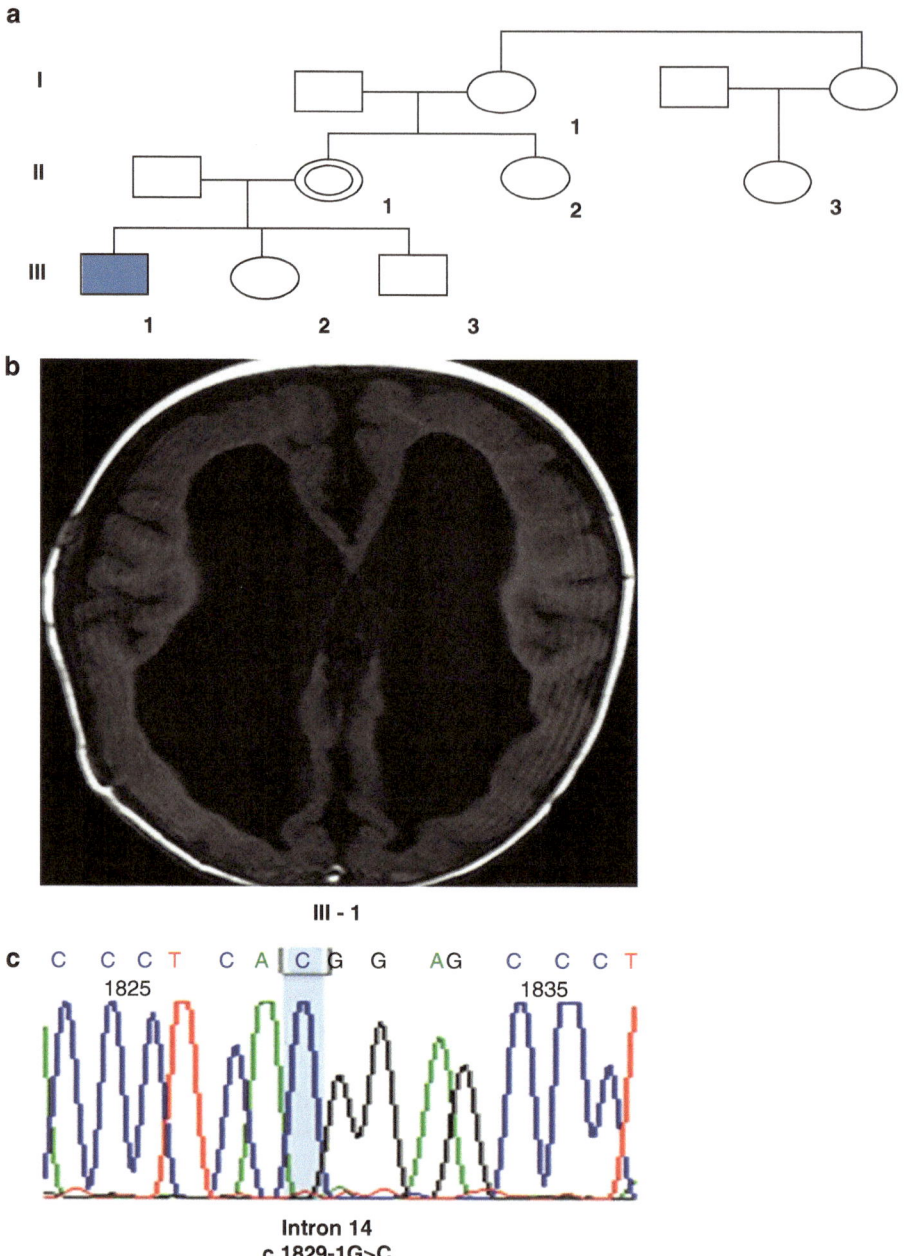

Fig. 4.3 (**a**) Pedigree of family HC27. (**b**) Postnatal CT scan of patient (III-1) shows bilateral severe ventricular dilatation. (**c**) Result of L1 gene analysis shows the position of the G to C transition in intron 14. This L1 mutation was reported previously [8]

Fig. 4.4 (**a**) Pedigree of family HC45. (**b**) Fetal MRI of fetus (II-1) at 20 weeks gestation shows bilateral severe ventricular dilatation. (**c**) Result of L1 gene analysis shows the deletion of ACC, which is a novel L1 mutation. *Sa* spontaneous abortion, *fd* fetal death due to twisting of the umbilical cord, *+TOP* termination of pregnancy

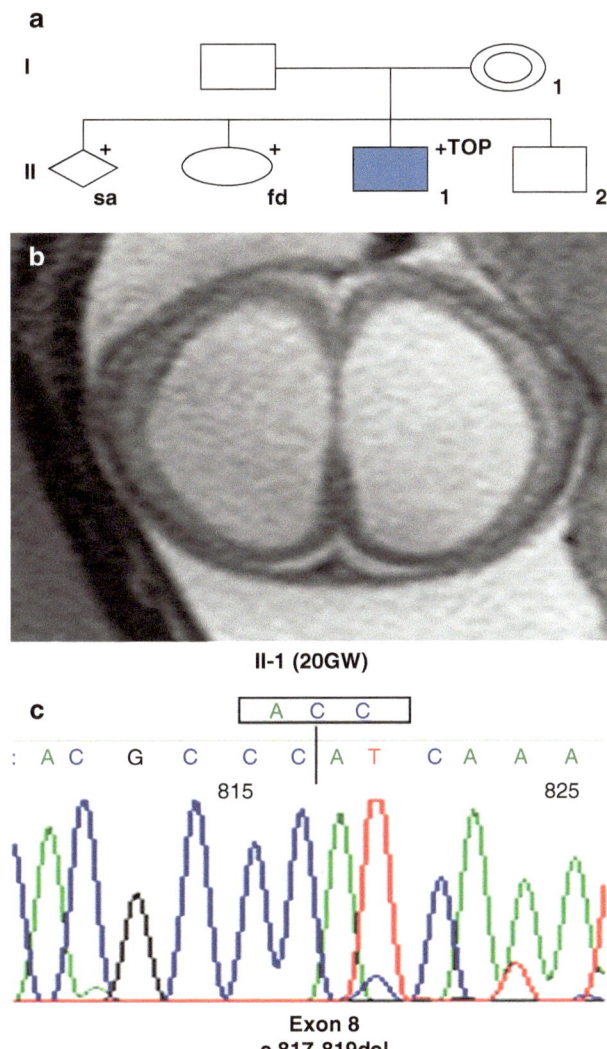

II-1 (20GW)

Exon 8
c.817-819del

4.5.3   Family HC53

At 24 weeks of gestation, severe ventricular dilatation was detected in a male fetus (II-1). The mother's family history was unremarkable. The boy was delivered weighing 3620 g at 37 weeks gestation. His thumbs were bilaterally adducted. As XLH was suggested, an LI gene analysis of the patient (II-1) and the mother (I-1) was performed. DNA sequencing revealed that the boy's L1 gene had a C1146 to A transition in exon 10. This transition would change tyrosine 382 to a stop codon and terminate the L1 protein at the forth fibronectin domain of L1. The mother was heterozygous for the same mutation. She became pregnant again and requested

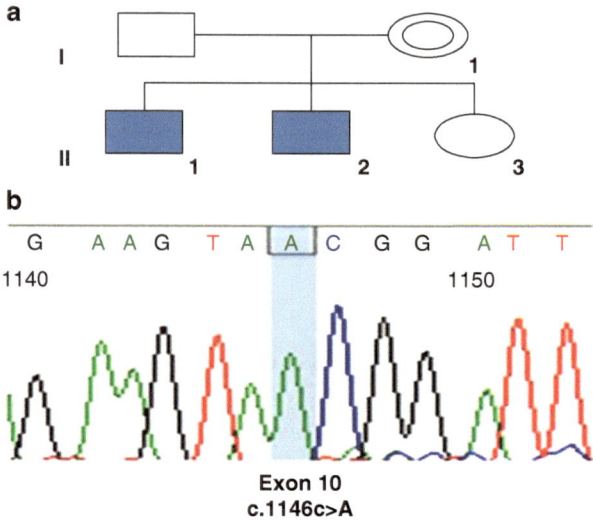

Fig. 4.5 (**a**) Pedigree of family HC53. (**b**) Result of L1 gene analysis shows the C to A transition, which is a novel L1 mutation. *TOP* termination of pregnancy

genetic analysis of the fetus (II-2) under genetic counseling. DNA was obtained by CVS at 12 weeks gestation, and the male fetus was determined to have an L1 gene mutation. The parents chose to terminate the pregnancy. Subsequently the mother (I-1) became pregnant and again requested prenatal genetic analysis. CVS was performed at 12 weeks gestation, and karyotype testing showed that the fetus was female (II-3). In accordance with Guideline 2004, genetic analysis was not performed. The mother continued her pregnancy and delivered a healthy girl (Fig. 4.5 shows case summary for HC53).

4.6   Discussion

In 2005, the Guideline for the Diagnosis and Management of Fetal Hydrocephalus was published in Japan. This is the first guideline related to the prenatal diagnosis of congenital disease to be published in Japan, and it is widely accepted by obstetricians, pediatrics, neurosurgeons, and especially patients' families. According to the data of a nationwide survey for congenital hydrocephalus, conducted by the Research Committee for Intractable Hydrocephalus in Japan, 55% of the congenital hydrocephalus cases have been detected prenatally as fetal hydrocephalus. However, fetal hydrocephaly varies widely in long-term outcome, largely because a wide variety of conditions may be responsible for this phenotype. Although other malformations may be excluded by fetal ultrasonography or fetal MRI, an initial finding of isolated ventriculomegaly may encompass XLH, corpus callosum agenesis, lissencephaly, chromosomal anomaly, hydrocephalus due to viral infection, or fetal intracranial hemorrhage. For prenatal counseling, precise diagnosis by fetal sonography, fetal MRI, and TORCH screening is very important. Prenatal L1 gene analysis plays an important role in obtaining an exact diagnosis when there is a finding of isolated ventriculomegaly.

To date, more than 200 mutations of the L1 gene have been identified in families with L1 syndrome. All known cases have showed severe retardation, both physical and mental without interfamilial variability [12]. In contrast to the phenomenon that mutations in SHH gene that is responsible for some types of holoprosencephaly (HPE) had a wide variety in phenotype from a symptomatic to severe phenotype of HPE, the method using precise serial observations by fetal sonography can detect the ventriculomegaly around 19 weeks gestation, but using prenatal L1 gene detection is able to diagnosis the fetus XLH or not in 13 weeks gestation. These 6 weeks gave the carrier mother not only the tremendous anxiety and severe physical stress in case of termination of pregnancy. Renier reported that XLH shows the poorest outcome of any prenatal noncommunicating hydrocephalus. From a mother carrying an L1 mutation, 50% of the male fetuses could have severe hydrocephalus, making prenatal molecular genetic diagnosis with genetic counseling beneficial for a family with XLH. The prenatal molecular genetic diagnosis of L1 syndrome was first reported by Jouet and Kenwrick in 1995 [17]. Since then, several case reports have been published [19]. This is the first report of systematically organized prenatal L1 gene analysis.

As previously reported, most L1 mutations are private, occurring within a single family. Within the 56 families in which we identified L1 gene mutations, only four types of L1 mutations were found in more than one family; three types were each found in two unrelated families, and one type was found in three unrelated families. Only half of the 56 families had a family history of hydrocephalus; the others were sporadic cases. Novel mutations were found in 70% of the cases. In another words, L1 mutations have no hot point.

Most western researchers have used SSCP (single-strand conformation polymorphisms) as their primary screening method [20]. We initially used SSCP as a screening method, but it failed to identify obvious L1 gene mutations in six families. Therefore, we switched to the direct sequencing method. DNA sequencing is now our standard technique for detecting genetic mutations, and it can examine all the L1 exons rapidly and reproducibly. The present results suggest that direct sequencing should be used to avoid false negatives and improve the accuracy of L1 genetic analysis. This is especially important in prenatal diagnosis, which requires perfect accuracy and prompts results if the parents are to be able to decide whether to continue the pregnancy.

The L1 gene is composed of 28 exons and a 3825 bp open reading frame. Therefore, direct sequencing of the entire gene requires at least 4 weeks. In the case of families in which the L1 gene mutation has previously been detected and analyzed, it only takes a week to analyze the few exons known to be involved. Therefore, prenatal gene analysis is limited to carriers already diagnosed with an L1 mutation. Performing L1 gene analysis in all cases of fetal hydrocephalus detected by ultrasonic echogram is not possible at present, but it will become more feasible as the cost and speed of analysis improve.

Conclusion

Prenatal L1 genetic testing with genetic counseling has been useful for families carrying L1 mutations.

Acknowledgments *Conflict of interest*: The author has no conflict of interest. The authors who are members of the Japan Neurological Society (JNS) have registered online Self-reported COI Disclosure Statement Forms through the website for JNS members.

References

1. Bickers DS, Adams RD. Hereditary stenosis of aqueduct of Sylvius as cause of congenital hydrocephalus. Brain. 1949;72:246–62.
2. Rosenthal A, Jouet M, Kenwrick S. Aberrant splicing of neural cell adhesion molecule L1 mRNA in a family with X-linked hydrocephalus. Nat Genet. 1992;2:107–12.
3. Jouet M, Rosenthal A, Armstrong G, MacFarlane J, Stevenson R, Paterson J, et al. X-linked spastic paraplegia (SPG1), MASA syndrome and X-linked hydrocephalus result from mutations in the L1 gene. Nat Genet. 1994;7:402–7.
4. SM G, Orth U, Zankl M, Schroder J, Gal A. Molecular analysis of the L1CAM gene in patients with X-linked hydrocephalus demonstrates eight novel mutations and suggests non-allelic heterogeneity of the trait. Am J Med Genet. 1997;71:336–40.
5. Finckh U, Schroder J, Ressler B, Veske A, Gal A. Spectrum and detection rate of L1CAM mutations in isolated and familial cases with clinically suspected L1-disease. Am J Med Genet. 2000;92:40–6.
6. Vos YJ, de Walle HE, Bos KK, et al. Genotype-phenotype correlations in L1 syndrome: a guide for genetic counselling and mutation analysis. J Med Genet. 2010;47:169–75.
7. Kanemura Y, Okamoto N, Sakamoto H, et al. Molecular mechanisms and neuroimaging criteria for severe L1 syndrome with X-linked hydrocephalus. J Neurosurg. 2006;105(5 Suppl):403–12.
8. Yamasaki M, Thompson P, Lemmon V. CRASH syndrome: mutations in L1CAM correlate with severity of the disease. Neuropediatrics. 1997;28:175–8.
9. Fransen E, Van Camp G, D'Hooge R, Vits L, Willems PJ. Genotype-phenotype correlation in L1 associated diseases. J Med Genet. 1998;35:399–404.
10. Kamiguchi H, Hlavin ML, Yamasaki M, Lemmon V. Adhesion molecules and inherited diseases of the human nervous system. Annu Rev Neurosci. 1998;21:97–125.
11. Bateman A, Jouet M, MacFarlane J, JS D, Kenwrick S, Chothia C. Outline structure of the human L1 cell adhesion molecule and the sites where mutations cause neurological disorders. EMBO J. 1996;15:6050–9.
12. Yamasaki M, Arita N, Hiraga S, Izumoto S, Morimoto K, Nakatani S, et al. A clinical and neuroradiological study of X-linked hydrocephalus in Japan. J Neurosurg. 1995;83:50–5.
13. Fransen E, D'Hooge R, Van Camp G, Verhoye M, Sijbers J, Reyniers E, et al. L1 knockout mice show dilated ventricles, vermis hypoplasia and impaired exploration patterns. Hum Mol Genet. 1998;7:999–1009.
14. Barami K, Kirschenbaum B, Lemmon V, Goldman SA. N-cadherin and Ng-CAM/8D9 are involved serially in the migration of newly generated neurons into the adult songbird brain. Neuron. 1994;13:567–82.
15. Yamasaki M, Nonaka M, Suzumori N, et al. Prenatal molecular diagnosis of a severe type of L1CAM syndrome (X-linked hydrocephalus). J Neurosurg Pediatr. 2011;8:411–6.
16. Renier D, Sainte-Rose C, Pierre-Kahn A, Hirsch JF. Prenatal hydrocephalus: outcome and prognosis. Childs Nerv Syst. 1988;4:213–22.

17. Jouet M, Kenwrick S. Gene analysis of L1 neural cell adhesion molecule in prenatal diagnosis of hydrocephalus. Lancet. 1995;345:161–2.
18. Kanemura Y, Takuma Y, Kamiguchi H, Yamasaki M. First case of L1CAM gene mutation identified in MASA syndrome in Asia. Congenit Anom (Kyoto). 2005;45:67–9.
19. Piccione M, Matina F, Fichera M, LoGiudice M, Damiani G, Jakil MC, et al. A noble L1CAM mutation in a fetus detected by prenatal diagnosis. Eur J Pediatr. 2010;169:415–9.
20. Michaelis RC, YZ D, Schwartz CE. The site of a missense mutation in the extracellular Ig or FN domains of L1CAM influences infant mortality and the severity of X linked hydrocephalus. J Med Genet. 1998;35:901–4.

Idiopathic Normal-Pressure Hydrocephalus Syndrome: Is It Understood? The Comprehensive Idiopathic Normal-Pressure Hydrocephalus Theory (CiNPHT)

5

Ahmed Ammar, Faisal Abbas, Wisam Al Issawi, Fatima Fakhro, Layla Batarfi, Ahmed Hendam, Mohammed Hasen, Mohammed El Shawarby, and Hosam Al Jehani

5.1   Introduction

Normal-pressure hydrocephalus (NPH), although an established clinical entity for more than 40 years, remains challenging for neurosurgeons, neurologists, and neuroscientists around the globe. The pathophysiology and etiology are multifactorial. Therefore, the management is controversial, and the outcome greatly varies between cases and centers. The original description of NPH (Hakim and Adams, 1965) was of hydrocephalus with normal CSF pressure in patients who were demented and who also exhibited psychomotor abnormalities [1–4]. A clinical triad of dementia, gait disturbance, and urinary incontinence was considered as the basis of the clinical diagnosis [1, 5–8]. Hakim's theory is based on the scientific equation (force = pressure × area). The increased CSF pressure over an enlarged ependymal and ventricular surrounding cerebral surface is significant and may alter the frontal and temporal cerebral functions causing the clinical presentations [3, 9–13]. This theory failed to

A. Ammar (✉) • F. Abbas • W. Al Issawi • L. Batarfi • A. Hendam • M. Hasen
Department of Neurosurgery, King Fahd University Hospital, Al Khobar, Saudi Arabia
e-mail: ahmed@ahmedammar.com

F. Fakhro
Defense Forces Hospital, Manama, Bahrain

M. El Shawarby
Department of Pathology, King Fahd University Hospital, Al Khobar, Saudi Arabia

H. Al Jehani
Department of Neurology and Neurosurgery, Montreal Neurological Institute and Hospital, McGill University, Montreal, QC, Canada

Department of Neurosurgery, King Fahd University Hospital, Al Khobar, Saudi Arabia

© Springer International Publishing AG 2017
A. Ammar (ed.), *Hydrocephalus*, DOI 10.1007/978-3-319-61304-8_5

explain all observed questions about iNPH such as why NPH is a progressive disease and why some patients do very well after shunt procedures and others don't [6, 8]. Recently, cerebrovascular diseases (CVS), small vessel disease, and subsequently, the regional cerebral blood flow (rCBF) are considered to be among the multifactorial etiologic causes of iNPH [14–19]. On the other hand, we observed, throughout our practice, changes in the cortices of patients with iNPH that could not be attributed solely to arterial insufficiency, and we speculate venous congestion or hypertension as the culprit in these patients. Normal-pressure hydrocephalus (NPH) so far has been categorized into two clinical forms, idiopathic NPH (iNPH) and secondary NPH (sNPH). iNPH is considered to be a syndrome of older adults, who presented with NPH triad. iNPH may occur in pediatric patients too in certain condition. Radiological examination by MRI shows cortical cerebral atrophy or narrowing and increasing the cortical subarachnoid spaces [20–24]. The ventricles are also dilated and surrounded by edema or transependymal transudation! sNPH affects patient surviving different central nervous system afflictions like SAH, stroke, head trauma, encephalitis, and meningitis and after brain surgery, which may occur at any age.

Idiopathic normal-pressure hydrocephalus is not a simple, straightforward disease, but rather a complex and a progressive syndrome [5, 25] with several pathophysiological derangements contributing to its development. It is progressive in nature as well. Understanding idiopathic normal-pressure hydrocephalus requires a new and wider vision with expanded thoughts as to the possible pathophysiology of this syndrome. None of previous proposed theories fully explain and elucidate all the ambiguity of the clinical and pathological findings of this syndrome [3, 6, 11, 18, 21, 26–35]. Careful studying of the previously published theories, hypothesis, and concepts of hydrocephalus allowed us to compose a new concept by merging most of these theories in new order based on the results of our research and clinical observations. This new theory is called "comprehensive normal-pressure hydrocephalus theory." This concept may provide wider and deeper understanding of this syndrome.

Several studies reported had explored the role of small blood vessels, venous congestion, ischemic (infarction) necrosis involving parenchyma and blood vessel walls, the gray and white matter metabolism, and recovery process [7, 12, 17, 26, 36–39].

5.2 Where Is the Pathology? A Closer Look on the Venous Side: The Research Methodology

5.2.1 Background

We observed changes in the cortical venous circulation. To examine the possibility of a venous derangement at play in these patients, we carried out an exploratory project by taking a biopsy from the cortex of patients with iNPH and correlating the results with radiological findings, clinical presentations, and progress of these patients. A cortical biopsy was taken from six patients with iNPH undergoing shunt surgery and examined for histopathological changes. TCD was used as a surrogate measure of ICP in these patients undergoing shunt procedure. The changes of the brain mantle or integrity (stiff brain) were also observed in these patients during shunt insertions [15, 21].

5.2.2 Research Objective and Aim

The aim of this research is to identify the cortical and deep cortical changes in cases of iNPH for better understanding the etiology and pathophysiology of this syndrome. The results of the histopathological examination correlated with the clinical presentation and the outcome after insertion a programmable valve.

5.2.3 Material and Methods

A total of six patients have been involved in this research study. We excluded patients with sNPH to harmonize our cohort. Informed consent was obtained from different six iNPH patients or surrogate decision-makers to enroll them in this study. The consent was taken for both the shunt insertion and the cortical biopsy. Six patients were previously worked up and were diagnosed with iNPH and were planned for shunting according to clinical and radiological criteria. Also, a second informed consent was taken for the shunt insertion as well as for the cortical biopsy. ICP was measured by both lumbar drain and TCD. The biopsies were examined in the pathology lab in King Fahd Hospital of the University (KFHU) in Khobar, Saudi Arabia, using H&E stain.

5.2.4 Results

Microscopic examination of cortical biopsies of the six patients revealed the presence of venous ischemic (infarctive) coagulative necrosis involving the brain parenchyma and blood vessel walls (Fig. 5.1). Reactive gliosis consistent with healed (old) ischemic changes with liquefactive necrosis which were seldom seen was also sometimes observed (Figs. 5.2 and 5.3). The presence of ischemic (infarctive coagulative) necrosis and gliosis involving the parenchyma and blood vessel walls clearly seen in six cases along with the observed changes in the cortical venous circulation leads to development of the concept that

> "venous thrombosis, if present, increases the superficial venous pressure, which may cause local cortical and subcortical venous ischemia/hypoxia, infarction, and gliosis. This, in turn, may result in alteration of CSF dynamics by reducing the CSF flow and absorption rate consequent to reduction of the brain integrity and pulsatile movements. This pathophysiological process runs in a vicious circle" [15, 18].

Fig. 5.1 Brain biopsy. Ischemic (infarctive) necrosis involving brain parenchyma and blood vessel walls H&E × 200

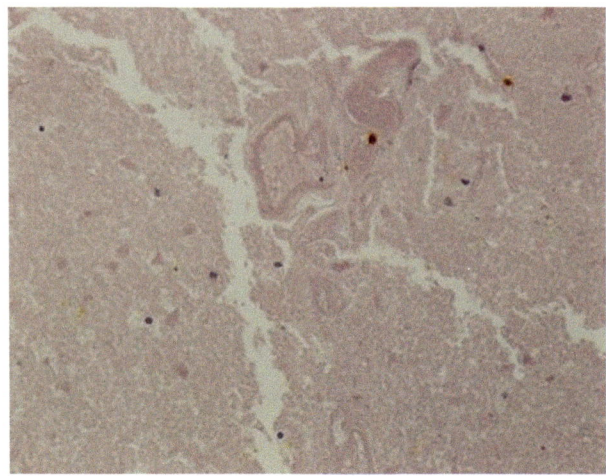

Fig. 5.2 Brain biopsy. Section showing reactive glial tissue H&E × 400

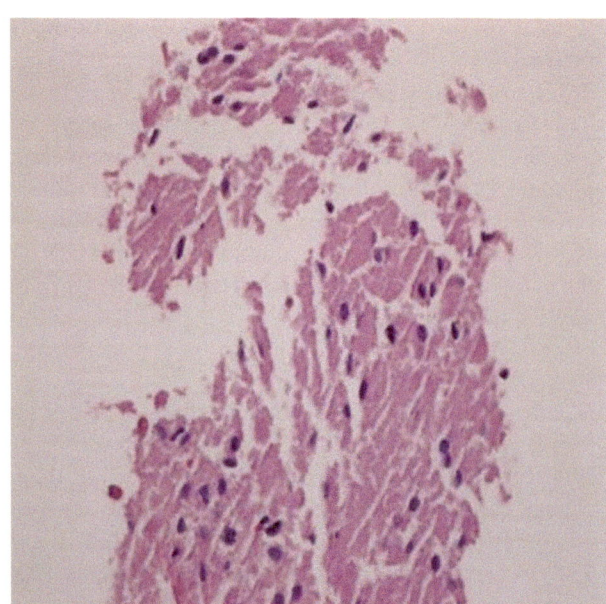

The results of this study clearly demonstrate the pertinent role of cortical and small subcortical veins in deep small blood vessels in the development of iNPH. Venous thrombosis and infarction increase the superficial venous pressure which, if transmitted to the subarachnoid space, may cause peripheral venous infarction or ischemia causing alterations of the CSF dynamics, that is, reduction

Fig. 5.3 Section showing reactive astroglial tissue with mild lymphocytic infiltration and focal calcification (right) H&E × 400

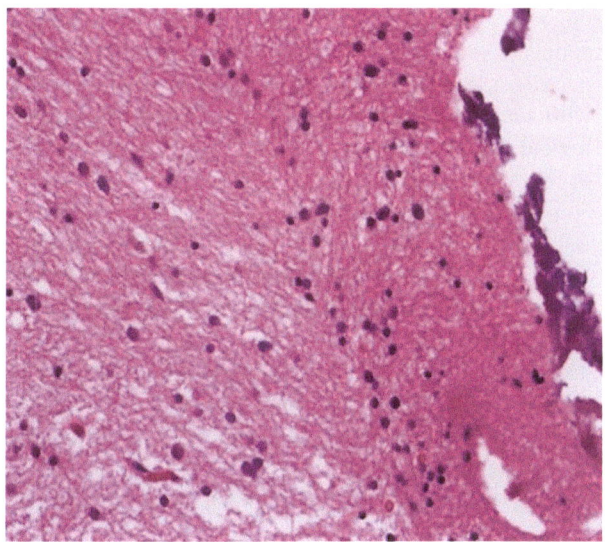

in the CSF flow toward the absorption point in the convexity and the impedance due to the reduction of the brain integrity resilience which is integral to the pulsatile nature of the block movements of CSF intracranially. Collectively, these abnormalities in the CSF hemodynamics are leading to decrease rate of CSF absorption, further maintaining this pathophysiological process runs in a vicious circle by maintaining a higher pressure in the milieu of the cortical surface and subarachnoid space. Results show that patients with what could be conveniently called "high-inflow NPH" show alterations, that is, reduction of the superficial venous compliance and a reduction in the blood flow returning via the sagittal sinus. The reduced venous pressure and the elevation of superficial venous pressure are considered to be among the primery events in the progress of primary iNPH with the other clinical and radiological spectrum of manifestation as and its consequences to that may occur and add to understanding the progress nature of iNPH.

Correlating the changes observed in the cortical venous circulation in patients with iNPH and the histopathological findings of venous ischemic (infarctive) necrosis involving parenchyma and blood vessel walls, these results are interpreted in a new concept, which is

"venous thrombosis and infarction increase the superficial venous pressure which may cause peripheral venous infarction or ischemia causing alteration of CSF dynamics by reducing the CSF flow due to reduction of the brain integrity and pulsatile and blocking the movements of CSF leading to decrease the rate of CSF absorption; this pathophysiological process runs in vicious circle."

These findings, when merged with other previous theories to present new theory, lead us to propose the "comprehensive idiopathic normal pressure hydrocephalus theory." This theory serves to provide a better understanding of this syndrome. It connects several pathological district processes into a unifying pathophysiological paradigm. Some isolated proven pathological facts together form new context, and with such a unifying theory, it can identify some critical pathological and physiological derangements that should be prevented or corrected to improve the outcome of this serious syndrome.

5.3 Comprehensive Idiopathic Normal-Pressure Hydrocephalus Theory (CiNPHT)

The value of this new concept is not to only explain the pathophysiology of iNPHs and its clinical presentations, but also it may suggest more efficient methods for managing these cases or at least abort the progressive mechanism. Algorithm 1 summarizes the theory.

This theory is based on:

(a) The clinical presentations and cumulative observation of iNPH patients over 38 years of experience in multiple centers
(b) The interesting novel histopathological findings in our cohort of patients from which brain biopsies were obtained
(c) The correlation of our findings with the existing results and implications of previous theories and hypothesis related to iNPH

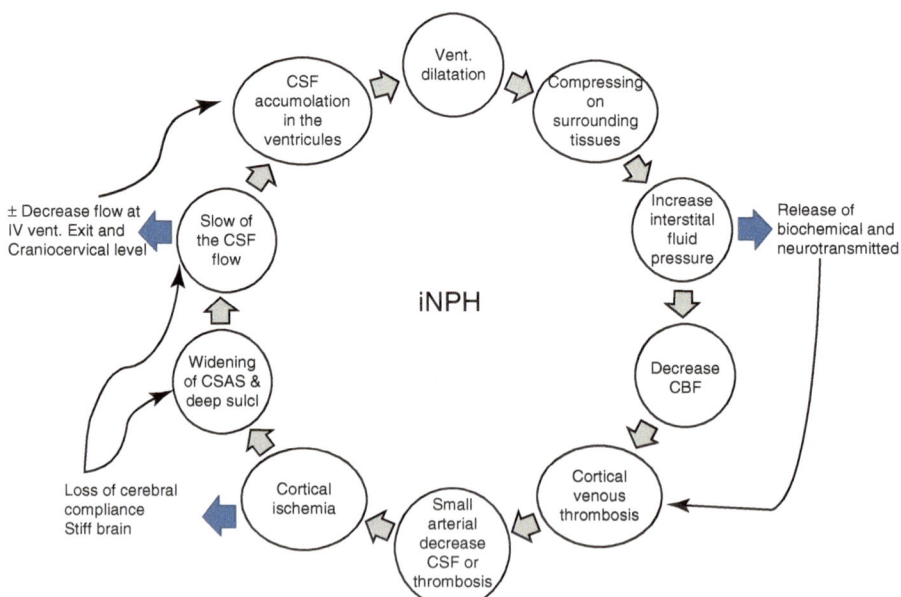

Algorithm 1 Comprehensive idiopathic normal-pressure hydrocephalus theory (CiNPHT)

5.3.1 Clinical Presentations and Observations

1. iNPH usually occurs in elderly patients.
2. Secondary normal-pressure hydrocephalus sNPH usually is a complication of CNS insult such as SAH, head trauma, or major cerebral surgery.
3. A clinical triad of dementia, gait disturbance, and urinary incontinence is very common; however, in some cases, cases vary in their presentation and proportion of severity of each disease; domain may vary.
4. iNPH is a progressive disease, the rate of which is variable among patients.
5. The ICP is changeable and exhibits a fluctuating course but tends to be reduced with/by time.
6. MRI of iNPH patients always invariably shows widening of cortical subarachnoid space (CSAS) in association with the ventricular dilatations.
7. The frontal horns of the lateral ventricular are surrounded by periventricular changes which are seen in MRI as high T2 intense signals which are usually interpreted as brain edema but could be due to transependymal transudation or other reasons such as demyelination or dissociation of the cerebral cortex syndrome. This is not frequently seen around the occipital horn as in the classical hydrocephalus.
8. There is no physical or organic obstruction of CSF flow from the lateral ventricles till the absorption site in the sagittal sinus via arachnoidal villi. The arachnoidal villi seem to be efficiently working in iNPH, unlike in the cases of sNPH, the cases after meningitis or SSAH, or meningitis.
9. Phase-contrast MRI quantifies CSF flow through the aqueduct and correlates with other images to show the CSF flow voids.
10. Different CSF dynamic studies by using isotopes and special MRI protocols show delay of the CSF flow from the ventricular system to the subarachnoid space which is delayed in the cases of NPH. This supports the CiNPHT as the increase in venous pressure and in perivenous activity lead to an increase resistance to CSF progress as it is not high enough to obstruct the flow yielding a communicating hydrocephalus pathophysiology.
11. According to the Hägen-Poiseuille equation, under steady-state pressure, the fluid will pass slower in the smaller channel compared to larger ones. If the venous congestion impared the CSF absorption, the CSF flow will be slow. We have learned from physics that the fluid flow in a wide space (tube) is slower than the flow in a smaller or thinner tube if the pressure in both cases remains the same.
12. The ventricular dilatation occurs gradually to accommodate not accounting to any increase in the ICP.
13. None of the iNPH patients died of brain herniation due to increased ICP, as far as we know. This is explained through CiNPHT, such that the derangement is downstream of all the CSF pathways, so the venous congestion will be uniformly distributed above and below the tentorium.
14. During insertion of the shunt, the brain looks pale and feels stiff more than the usual ordinary brain.

5.3.2   Different NPH Theories and Hypothesis That Merged as Comprehensive Theory

1. Hakim-Adam theory (Hakim S., Adams RD., 1965) [3]
2. Transmantle pressure gradient (Hoff J. and Barber R., 1974) [10]
3. Restricted arterial pulsation hydrocephalus (Greitz D., 1993) [20]
4. Unifying theory for definition and classification of hydrocephalus (Raimondi AJ, 1994) [33]
5. Bulk flow theory (Rekate H, 1994) [21]
6. Evolution theory in cerebrospinal fluid dynamics and minor pathway hydrocephalus (Oi S., Di Rocco C., 2006) [40]
7. Hemodynamic theory of venous congestion (Bateman GA, 2004) [27]
8. Importance of cortical subarachnoid space in understanding hydrocephalus (Rekate H, 2008) [21]
9. Reassessing CSF hydrodynamics and novel hypothesis (Chikly B. and Quaghebeur J., 2013) [28]
10. Pulsatile vector theory (Preuss M. et al. 2013) [18]
11. Intimate exchange between cerebrospinal fluid and interstitial fluid. (Matsumae M. et al. 2016) [41]

The clinical presentations and observations in iNPH patients cannot be explained by only one theory but will be better understood if we combine several theories together (comprehensive iNPH pathophysiology theory).

5.3.3   Important Aspects of Comprehensive Theory

1. This is not a new theory but rather a practical constructive integration of several aspects of several of the previously mentioned theories and hypotheses. This new concept is comprehensive in the sense that it includes attributes of most previous theories, which were carefully analyzed and merged to yield a practical understanding of this complex entity.
2. We interpreted in a new analogue (context) to construct the comprehensive iNPH theory. Most of these theories have been studied carefully, analyzed, and merged together into a new concept.
3. The dynamics of CSF is put simplistically as that governed by the understood factors affecting the flow of CSF in eight (8) consecutive compartments, namely, (1) lateral ventricles (via the foramen of Monro to the third ventricle), (2) third ventricle, (3) the aqueduct of Sylvius, (4) fourth ventricle, (5) exit to subarachnoid space (via the foramina of Magendie and Luschka), (6) the cortical subarachnoid space (CSAS), (7) the spinal subarachnoid space (SSAS), and (8) the spinal canal space (SCS). This understanding does not take into consideration that each of these compartments plays a very important and possibly an independent or complementary roles in maintaining the normal CSF dynamics. Any alteration in the flow of CSF in one or more of these compartments might be

compensated or compromised by reactionary change in the behavior of the compartment upstream or downstream to that suffering the derangement. These could also behave in a simple way of allowing the flow through them without modification into the subsequent path or follow the path by other compartments. We propose that; failure of intra-compartmental interaction might result of the failure of the compensatory mechanisms producinh NPH or pseudotumor cerebri and any other type of what now can be considered the "continuum" hydrocephalus or other types of hydrocephalus.

4. The vicious circular nature of iNPH pathophysiological process can now be practically and tangibly appreciated.

5.4 Brief Explanation of the Comprehensive NPH Pathophysiology Theory as Illustrated in Algorithm 1

(a) Ventricular dilatation occurs due to the increased accumulation of CSF inside the ventricles due to slowdown and delay of CSF outflow from the ventricular systems to the cortical subarachnoid space. The ventricles become gradually dilated due to the following reasons: (1) slow CSF flow out of the fourth ventricles to the subarachnoid space and (2) the active production of CSF is not reduced so that the production will exceed the absorption. The progressive ventricular dilatation increases the pressure load (mechanical/physical stress) on the periventricular white matter, causing destruction of periventricular white matter axons through different biochemical actions, hypoxia, and ischemia; (3) with chronic dilatation and pressure, the change in the cerebral tissues progresses including the ependymal layer. With several attempts of healing in face of progressive dilatation, it will lead to the decrease brain compliance around the ventricle, which results in "stiff ventricle/brain" due to ischemia and increase interstitial fluid. With time, the ependymal wall loses its plasticity, and pulsatility will be reduced significantly which in turn abolishes the "pulsatile vector" or reduces the ability to pulsate (pulsatile vector theory) [18], and that is proposed to push the CSF through outlet of the ventricular cavities through the outlet foramina of the ventricles (bulk theory) [21] out the ventricular cavity. The periventricular hypodensity on CT or hyperintensity on MRI white signals may be signs of disruption of the white matter and demyelination caused by ventricular dilatation and hypoxia-ischemia manifestation rather than simply transependymal flow of CSF resulting in periventricular edema and transependymal transudation to a varying degree as well. All these findings are indicating that a major part of the pathology occurs in the periventricular areas.

(b) The expansion of the ventricles will cause pressure on a larger surface area of the brain (Hakim theory) to produce the NPH triad. We believe NPH is a syndrome, not a simple disease. The triad and other clinical presentations are due to the subcortical disconnection syndrome. The increased ICP and paraventricular reduction of CBF, CPP, hypoxia, and ischemia lead to metabolic, biochemical, and molecular disruptions. Subsequently, oligodendroglia and axons are gradually damaged with demyelination occurring as a final result. When dysregulation is significant, calcium channel-related proteolysis is activated, and the risk of apoptotic cell death is expected. Thus, early diagnosis and new potential therapeutic drug interventions may reverse or at least hasten this serious pathophysiological process.

(c) The interstitial fluid accumulates [26, 41], and subsequently, the interstitial transmantle pressure is increased possibly leading to (a) cortical and deep cerebral deep venous congestion and thrombosis, (b) hindrance or blocking of the alternative intraparenchymal CSF pathways, and (c) progressive worsening of the transperiependymal transudation of CSF accumulation in face of ongoing transependymal transudation mechanism. The increase of the interstitial fluids and pressure will in turn cause further injury or damage to neural and glial cells and alteration in the neurotransmitters and releases some chemicals like CA which may cause spasm and thrombosis of small blood arteries and arterioles subsequently leading to ischemia and hypoxia of vessels. Oxidative stress will ensue leading to further damage of the niche of precursor oligodendroglia precursor cells (PreOL), which are essential for remyelination and healing process. This scarring will lead to elevation of the protein level in CSF, which might render the CSF difficult to absorb further advancing iNPH progression. This proposed hostile environment is supported by the finding which may contribute to CSF reduction of absorption subsequently causing further advancing the iNPH syndrome. The subcortical disconnection syndrome is manifested, and glial scars are developed that contain high level of CD44 cells, which are known to block PreOL differentiation, hindering the remyelination and recovery process. With this, the vicious circle of damages enters a "maintenance phase" with permanent advancement of the disease despite any attempt at treatment and syndrome is advanced.

(d) The ependymal wall itself is not immune to these hostile interactions as it is exposed to the same mechanical stress of ventricular dilatation, hypoxia, ischemia, molecular changes, and calcium proteolysis causing death of some of its cells. Reactive astrocytes may replace the dead cells at the ventricular border. The reactive astrocytes may produce tumor necrosis factor-a (TNFa), which causes hippocampal and neocortical dysfunction with significant decline in cognitive abilities of patients. TNFa and transforming growth factors—β113,51,73,137—are a useful biomarkers for detection of the progress of NPH. Further ventricular dilatation has serious consequences such as elevation of ICP, accumulation of interstitial tissue fluid, elevation of interstitial pressure, and deep venous and arteriolar ischemia, and the CBF is reduced.

(e) Congested or thrombosed deep and cortical veins lead to a decreased venous drainage of the brain leading to the accumulation of the above-mentioned toxic factors in the cerebral mantle, further intensifying the derangement of iNPH. The draining of the blood is a result of released chemicals of damaged cells or the increased interstitial fluid's pressure.

(f) The small arteriolar spasm or occlusion occurs as well.

(g) Occlusion of, congestion of, or compromised small cortical blood vessels (venules and arterioles) will cause cortical ischemia, as some of these blood vessels get thrombosed as the ischemia is advanced. As the cortical tissue changes in the form of loss of integrity, gliosis develops, and the cortical tissues became stiff with loss of compliance and transmission of shock waves or pulsatile waves. The deep cortical ischemia also affects the ependymal wall which significantly reduces its compliance.

(h) The cortical subarachnoid spaces increase in width, depth, and volume (as usually demonstrated by MRI) [21].

(i) The pressure gradient between the ICP inside the ventricular and the cortical subarachnoid space (CSAS) is altered. The resulting positive pressure gradient between the ventricular systems and the cortical subarachnoid space causes further dilatation of the ventricles (transmantle pressure gradient theory) [10]. Therefore, NPH is progressive in nature.

(j) As the CSF is produced at the same rate, accumulation is increased inside the ventricles, and with delayed downflow to the subarachnoid space, the ventricles dilate more, and the pressure increases on the surrounding tissue causing cell damage and ischemia, which consequently widens and deepens the cortical subarachnoid space, which delays the CSF flow and absorption increasing CSF accumulation in the ventricles. This is the vicious circle of iNPH. It should be disrupted in different points in order to improve the outcome of this syndrome.

5.5   Evidence to Support the Concept of Comprehensive iNPH Theory

A study of cerebral blood flow at rest in patients with NPH compared to normal controls was conducted [29] using O-labeled water positron emission tomography (PET) with co-registration of magnetic resonance images (MRI) to investigate the role of the cerebral vasculature in NPH. It was demonstrated that the mean CBF was decreased in the cerebrum and cerebellum of patients with idiopathic NPH compared to normal healthy controls. This study supports the concept that apart from being a disorder of CSF circulation, there are cerebrovascular factors involved in the pathophysiology of NPH and that these factors may be related to the gait disorder of NPH [40].

A patient with clinical features of idiopathic normal-pressure hydrocephalus, who responded dramatically to shunting, was found at necropsy to have a severe hypertensive and arteriosclerotic vasculopathy with multiple lacunar infarcts. There was no pathological evidence of thickened leptomeninges, fibrosis of the arachnoid villi, or Alzheimer's disease. An abnormal absorption mechanism was demonstrated with cisternography and by an increase in the concentration of homovanillic acid in the cerebrospinal fluid. It is suggested that vascular changes may play an important role in the pathophysiology in some cases of normal-pressure hydrocephalus [5].

A high prevalence of CVD and Alzheimer's disease was found in patients shunted for INPH, which was reflected, although less commonly, by similar neuropathological biopsy findings. No significant correlation was found between the presence of comorbidity and shunt outcome. The findings support the perception of INPH as a multi etiological clinical entity, possibly overlapping pathophysiologically with CVD and Alzheimer's disease [25].

Fundamental issues in NPH, which also represent deficiencies in knowledge, are those of pathogenesis and pathophysiology. In general, hydrocephalus is due to obstruction present with ventricular enlargement and raised CSF pressure. However, the apparent paradox of ventricular enlargement with normal CSF pressure is more difficult, and some mechanisms have been proposed [3, 30]. Narrowed periventricular arterioles due to hyaline vessel wall degeneration in hypertensive patients might be another precipitating factor and would explain why the prevalence of vascular risk factors in NPH is increased [5, 25]. Vascular compliance is significantly different in the brains of healthy subjects as compared with that in patients with ischemia/atrophy or NPH [3, 15, 30].

The key to understanding the CSF dynamics and pathophysiology of iNPH in particular and hydrocephalus, in general, is to understand the factors that control the flow of CSF through the cortical subarachnoid space. Cortical subarachnoid space is and should be considered as one of the main compartments of CSF dynamics. CSF dynamics will never be completely understood unless the flow of the CSF in the subcortical space is understood. The changes in the size and volume of cortical subarachnoid space and the underneath pulsatile forces of cerebral cortex may play an important role in CSF dynamics. Rekate H. et al. were very right when they wrote that "the important role of the CSAS in the pathophysiology of various forms of hydrocephalus has been largely ignored [21]. Attention to the dynamics of CSF in this compartment will improve understanding of enigmatic conditions of hydrocephalus and improve selection criteria for treatment paradigms such as ETV. These concepts lead to clearly defined problems that may be solved by the creation of a central database to address these issues. We share Dr. Rekate's request to have international cooperation and collaboration to clinical experiences, scientific experiences, and thoughts, and all related data are very valid and badly needed.

5.6   Discussion

There has been and still ongoing sincere scientific strives to find answers of yet unanswered questions related to the diagnosis, causes, pathophysiology, and the best method of treatment and outcome of iNPH.

Many theories had been formulated trying to explain iNPH phenomena including the structural, cerebral blood flow and CSF flow, but none could explain it all! Each theory is good in explaining part of the syndrome, but fails to explain the rest of the problems. We consider iNPH as a syndrome not a disease, caused and triggered by different factors and developed and progressed through a vicious cycle of biological and pathological events. Therefore, the clinical presentation varies and progressed some times.

Further explanation of our new theory is based on the pathology reports from the six patients, literature reviews on CSF hydrodynamics, and cerebral circulation and the alterations of ICP. In CSF hydrodynamics, there are three phases in which the disturbance in any one of these might be the pathophysiologic cause of NPH: (1) CSF production, (2) CSF flow, and (3) CSF absorption. However, none of these processes can alone explain the whole clinical findings and sequelae of the syndrome. For example, the hydrodynamic theory cannot explain why there are changes in the brain parenchyma. Nevertheless, the hydrodynamic theory is very useful and valid to explain CSF absorption. Recently, new theories suggest vascular changes might be the pathophysiologic cause of iNPH. Others suggested that both a disturbance in CSF hydrodynamics and vascular changes might be the cause. However, none of these theories were proved before, nor the mechanisms by which they cause the clinical manifestation were explained. Furthermore, cerebral vessel diseases were studied as a separate entity. Cerebral arterial occlusion causes ischemia and subsequently infarction in the areas they supply. Similarly, it was found from studies on animal models that venous occlusion causes ischemia and subsequently infarction of the draining areas. The differences are: (1) venous occlusion takes a longer time to cause ischemia than arterial occlusion, (2) venous congestion and occlusion occur at a much lower pressure than that affecting arterial vessels, and (3) the occurrence and extent of ischemia depend on the presence or absence of collateral venous outflow, which is not clearly demarcated whether the occlusion occurred acutely or chronically. Based on the result of our research and the observation of changes in cortical venous circulation and pathological findings of venous ischemic (infarction) coagulative necrosis involving the parenchyma and blood vessel walls in the six patients undergoing shunt insertion for iNPH, we concluded that venous congestion, thrombosis, and infarction increase the superficial venous pressure which may cause peripheral venous infarction and gliosis resulting in alteration of CSF dynamics by reducing the CSF flow due to reduction of the brain integrity and pulsatility and blocking the movements of CSF leading to decreased rate of CSF absorption, and this pathophysiological process runs in vicious cycle. Venous obstruction will increase the superficial venous pressure, which will impede blood drainage, thereby decreasing tissue perfusion and oxygenation with consequent peripheral ischemia and infarction.

Vascular disease in the form of venous obstruction will lead to increase in the superficial venous pressure. As a result, it will affect CSF hydrodynamics by decreasing CSF absorption. Venous obstruction will also increase the superficial venous pressure, which will impede blood drainage, decreasing tissue perfusion and oxygenation causing peripheral venous infarction or ischemia. Since venous obstruction is mostly subclinical, most of these initial events will pass unnoticed. Moreover, if the obstruction continues, this leads to a chronic increase in superficial venous pressure and decreased CSF absorption. As a consequence, this will result in the accumulation of CSF in the ventricles, and because this process occurs slowly and over a long interval, there is time for the ventricles to accommodate the increase in volume and pressure called ventriculomegaly; CSF pressure will plateau after that and will end up with ventriculomegaly without an increase in CSF pressure. Furthermore, venous obstruction will lead to the following consequences in a sequence: (1) increased dural sinuses venous pressure, (2) increased CSF production and decreased CSF absorption, (3) rupture of the venous structures (hematoma), (4) cortical ischemia (infarction) with subsequent atrophy of the area involved, (5) reduced capillary perfusion pressure, (6) development of interstitial edema, and (7) blood-brain barrier disruption ending to decreased cerebral blood flow. These changes will lead to the parenchymatous changes and manifestations associated with NPH. This is supported by other diseases found to be associated with iNPH such as Alzheimer's disease having similar cortical ischemia. Finally, venous occlusion is reversible if detected early, and intervention took place in the initial stage of the disease, unlike arterial occlusion. Therefore, early detection of iNPH in patients and early management improve the clinical manifestations as well as the outcome. Deep and small vessels involvment is resulting in hypoxia and ischemia and subsequently infarction and loss of cerebral tissues integrity, which is felt during surgery as cerebral parenchyma is getting harder or stiff as appointed by others. These findings do not explain all the pathophysiological process alone, but it is a part of a larger process, which we collected and called "comprehensive iNPH theory" (CiNPHT).

Conclusion

Normal-pressure hydrocephalus (NPH) should be clearly divided into iNPH syndrome and sNPH disease. They are completely two different identities, and the pathophysiology, etiology, and clinical outcomes are different. Therefore, iNPH syndrome and sNPH disease should not amalgamate together. Idiopathic normal-pressure hydrocephalus is a complex syndrome and is progressive in nature. The "comprehensive idiopathic normal pressure theory" (CiNPHT) provides a wider and deeper understanding of the different related issues, phases, and aspects of iNPH such as the causes and factors contributing to the alteration of normal CSF dynamics. There are several factors contributing to the nature, etiology, and progress of NPH, which are the cortical blood vessels being small, changes in cerebral integrity (stiffness and pulsatility), alteration of CSF dynamics, delay of CSF flow in the cortical subarachnoid space, widening of the cortical subarachnoid space and ventricular dilatation, and changes in the ependymal wall. Ischemia in the deep venous territory is not a prerequisite for NPH. Patients with high-inflow NPH show alterations in superficial venous compliance and a reduction in the blood flow returning via the sagittal sinus. These changes together suggest that an elevation in superficial venous pressure may occur in NPH. Although the CSF hydrodynamic theory can describe some aspects of the pathophysiology of NPH, it cannot explain the vascular changes observed in the cortical veins as well as the clinical manifestations associated with this disease. These changes, together with the pathological changes and careful review and understanding of previous theories, have allowed us to produce the comprehensive new concept and hypothesis.

References

1. Bech-Azeddine R, Høgh P, Juhler M, Gjerris F, Waldemar G. Idiopathic normal-pressure hydrocephalus: clinical comorbidity correlated with cerebral biopsy findings and outcome of cerebrospinal fluid shunting. Neurol Neurosurg Psychiatry. 2007;78(2):157–61.
2. Gallia GL, Rigamonti D, Williams MA. The diagnosis and treatment of idiopathic normal pressure hydrocephalus. Nat Clin Pract Neurol. 2006;2(7):375–81.
3. Hakim S, Adams RD. The special clinical problem of symptomatic hydrocephalus with normal cerebrospinal fluid pressure. Observations on cerebrospinal fluid hydrodynamics. J Neurol Sci. 1965;2:307–27.
4. AM K, Alafuzoff I, Savolainen S, Sutela A, Rummukainen J, Kurki M, Jääskeläinen JE, Soininen H, Rinne J, Leinonen V, Kuopio NPH Registry. Poor cognitive outcome in shunt-responsive idiopathic normal pressure hydrocephalus. Neurosurgery. 2013;72(1):1–8.
5. Koto A, Rosenberg G, Zingesser LH, Horoupian D, Katzman R. Syndrome of normal pressure hydrocephalus: possible relation to hypertensive and arteriosclerotic vasculopathy. J Neurol Neurosurg Psychiatry. 1977;40(1):73–9.
6. Malm J, Eklund A. Idiopathic normal pressure hydrocephalus. Pract Neurol. 2006;6:14–27.
7. Owler BK, Fung K, Czosnyka Z. Importance of ICP monitoring in CSF circulation disorders. Br J Neurosurg. 2001;15(5):439–40.
8. Udvarhelyi GB, Wood JH, James AE, Bartlet D. Results and complications in 55 shunted patients with normal pressure hydrocephalus. Surg Newrvl. 1975;3:271–5.
9. Daniel RT, Lee GYF, Halcrow SJ. Low-pressure hydrocephalic state complicating hemispherectomy: a case report. Epilepsia. 2002;43:563–5.
10. Hoff J, Barber R. Transcerebral mantle pressure in normal pressure hydrocephalus. Arch Neurol. 1974;31:101–10.
11. Johanson CE, Szmydynger-Chodobska J, Chodobski A, Baird A, McMillan P, Stopa EG. Altered formation and bulk absorption of cerebrospinal fluid in FGF-2-induced hydrocephalus. Am J Phys. 1999;277(1 Pt 2):R263–71.
12. Owler BK, Jacobson EE, Johnston IH. Low pressure hydrocephalus: issues of diagnosis and treatment in five cases. Br J Neurosurg. 2001;15:353–9.
13. Toma AK, Holl E, Kitchen ND, Watkins LD. Evans' index revisited: the need for an alternative in normal pressure hydrocephalus. Neurosurgery. 2011;68(4):939–44.
14. Chang CC, Asada H, Mimura T, Suzuki S. A prospective study of cerebral blood flow and cerebrovascular reactivity to acetazolamide in 162 patients with idiopathic normal-pressure hydrocephalus. J Neurosurg. 2009;111(3):610–7.
15. Grant A. Bateman: vascular compliance in normal pressure hydrocephalus. AJNR Am J Neuroradiol. 2000;21:1574–85.
16. Johansson E, Ambarki K, Birgander R, Bahrami N, Eklund A, Malm J. Cerebral microbleeds in idiopathic normal pressure hydrocephalus. Fluids Barriers CNS. 2016;13:4.
17. Owler BK, Pickard JD. Cerebral blood flow and normal pressure hydrocephalus. A review. Acta Neurol Scand. 2001;104:325–42.
18. Preuss M, Hoffman K-T, Reiss-Zimmermann M, Hirsh W, Merkwnschlager A, meixensberger J, Dengi M. Updated physiology and pathophysiology of CSF circulation—the pulsatile vector theory. Childs Nerv Syst. 2013;29:1811–23.
19. Sharma AK, Gaikwad S, Gupta V, Garg A, Mishra NK. Measurement of peak CSF flow velocity at cerebral aqueduct, before and after lumbar CSF drainage, by use of phase-contrast MRI: utility in the management of idiopathic normal pressure hydrocephalus. Clin Neurol Neurosurg. 2008;110(4):363–8.
20. Greitz D. Cerebrospinal fluid circulation and associated intracranial dynamics. A radiologic investigation using MR imaging and radionuclide cisternography. Acta Radiol Suppl. 1993;386:1–23.
21. Rekate HL, Nadkarni TD, Wallace D. The importance of the cortical subarachnoid space in understanding hydrocephalus. J Neurosurg Pediatr. 2008;2:1–11.

22. Shiino A, Nishida Y, Yasuda H, Suzuki M, Matsuda M, Inubushi T. Magnetic resonance spectroscopic determination of a neuronal and axonal marker in white matter predicts reversibility of deficits in secondary normal pressure hydrocephalus. J Neurol Neurosurg Psychiatry. 2004;75(8):1141–8.
23. U S, Kondziella D. Neuronal glial interaction in different neurological diseases studied by ex vivo 13C NMR spectroscopy. NMR Biomed. 2003;16(6–7):424–9.
24. Yamashita F, Sasaki M, Takahashi S, Matsuda H, Kudo K, Narumi S, Terayama Y, Asada T. Detection of changes in cerebrospinal fluid space in idiopathic normal pressure hydrocephalus using voxel-based morphometry. Neuroradiology. 2010;52(5):381–6.
25. Casmiro M, D'Alessandro R, Cacciatore FM, Daidone R, Calbucci F. Risk factors for the syndrome of ventricular enlargement with gait apraxia (idiopathic normal pressure hydrocephalus): a case-control study. J Neurol Neurosurg Psychiatry. 1989;52:847–52.
26. Agre P, Nielsen S, Ottersen OP. Towards a molecular understanding of water homeostasis in the brain. Neuroscience. 2004;129(4):849–50.
27. Bateman GA. Idiopathic intracranial hypertension: priapism of the brain? Med Hypotheses. 2004;63:549–55.
28. Chikly B, Quaghebeur J. Reassessing cerebrospinal fluid (CSF) hydrodynamics: a literature review presenting a novel hypothesis for CSF physiology. J Bodyw Mov Ther. 2013;17:344–54.
29. Hebb AO, Cusimano MD. Idiopathic normal pressure hydrocephalus: a systematic review of diagnosis and outcome. Neurosurgery. 2001;49(5):1166–84; discussion 1184–1166.
30. Levine DN. Intracranial pressure and ventricular expansion in hydrocephalus: have we been asking the wrong question? J Neurol Sci. 2008;269:1–2):1–11.
31. Linninger AA, Sweetman B, Penn R. Normal and hydrocephalic brain dynamics: the role of reduced cerebrospinal fluid reabsorption and ventricular enlargement. Ann Biomed Eng. 2009;73:1434–47.
32. Momjian S, Bichsel D. Nonlinear poroplastic model of ventricular dilation in hydrocephalus. J Neurosurg. 2008;109:100–7.
33. Raimondi AJ. A unifying theory for the definition and classification of hydrocephalus. Childs Nerv Syst. 1994;10(1):2–12.
34. Ishikawa M. Progress in diagnosis of and therapy for idiopathic normal-pressure hydrocephalus—classical view of cerebrospinal production, absorption and bulk flow and its criticism. Rinsho Shinkeigaku. 2014;54(12):1184–6.
35. Sá Santos S, Sonnewald U, Carrondo MJ, Alves PM. The role of glia in neuronal recovery following anoxia: in vitro evidence of neuronal adaptation. Neurochem Int. 2011;58(6):665–75.
36. Akins PT, Guppy KH. Sinking skin flaps, paradoxical herniation, and external brain tamponade: a review of decompressive craniectomy management. Neurocrit Care. 2008;9:269–76.
37. Fujimura M, Onuma T, Kameyama M, Motohashi O, Kon H, Yamamoto K, Ishii K, Tominaga T. Hydrocephalus due to cerebrospinal fluid overproduction by bilateral choroid plexus papillomas. Childs Nerv Syst. 2004;20(7):485–8.
38. Kondziella D, Sonnewald U, Tullberg M, Wikkelso C. Brain metabolism in adult chronic hydrocephalus. J Neurochem. 2008;106(4):1515–24.
39. Kondziella D, Eyjolfsson EM, Saether O, Sonnewald U, Risa O. Gray matter metabolism in acute and chronic hydrocephalus. Neuroscience. 2009;159(2):570–7.
40. Oi S, Di Rocco C. Proposal of "evolution theory in cerebrospinal fluid dynamics" and minor pathway hydrocephalus in developing immature brain. Childs Nerv Syst. 2006;22:662–9.
41. Matsumae M, Sato O, Hirayama A, Hayashi N, Takizawa K, Atsumi H, Sorimachi T. Research into the physiology of cerebrospinal fluid reaches a new horizon: intimate exchange between cerebrospinal fluid and interstitial fluid may contribute to maintenance of homeostasis in the central nervous system. Neurol Med Chir (Tokyo). 2016;56(7):416–41.

Post-infection and Multiloculated Hydrocephalus (Complex Hydrocephalus)

6

Ahmed Ammar, Fatima A. Fakhroo, Ahmed Abdelfattah, and Mohammed Shawarby

6.1 Definitions and Nomenclatures

Hydrocephalus is one of the most serious and important issues in neurosurgery and in particular in pediatric neurosurgery. Loculated hydrocephalus or complex hydrocephalus is a rare form of hydrocephalus in which discrete fluid-filled compartments trapped or formed in the ventricular system of the brain [1]. Septated hydrocephalus and complex hydrocephalus are synonyms for loculated hydrocephalus. Loculated hydrocephalus is divided into two subdivisions based on the number of compartments formed: (1) uninoculated hydrocephalus and (2) multiloculated hydrocephalus. Uninoculated hydrocephalus is defined as cystic dilatation of one segment of the normal ventricular system, e.g., obstruction of one of the foramina of Monro, or an isolated fourth ventricle due to an obstruction of the foramen of the aqueduct and fourth ventricle outlets [1]. On the other hand, multiloculated hydrocephalus is defined as multiple separated cystic cavities or spaces located in or about the ventricular system, filled with fresh or altered CSF, and isolated by multiple intraventricular septations [2, 3]. Other terms used in literature to describe multiloculated hydrocephalus are "multicystic" and "multicompartment." Thus, uninoculated and multiloculated hydrocephalus are two different disease entities, with different morphologic and radiologic findings, most common etiologic causes, pathological changes, treatment options, and prognosis. Unfortunately, in literature,

A. Ammar (✉) • A. Abdelfattah
Department of Neurosurgery, King Fahd University Hospital, Imam Abdulrahman Bin Faisal University, Al Khobar, Saudi Arabia
e-mail: ahmed@ahmedamma.com

F.A. Fakhroo
Military Defense Hospital, West Rifaa, Kingdom of Bahrain

M. Shawarby
Department of Pathology, King Fahd Hospital of the University, Imam Abdulrahman Bin Faisal University, Al Khobar, Saudi Arabia

© Springer International Publishing AG 2017
A. Ammar (ed.), *Hydrocephalus*, DOI 10.1007/978-3-319-61304-8_6

these two terms are used interchangeably, which lead in some circumstances to the inability to compare the results of different studies because of the misuse of the correct terms.

6.2   Etiology and Risk Factors

Multiloculated hydrocephalus is a complication and sequela of different pathological causes or iatrogenic factor. There is a report of an intrauterine case of multiloculated hydrocephalus [4]. It may be caused as a sequel of intraventricular hemorrhage. The most frequent causes of multiloculated hydrocephalus are:

1. Meningitis and encephalitis
2. Intraventricular hemorrhage
3. Post-shunt infection [5]
4. Shunt over drainage
5. Multiple neuroepithelial cysts [1]

   Predisposing factors are:

1. Low birth weight
2. Preterm birth, especially before 32 weeks which increase the incidence of intraventricular hemorrhage
3. Perinatal complications
4. Congenital central nervous system malformations [2]

6.3   Pathophysiology

The definite pathophysiology of multiloculated hydrocephalus is not very well understood. Therefore, we carried out a study to understand, reveal, document the different stages of septum formation, and confirm the progressive nature of intraventricular septae (multiloculated hydrocephalus).

6.3.1   Illustrative Case

A 1-day-old premature boy born at 30GW was referred to our department. The Apgar score was low, and he had tense anterior fontanelle. CT scan showed intraventricular hemorrhage. Therefore, EVD was inserted. The infant went through a stormy course as he was operated several times for insertion of VP shunt, shut infection, EVD, and shunt revision. We decided to carry on this research as soon as we noticed the early formation of intraventricular septum. We decided to take a biopsy

Fig. 6.1 (**a**) Brain CT of the head shows the early stage of intraventricular septae. (**b**) H&E histology of the septum showing fibrin entangling inflammatory cells

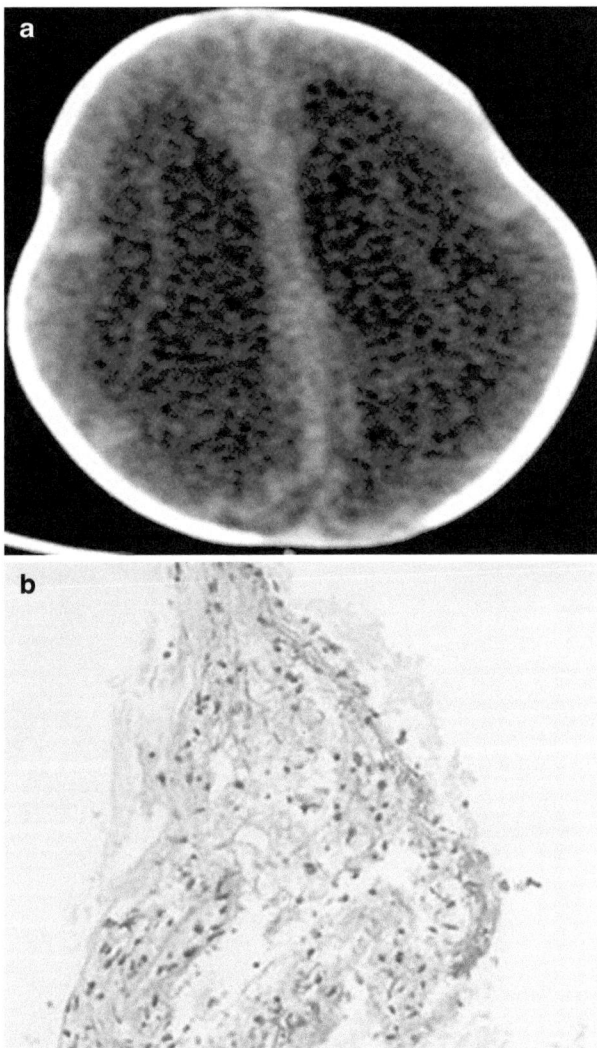

of the septum by the endoscope and correlate the results with the CT scan findings. The histopathology of the tissues of septum was carefully studied every time there is a need for shunt revision. The results of histopathology were correlated with the CT scan findings. The early septum appeared as a membrane consisting of glial or fibroglial tissue with variable infiltration by lymphocytes sometimes accompanied by plasma cells and foamy histiocytes (Fig. 6.1a, b and 6.2a, b). Edema and perivascular fibrosis were sometimes also noted (Fig. 6.3a, b). Eventually, the septum

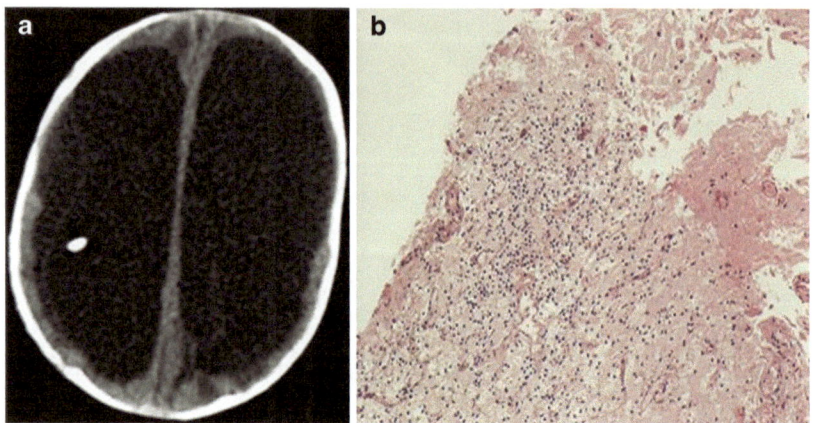

Fig. 6.2 (**a**) Three months later, more fine septae can be seen within both lateral ventricles and more brain atrophy. (**b**) Histology showing numerous mixed inflammatory cells with foamy macrophages

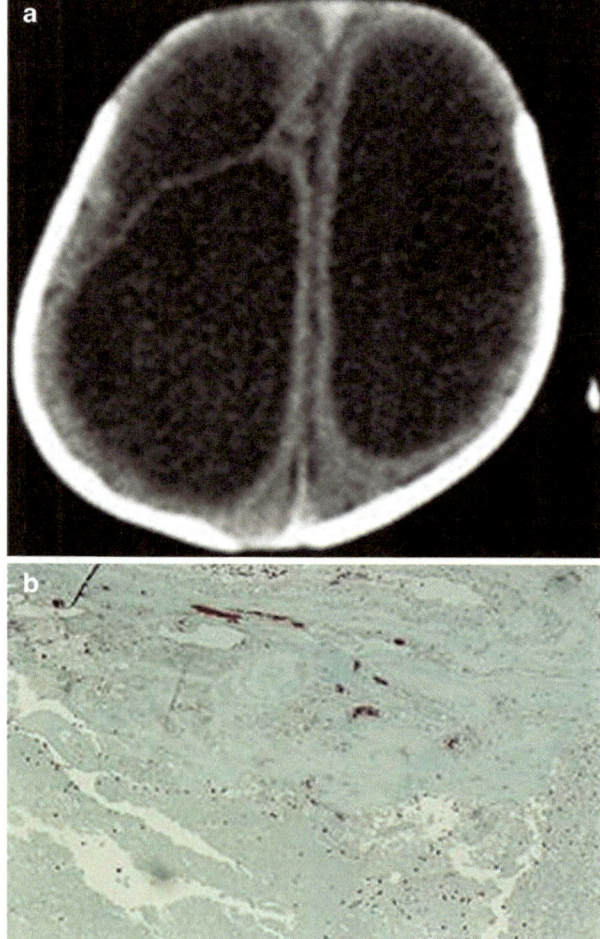

Fig. 6.3 (**a**) Further, 4 months later, axial CT showed a septum across the right lateral ventricle. (**b**) Histology of the septum: collagenous fibrous tissue originating from thickened blood vessel wall (Masson trichrome stain)

Fig. 6.4 (**a**) Nonenhanced axial brain CT shows development of thick septum crossing from one ependymal surface to the opposite side within the right lateral ventricle (A,C,D); brain atrophy was evident. (**b**) Histology of the septum showing collagenous fibrosis

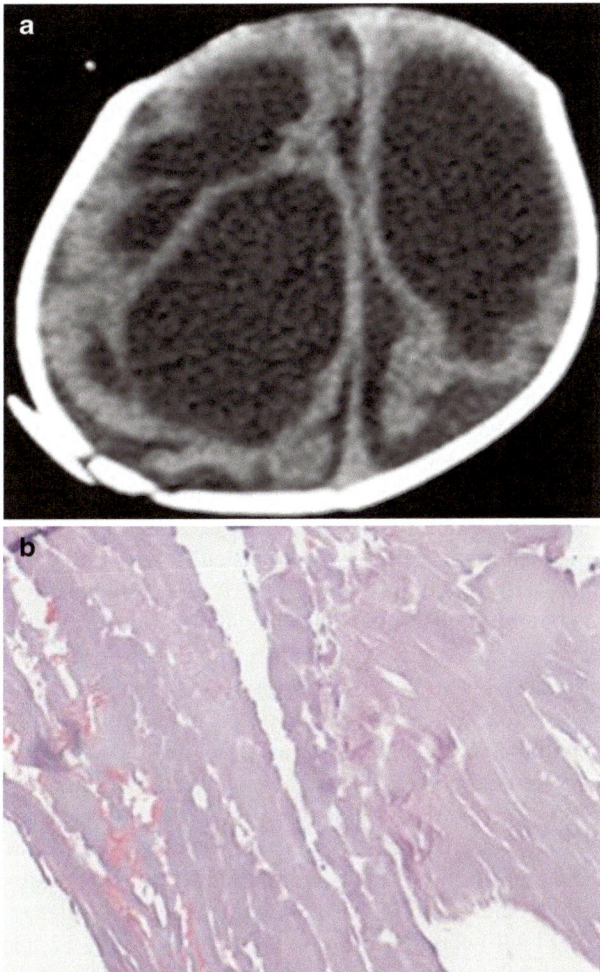

showed collagenous fibrosis (Fig. 6.4a). The meninges also showed collagenous fibrosis or inflammation with fibrin deposition and infiltration by mononuclear inflammatory cells (lymphocytes, plasma cells) and neutrophils (Fig. 6.4b). At the final stage, CT and MRI show dilatation and loss of configurations of the ventricles. Progressive brain atrophy is common as brain melts out and is replaced by multiple cysts, and this is likely irreversible. Sometimes it is difficult to differentiate between ventricles and periventricular cysts [6].

6.3.1.1 The Stages of Ventricular Loculation and Septum Formation

The results of this research prove that disease is a progressive one. The stages of formation of multiloculated hydrocephalus and septum formation can be summarized in four stages which are:

Stage 1: Formation of fibrinous, intraventricular membranes
Stage 2: Increasing infiltration of membranes by inflammatory cells (lymphocytes sometimes accompanied by plasma cells and foamy histiocytes)
Stage 3: Gliosis and early (perivascular) fibrosis of membranes
Stage 4: Diffuse collagenous fibrosis of membranes with eventual transformation of membranes into dense, fibrotic septa with softening of the brain, loss of integrity, ventricular dilatation, and low ICP

Intraventricular septae can be congenital or acquired and can be true or false septa. These septations will lead to the development of multiloculated hydrocephalus. The timing of the start of development of ventricular septations varies with an average of 2–4 months following ventriculitis [6–8]. The combination of bacterial infection especially gram-negative and intraventricular hemorrhage will lead to the most severe form of multiloculated hydrocephalus. Intraventricular septations occur as a result of fibrous adhesions developing within the ventricles and may contribute to the development of multiloculated hydrocephalus. These membranes appeared as filmy, translucent structures. Schultz and Leeds (2010) characterize the pathological findings of ventriculitis as small areas of denuded ependyma, with glial tufts extending through the fragmented ependyma into the ventricular lumen [9]. According to their description, the intraventricular septum consists of fibroglial elements with round cells and some polymorphonuclear cells [10]. Berman and Banker [11] studied meningitis series; there were nine who also had parenchymal infarction, microabscesses, or periventricular leukomalacia. Meningitis is frequently associated with purulent ventriculitis. They also observed that it causes an inflammatory response in the ependyma with patchy or diffuse denudation of this cell layer and development of subjacent gliosis and gliotic tufts at the sites of ependymal disruption. They state further that glial tufts frequently projected into the exudate from the subependymal tissue [12]. Septations can be true or pseudo septae depending on the origin either in the ventricles or in periventricular territories that later become ventriculized. These need to be observed early during their formation because later they cannot be differentiated. True septa may span the ventricular walls or float in the lumen. They can be delicate or coarse and resemble cobwebs or thin veils. They may be rather complex, occupying the entire ventricle, or focal with only solitary strands noted [13]. In a study involving five cases by L. Sagan et al. in 2008, biopsy of the intraventricular septum was taken for analysis, and the findings were observed. Ependymal breakage and inflammatory process reactivate fetal mechanisms of germinal matrix glial cells migration. These cells are behaving as fetal radial glia-forming intraventricular septae, leading to the compartmentalization of ventricular system. The central compartment is not a uniform entity [12].

The core underlying pathology for multiloculated hydrocephalus is *inflammatory process*, and the inflammatory triggering might be either infectious or chemical. The trigger to this disease is an injury by inflammatory reactions to the ependymal cells lining the ventricular system of the brain. Other disease insults induce these inflammatory reactions to the central nervous system such as meningitis mostly bacterial meningitis, intraventricular hemorrhage, and shunt-related infections. The first two accounts for most cases, respectively. When there is an inflammation occurring at the ependymal surface, this will destruct the ependymal cells lining the ventricular system and will expose the underlying subependymal layer. Furthermore, the inflammatory process reactivates fetal mechanisms of germinal matrix glial cell migration. Glial cell proliferation continues growing to form a nidus for the formation of septations that span the ventricles. Septations are also formed by the accumulation of inflammatory exudates and debris. Subsequently, isolated, the ventricular cavity is divided into different isolated compartments. The cerebrospinal fluid flow is altered, leading to the accumulation of cerebrospinal fluid within a loculated cavity, *resulting in obstructive hydrocephalus*, with progressive dilatation and mass effect on the adjacent part of the brain. Macroscopically [14], the membranes are completely extending through the ventricle to the other end or incompletely extending to part of the lumen [10]. The membranes may be transparent, thin, and avascular; on the other hand, they may be thick and highly vascularized. Microscopically the septations are membranes composed of fibroglial tissues and round and polymorphonuclear cells. Features of chronic ventriculitis usually present in the form of subependymal gliosis, glial tufts extending through the destructed ependyma into the ventricular lumen [15, 16]. An attempt to prevent the formation of intraventricular septations was done by Schulz et al. by the installation of streptokinase-streptodornase with or without administration of steroids. This trial was unsuccessful [17]. We studied and confirmed the progressive nature of intraventricular septae pathology in his patient by taking an endoscopic biopsy from septae every time there is a need for revision.

In a study involving five cases, a biopsy of intraventricular septa was taken for analysis, and the findings were observed. Ependymal breakage and inflammatory process reactivate fetal mechanisms of germinal matrix glial cells migration. These cells are behaving as fetal radial glia form intraventricular septa, leading to the compartmentalization of ventricular system. The central compartment is not a uniform entity [18].

6.4   Classifications

The value of calcification has unified the type and stage of pathology, dealing with and find out the best method of treating each type. Classification is essential for good and reliable data collection and research. The first classification is the one suggested by Spenato et al. in 2004, in which multiloculated hydrocephalus was subdivided into four forms according to the sites of obstruction, anatomic, and imaging appearance [15]. However, this classification does not consider the pathophysiologic cause. Also it does not provide a base for treatment of patient according to the type [12]. Kelsbeck JE et al., in 1980, presented the following classification:

"(1) multiple intraventricular septations, resulting from the presence of membranes or septa that encroach the ventricular system. (2) Isolated lateral ventricle and unilateral hydrocephalus, secondary to obstruction of one Monro foramen, leading to accumulation of fluid in only one lateral ventricle, while the remaining ventricular system is of normal size. (3) Entrapped temporal horn, resulting from adhesions in the region of the trigone, leading to isolation of the temporal or temporooccipital horn containing choroid plexus. (4) Isolated the fourth ventricle, caused by obstructions of the aqueduct of Sylvie and the foramina of Luschka and Magendie" [8].

There are other classifications [12].

6.5   Clinical Features

Patients with multiloculated hydrocephalus usually present with symptoms of increased intracranial pressure that differ in infants and older children, along with focal neurological deficits. Infant's symptoms include irritability, impaired conscious level, vomiting, and macrocephaly. On the other hand, in older children symptoms include a headache, deterioration of conscious level, and vomiting. Other symptoms include seizures, gait ataxia, hemiparesis, and developmental delay.

On physical examination, patients with multiloculated hydrocephalus might present with signs of increased intracranial pressure. These are tense anterior fontanelle, increased skull circumference, "cracked pot" sound on skull percussion, thin scalp with dilated veins, and "setting the sun" appearance which includes lid retraction and impaired gaze upward. Other signs might be found: cognitive deficits, psychomotor retardation, gait ataxia, and seizures.

6.6   Diagnostic Studies

Radiological studies diagnose multiloculated hydrocephalus. During the last 40 years with the advancement in neuroradiology, multiloculated hydrocephalus is diagnosed at an earlier stage, which positively affected the prognosis of the disease [2, 19] (Table 6.1).

Table 6.1 Summary of the most common methods of diagnosis and the advantages and disadvantages of each method

	Description	Advantages	Disadvantages
Ultrasound	– Used in the evaluation of neonates and infants with opened fontanelles	– Noninvasive	– Operator dependent
	– It is useful in evaluating patients with multiloculated hydrocephalus since it can demonstrate the cyst walls and reveal any compartmentalization that has occurred	– No ionizing radiation	
CT scan	– Used for screening	– No sedation is needed	– Exposure to ionizing radiation
	– Before the era of MRI, it was considered to be the diagnostic method of choice	– Noninvasive	– Difficult to visualize the cyst walls accurately
			– Unable to identify whether communications between cavities are present or absent
			– No multiplanar view as MRI??
Contrast CT ventriculography	– Before the era of MRI, it was superior to plain CT scan	– Accurately defines the margins of the cysts	– Invasive
		– Absolute confirmation of noncommunicating loculations	– Exposure to ionizing radiation
		– Immediate visualization of sequestrated ventricular compartments	– In multiloculated hydrocephalus, multiple punctures for different compartments to show the presence or absence among compartments
Magnetic resonance imaging with gadolinium	– Currently the diagnostic method of choice	– Noninvasive	– Doesn't define the CSF flow
	– MRI CISS is the one used	– No ionizing radiation	– Unable to identify whether communications between cavities are present or absent
		– More sensitive in revealing septations	
		– Multiplanar view with detailed picture in three different planes	

6.7   Treatment

The only definitive treatment for multiloculated hydrocephalus is surgical. Although medical treatment might be indicated in some cases for preexisting comorbidities like central nervous system infections. The goals of surgical treatment are (1) to drain the compartments mainly by ventricular shunting and (2) creating fenestrations between adjacent compartments to create a window for cerebrospinal circulation, thus decreasing the need for multiple shunt system [10, 14, 20–23].

The oldest and traditional surgical treatment was the insertion of multiple shunts each in the isolated compartment. The incidence of failure and complications was high. Also, the need for multiple shunt revision was indicated for many cases, which is associated with high morbidity and mortality rates.

The microsurgical techniques were considered previously to be the treatment of choice. Access to the ventricular system is done either by transcallosal or transcortical. However, both techniques produced serious complications. Nevertheless, the advantage of this technique is the direction of the intraventricular septations under direct visualization. Multiple wide fenestrations can be done, and bleeding is controlled. Rhoton AL and Gomez MR, in 1972, reported their success in treating two patients by direct intraventricular surgery converting it into the unilocular system [24].

Currently, the neuroendoscopic technique is the method of choice for treating such patients as this surgical technique allows the fenestration of different septae and turning the isolated multiloculate ventricles (cysts) into one connected dilated ventricle, to permit insertion of a single shunt [1, 3, 10, 14, 16, 20–23, 25–27]. The studies using neuroendoscope showed a reduction of the shunt revision's rate per year following fenestration [1, 10, 25]. Lewis et al., in 1995, reported a reduction in the shunt revision rate from 3.04 to 0.25 revisions per year [25]. However, the difficulty may sometimes appear as the neurosurgeon may loose the orientation of exact location of septum during the surgery. Therefore, coupling the neuronavigation technique with neuroendoscope proved to be very useful [17]. We consider this method of surgery as the method of choice, and we adapt it with great success to fenestrae all the intraventricular septums for years.

6.8   Prognosis

Multiloculated hydrocephalus is considered so far, to be one of the diseases with unfavorable prognosis due to the progressive nature of the disease. There is hope in the future to stop the progression of this disease by using pharmacological agents to inhibit the underlying inflammatory pathogenesis of this disease. However, there is a need for preclinical and clinical trials to prove the efficacy and safety of this new modality of management.

The prognosis of multiloculated hydrocephalus depends on the extent of intraventricular septations, the surgical procedure done, and the presence or absence of previous neurological insults and morbidities. The outcome is evaluated by the following: (1) the incidence of improvement of hydrocephalus in postoperative MR imaging (2) avoiding or eliminating the need for shunting, (3) simplifying complex shunt system, and (4) reducing shunt revision rate. The early discovery of shunt complication, meningitis, and IVH may help to reduce the risk of such serious disease. Once the disease is diagnosed, aggressive antibiotic treatment, which may include intrathecal antibiotic injection, and daily washing the ventricles with saline may help to abort the progressive sequale of the disease.

6.9   Discussion

The incidence and prevalence of multiloculated hydrocephalus are increasing probably due to increasing survival rates of children and neonates who suffered intraventricular hemorrhage and meningitis [14]. Shunt infection remains to be one of the most frequent causes of such serious disease. Therefore, every effort should be done to avoid shunt infection. There is no international consensus about the incidence of multiloculated hydrocephalus; the incidences vary between 7 and 30% or more. Lorber and Pickering [28] estimated the incidence of multiloculated hydrocephalus in cases of men ingitis to be more than 30% [22]. However, Reinprecht et al. [29], reported a long-term follow-up of posthemorrhagic hydrocephalus infants and revealed that intraventricular septations occurred only in 7% of cases [25]. Cipri S and Gambardella G found out that the prevalence of multiloculated hydrocephalus was 20% in patients with hydrocephalus admitted in their department [11]. A unanimously agreed upon classification, definition, and terminology are vital in finding out the real incidence of this disease and performing good research [3, 16, 28, 30]. It is noticeable that there is occasional mix-up between the terms multiloculated hydrocephalus and uninoculated hydrocephalus. They are two different disease entities with different radiologic findings, treatment, and prognosis. However, in some occasions, uninoculated hydrocephalus may progress to the multiloculated status.

Inflammation is the reaction of the host to tissue damage due to biological, chemical, or physical agents. It is a defensive process which in most cases resolves spontaneously. Unresolved or "chronic" inflammation, however, is a pathologic process which is harmful to the host. The inflammation is out of control, and it is no longer possible to differentiate between the damage caused by the injurious agent and that resulting from the inflammation itself.

In the reaction to an infection, cells recruit and activate other cells using multiple stimuli. Pro-inflammatory stimuli lead to the following processes: macrophages secrete chemotactic factors which attract neutrophils. Macrophages and neutrophils release ROIs (reactive oxygen intermediates), RNIs (reactive nitrogen intermediates), and hyaluronidase which decomposes the hyaluronic acid of the extracellular matrix. The resulting hyaluronic fragments, in turn, act as signals of damage: through CD44 macrophages, they induce the further release of chemotactic factors and MMPs (matrix metalloproteinases). MMPs break down collagen, proteoglycans, and fibronectin. TNF from macrophages activates neutrophils; these release more elastase, and the chain reactions are increased, in that elastase and ROIs, in turn, activate the MMPs. MMPs activate TGF-β from macrophages, which are the most powerful chemotactic factor. Elastase also breaks down the TGF-β-binding protein and thus supports the activation of TGF-β. The acute phase of the inflammation is now complete [31].

However, as soon the switchover is made from tissue damaging to healing, the cell and cytokine ensemble is regrouped. In this connection, it can be seen that certain molecules (e.g., TNF, IFN-γ, TGF- β, PGE2) can have pro- or anti-inflammatory effects, depending on the timing and the context. The drama of the healing of the tissue and the anti-inflammatory process begins when the complement system, the neutrophils, and the macrophages start to kill the infecting agent and digest cell fragments. The macrophages secrete protease inhibitors (SLPI—secretory leukocyte protease inhibitor—among others), which have anti-inflammatory and wound-healing effects. SLPI also suppresses the further release of elastase and ROIs by TNF-stimulated neutrophils, inhibits the already released elastase, but protects the TGF-β, and, through synergistic action, deactivates the neutrophils. CD44-positive macrophages decompose the hyaluronic fragments, and the chemotactic damage signal is thus eliminated; after that no fresh neutrophils are recruited. For their part, the neutrophils that are present trigger the apoptosis. Macrophages consume the dead neutrophils and break down their elastase reserves; as a result of which, they release more TGF-β. TNF stimulates the macrophages to release IL-12, which induces the release of IFN-γ in the lymphocytes. IFN-γ now suppresses the production of chemokine, and the TGF-β that is already present promotes fibrosis [31].

The exact incidence of postinfectious hydrocephalus is not known, but it was reported by some to be 30–40% and up to 60% [27, 28, 32]. The criteria for post-infective hydrocephalus for infants are (1) infants born with normal-sized heads with subsequent development of hydrocephalus, (2) history of febrile illness after birth, and (3) CSF cytology and biochemistry suggestive of post-infection sequelae [33]. The commonest cause of bacterial meningitis in patients over the age of 2 months is gram-negative encapsulated organisms either *H. influenzae, N. meningitidis*, or *Str. pneumoniae. Escherichia coli* is the most common infecting agent in neonatal meningitis [34]. Bacteria may reach the CSF by one of the three major pathways: hematogenous spread from a contiguous structure or by direct implantation within the CSF. The main organisms involved are enteric gram-negative bacteria. These organisms are encapsulated and possess endotoxin which may facilitate the breakdown of the blood-brain barrier and then the entry of organisms into the brain, in addition to its direct neuronal damaging effect [35]. The most obvious gross pathological change in bacterial meningitis is the exudate in the subarachnoid space (SAS) which is more marked around the area of the basal cistern and the cisterna magna. This is part of the host defense and may take considerable periods of time to be resorbed. It results in blockage of normal CSF flow in the SAS, resulting in hydrocephalus [5, 34, 36] and pial cell necrosis and displacement into the SAS [13, 37]. In severe cases, arachnoid necrosis may occur with coexistent subdural empyema and encephalitis. Ventriculitis mainly seen in the lateral ventricles may also occur. So hydrocephalus is the result of exudates obstructing the CSF pathways (e.g., in the aqueduct, foramina of Luschka or Magendie, cortical SAS, or at the arachnoid villus) directly or by a later fibrotic gliotic reaction. This exudate also surrounds cranial nerves and may lead to cranial nerve palsies. Intracranial blood vessels are involved by arteritis and phlebitis of small vessels and may lead to secondary parenchymal changes such as brain softening, infarction, hemorrhage, or abscess [2, 12]. Later in the disease course, major dural sinus thrombosis may occur. Cerebral edema, due to leuko or other toxic factors, vessel involvement, and encephalitis may then lead to increased intracranial pressure (ICP) with the hazard of a life-threatening herniation of the brain, venous stasis, venous thrombosis and infarction, worsening brain edema, and a rise in ICP. Also, hydrocephalus may lead to further increase in ICP. The cellular element of the exudate consists of polymorphonuclear leucocytes in the first 48 h, followed by lymphocytes and macrophages, and then by plasma cells and fibroblasts. The arachnoid villi also are affected by fibrosis and adhesions. Intraventricular septations occur as a result of fibrous adhesions developing within the ventricles and may contribute to the development of multiloculated hydrocephalus [17]. Tuberculous meningitis leads to adhesion formation, an obliterative vasculitis, and an encephalomyelitis. The exudate is usually basal and fibrinous, heals by fibrous scar, and is responsible for hydrocephalus [38–40]. In TB meningitis, CSF diversion is not always required in the presence of ventriculomegaly since this is mainly due to brain softening [39, 40]. Infection is a well-documented complication of cerebrospinal fluid shunts and accounts for 5–10% of cases. Early reports stated that neonatal meningitis was the common cause of multiloculated hydrocephalus accounting for up to 75% of cases [7, 31, 32]. Multiloculated hydrocephalus refers to the presence of an isolated CSF compartment or compartments within the ventricular system that may progressively enlarge despite a functioning shunt system [15]. Pathological findings in postmeningitic hydrocephalus showed components of both communicating as well as obstructive hydrocephalus. The pathogenesis of septation is still unclear but believed to be secondary to inflammatory changes and gliosis after ventriculitis, especially caused by gram-negative organisms [8]. The inflammatory stimulus can be infectious or chemical. It is believed that the septations giving rise to multiple loculations probably represent the organization of intraventricular exudates and debris produced by ventriculitis. An inflammatory response at the ependymal surface could encourage the proliferation of subependymal glial tissue, upon which exudates and debris organize and serve as a nidus for the formation of septations that span the ventricles [21]. The septations alter the ventricular anatomy and disrupt the normal flow of cerebrospinal fluid leading to its accumulation within a loculated cavity with progressive dilatation and mass effect [41]. Intraventricular septations are usually the result of meningitis during the neonatal period. The inflammation and destruction of the ependyma allow glia to project into the lumen, bridging crucial areas and serving as a nidus for the formation of intraventricular septations [17]. Grossly the membranes appear translucent and vary in thickness. Microscopically subependymal gliosis, small areas of denuded ependyma, and glial tufts extending through the denuded ependyma into the ventricular lumen indicate the presence of chronic ventriculitis [21]. In a study involving five cases, a biopsy of intraventricular septa was taken for analysis, and the findings were observed. Ependymal breakage and inflammatory process reactivate fetal mechanisms of germinal matrix glial cells migration. These cells are behaving as fetal radial glia form intraventricular septa, leading to the compartmentalization of ventricular system. The central compartment is not a uniform entity [18]. Cushing reported the first case of loculated hydrocephalus in 1908 [42]. The first case of multiloculated hydrocephalus was reported by Rhoton in 1972 [24]. The early cases of loculated hydrocephalus were described based on autopsy studies, then X-ray imaging, and since the 1970s by using CT and later MRI scans. High-resolution MR imaging often helps to identify multiloculated hydrocephalus at an early stage. Multilocular hydrocephalus appears to be a progressive disease that leads to death or severe psychomotor retardation. CSF over-drainage by ventricular shunt system can cause morphological changes in the CSF pathways possibly leading to the isolation of compartments. Infections may be associated with the development of multiloculated hydrocephalus, particularly those with gram-negative organisms and concomitant intraventricular hemorrhage. Ventricular tap and CSF study in neonates with hydrocephalus has resulted in a large number of infants being diagnosed with post-infective hydrocephalus [8]. CSF should be cleared first using Ommaya reservoir or ventricular-subgaleal shunt temporary then performing endoscopic third ventriculostomy with or without lysis of septations if present [10]. Hydrocephalus may be of the arrested type, but if ventriculoperitoneal shunts are needed, CSF should be sterilized. If it is multiloculated hydrocephalus endoscopy is a very useful tool to make a unique cavity trying to use a single shunt and reducing revision rate. Fenestrations should be large to avoid their closure [39]. For patients with multiloculated hydrocephalus, the prognosis is generally poor compared to those with the unilocular type.

Despite the different and multiple treatments proposed, the neurological outcome is extremely poor in almost all patients. Compartmentalization appears to be a progressive disease that leads to death or severe psychomotor retardation [34]. In the past decade, major advances in neonatal intensive care have allowed the survival of a large number of low-birth-weight premature infants. Intraventricular hemorrhage arising from the germinal matrix is frequent in these small babies, and pan ventricular enlargement occurs in about half of them. This may be a transient phenomenon, not requiring treatment, or may be progressive with signs and symptoms of hydrocephalus, commonly because of the obstruction at the foramen of Monro or the Sylvian aqueduct by the clot. The role that neonatal meningitis plays in the development of MH has been described [34, 35]. Gram-negative bacterial meningitis has been reported to be a frequent cause of lateral ventricle entrapment, as well as CSF shunt-related infection. The commonly accepted causes of ventriculitis are neonatal meningitis, shunt infection, and intraventricular hemorrhage. It is thought that intraventricular septations probably represent the organization of exudates and debris produced by ventriculitis, regardless of whether it is of chemical or infectious nature [36]. The inflammatory response of the ependymal surface could lead to the formation of the fibroglial septa. The septation not only alters the ventricular anatomy but disrupt the normal flow of CSF. In the end, production and accumulation of CSF within a loculated cavity can lead to progressive dilatation and mass effect [4].

Microscopically, the septa are composed of fibroglial elements, with some round and polymorphonuclear cells [36]. According to Oi et al., the excess drainage of CSF via a ventricular shunt system can cause morphological changes in the CSF pathways [5, 37] and possibly lead to the isolation of compartments. The obstruction of the foramen of Monro in isolated unilateral hydrocephalus and aqueductal obstruction in isolated fourth ventricle after shunt placement occurs in a previously communicating ventricular system. In both instances, a reduction in the ventricular size is initially seen after shunting. Isolation then gradually develops, and re-enlargement of the isolated compartment is observed. Various types of isolation may then develop, depending upon the side of occlusion [13, 37]. There are other conditions that may play important roles as etiological factors, such as shunt-related infection [4, 13], direct ependymal trauma during catheter insertion, head injury [38], and intracranial surgery [39]. The septa sometimes are thin and translucent and, in some cases, are thick and very vascularized, consequently requiring a careful coagulation before being perforated. Regarding the diagnosis of MH, CT or MRI is usually utilized, but these examinations may fail to disclose communication between cavities accurately. Pneumoventriculography and contrast ventriculography are currently carried out in selected cases only [6]. Constructive interference in steady-state, three-dimensional Fourier transformation (CISS) allows demonstrating intracystic and intraventricular septa that cannot be visualized in the MH by using T2-weighted imaging [40]. As the MH is an evolutionary process, we need to follow the patients periodically with images. In newborns, we prefer to perform follow-up with sonography to avoid radiation. The use of CT scanning for follow-up purposes in neonatal meningitis patients will facilitate earlier diagnosis of compartmentalization of the lateral ventricles, document progression of hydrocephalus, and probably help in the earlier diagnosis of shunt malfunction [34].

Conclusion

Prevention of the formation of multiloculated hydrocephalus is the best treatment. It is, therefore, necessary to find ways out of this dilemma and to modulate the inflammatory reaction at various sites so that the acute reaction is not stopped too soon, while on the other hand, a stop signal that has been initiated is increased accordingly so that the healing phase can begin. In this chapter, we presented a classification which may explain how intraventricular septations may develop and the pathophysiological stages of developing this serious disease.

References

1. Fritsch MJ, Mehdorn M. Endoscopic intraventricular surgery for treatment of hydrocephalus and loculated CSF space in children less than one year of age. Pediatr Neurosurg. 2002;36: 183–8.
2. Cairns H, Russel DS. Cerebral arteritis and phlebitis in pneumococcal meningitis. J Pathol Bacteriol. 1946;58(4):649–65.
3. Gangemi M, Maiuri F, Donati P, Sigona L, Iaconetta G, de Divitiis E. Neuro-endoscopy. Personal experience, indications and limits. J Neurosurg Sci. 1998;42:1–10.

4. Carpenter RR, Petersdorf RG. The clinical spectrum of bacterial meningitis. Am J Med. 1962;33:262.
5. Dodge PR, Swartz MN. Bacterial meningitis: a review of selected aspects II. Special neurological problems, postmeningitis complications and clinicopathological correlations. N Engl J Med. 1965;272:954.
6. Albanese V, Tomasello F, Sampaolo S. Multiloculated hydrocephalus in infants. Neurosurgery. 1981;8:641–6.
7. Jamjoom AB, Mohammed AA, Al-Boukai A, et al. Multiloculated hydrocephalus related cerebrospinal fluid shunt infection. Acta Neurochir. 1996;138:714–9.
8. Kalsbeck JE, De Sousa AL, Kleiman MB, et al. Compartmentalization of the cerebral ventricles as a sequela of neonatal meningitis. J Neurosurg. 1980;52:547–52.
9. Schultz P, Leeds NE. Intraventricular septations complicating neonatal meningitis. J Neurosurg. 1973;38:620–6.
10. Oi S, Abbott R. Loculated ventricles and isolated compartments in hydrocephalus: their pathophysiology and the efficacy of neuroendoscopic surgery. Neurosurg Clin N Am. 2004;15(1):77–87.
11. Berman PH, Banker BQ. Neonatal meningitis. A clinical and pathologic study of 29 cases. Pediatrics. 1966;38:6–24.
12. Andresen M, Juhler M. Multiloculated hydrocephalus: a review of current problems in classification and treatment. Childs Nerv Syst. 2012;28(3):357–62.
13. Nelson E, Blinzinger K, Hager H. An electron microscopic study of bacterial meningitis. I. Experimental alterations in the leptomeninges and subarachnoid space. Arch Neurol. 1962;6:390–403.
14. Oi S, Hidaka M, Honda Y, Togo K, Shinoda M, Shimoda M, Tsugane R, Sato O. Neuroendoscopic surgery for specific forms of hydrocephalus. Child Nerv Syst. 1999;15:56–68.
15. Spenato P, Cinalli G, Carannante G, Ruggiero C, Del Basso de Caro ML. Multiloculated hydrocephalus. In: Cinalli G, Maixner WJ, Sainte-Rose C, editors. Pediatric hydrocephalus. Milano: Springer; 2004. p. 219–44.
16. Spennato P, Cinalli G, Ruggiero C, Aliberti F, Trischitta V, Cianciulli E, Maggi G. Neuroendoscopic treatment of multiloculated hydrocephalus in children. J Neurosurg. 2007;106(1 Suppl):29–35.
17. Schulz M, Bohner G, Knaus H, Haberl H, Thomale UW. Navigated endoscopic surgery for multiloculated hydrocephalus in children. J Neurosurg Pediatr. 2010;5(5):434–42.
18. Sagan L, Hnatyszyn G. Natural history of multiloculated hydrocephalus. Neurology Volume 2008;S67–S68.
19. Nathan C. Points of control in inflammation. Nature. 2002;420:846–52.
20. Cipri S, Gambardella G. Neuroendoscopic approach to complex hydrocephalus, personal experience and preliminary report. J Neurosurg Sci. 2001;45:92–6.
21. El-Ghandour NMF. Endoscopic cyst fenestration in the treatment of multi-loculated hydrocephalus in children. J Neurosurg Pediatr. 2008;1:217–22.
22. Etus V, Kahilogullari G, Karabagli H, Unlu A. Early endoscopic ventricular irrigation for the treatment of neonatal posthemorrhagic hydrocephalus. A feasible treatment option or not ? -a multi center report. Turk Neurosurg. 2016.
23. Zuccaro G, Ramos JG. Multiloculated hydrocephalus. Childs Nerv Syst. 2011;27(10):1609–19.
24. Rhoton AL, Gomez MR. Conversion of multilocular hydrocephalus to unilocular. Case report. J Neurosurg. 1972;36:348–50.
25. Lewis AI, Keiper GL Jr, Crone KR. Endoscopic treatment of loculated hydrocephalus. J Neurosurg. 1995;82:780–5.
26. Eshra MA. Endoscopic management of septated multiloculated hydrocephalus. Alexandria J Med. 2014;50:123–6.
27. Warf BC. Hydrocephalus in Uganda: the predominance of infectious origin and primary management with endoscopic third ventriculostomy. J Neurosurg. 2005;102(1):1–15.
28. Lorber J, Pickering D. Incidence and treatment of postmeningitic hydrocephalus in new-born. Arch Dis Child. 1966;41:44–50.

29. Reinprecht A, Dietrich W, Berger A. Posthemorrhagic hydrocephalus in preterm infants: long term follow up and shunt related complications. Childs Nerv Syst. 2001;17:663–9.
30. Akbari SH, Holekamp TF, Murphy TM, Mercer D, Leonard JR, Smyth MD, Park TS, Limbrick DD Jr. Surgical management of complex multiloculated hydrocephalus in infants and children. Childs Nerv Syst. 2015;31(2):243–9.
31. Gandhoke GS, Frassanito P, Chandra N, Ojha BK, Singh A. Role of magnetic resonance ventriculography in multiloculated hydrocephalus. J Neurosurg Pediatr. 2013;11:697–703.
32. Handler LC, Wright MG. Postmeningitic hydrocephalus in infancy. Ventriculography with special reference to ventricular septa. Neuroradiology. 1978;16:31–5.
33. Chatterjee S, Chatterjee U. Overview of post-infective hydrocephalus. Childs Nerv Syst. 2011;27:1693–8.
34. Feigin RD. Bacterial meningitis in the newborn infant. Clin Perinatol. 1977;4:103.
35. Ducker TB. The pathogenesis of meningitis Systemic effects of meningococcal endotoxin within the cerebrospinal fluid. Arch Neurol. 1968;18:123.
36. Alon V, Naveh V, Gardos M, Freedman A. Neurological sequelae of septic meningitis. Isr J Med Sci. 1979;15(6):512–7.
37. Waggener JD. The pathophysiology of bacterial meningitis and cerebral abscesses: an anatomical interpretation. Adv Neurol. 1974;6:1–17.
38. Dastur DK, Manghani DK, Udani PM. Pathology and pathogenetic mechanisms in neurotuberculosis. Radiol Clin N Am. 1995;33:733–52.
39. Palur R, Rajsekhar V, Chandy MJ, et al. Shunt surgery for hydrocephalus in tuberculous meningitis: a long term follow-up study. J Neurosurg. 1991;74:64–9.
40. Sil K, Chatterjee S. Shunting in tuberculous meningitis: a surgeon's nightmare. Childs Nerv Syst. 2008;24(9):1029–32.
41. Nida TY, Haines SJ. Multiloculated hydrocephalus: craniotomy and fenestration of intraventricular septations. J Neurosurg. 1993;78(1):70–6.
42. Cushing H. Surgery of the head. In: Keen WW, editor. Surgery, its principles and practice, vol. III. Philadelphia: Saunders; 1908. p. 17–276.

Part III

Clinical Presentation of Hydrocephalus

Hydrocephalus in Adolescence

Dominic Venne

7.1 Etiologies of Hydrocephalus in Adolescence

As for many patients regardless of their age, hydrocephalus can be caused by several conditions. However, adolescents being between the pediatric and the adult ages, the etiological spectrum is wider and encompasses both age groups.

7.1.1 De Novo Hydrocephalus (Primary Hydrocephalus)

(a) *Obstructive hydrocephalus*: Compared to adults, in this population group, intra-axial lesions are more frequent conditions than extra-axial ones. After the age of 10 years, intra-axial tumors are mainly supratentorial, and they include optic apparatus, brain stem and cerebral gliomas, as well as other lesions that can be associated with syndromic conditions (e.g., giant cell astrocytomas associated with tuberous sclerosis). Extra-axial lesions include colloid cyst of the third ventricle, arachnoid cyst, and dermoid and epidermoid cysts, to name a few of them. Infratentorial tumors include cerebellar astrocytomas, medulloblastomas, ependymomas and brainstem gliomas. Non-tumoral aqueductal stenosis is also seen, but usually this condition has been diagnosed and treated earlier in life [1].

(b) *Communicating hydrocephalus* is also relatively frequent in this age group. It is possible that many of these patients had latent or "compensated" hydrocephalus. Compensated hydrocephalus might be secondary to childhood or even

D. Venne
Department of Neurosurgery, Cleaveland Clinic, Abu Dhabi, United Arab Emirates
e-mail: dominic_venne@yahoo.com

© Springer International Publishing AG 2017
A. Ammar (ed.), *Hydrocephalus*, DOI 10.1007/978-3-319-61304-8_7

neonatal hemorrhages or infections (see below arrested hydrocephalus). Idiopathic intracranial hypertension (IIH) can also be seen in these young patients. The association of IIH with obesity, endocrine disorders, usage of retinoid acid products, acne medications, and oral contraceptive reflects the poor understanding of this entity [2]. Cerebral sinus venous stenosis or obstruction is nowadays frequently diagnosed and might represent a more plausible etiology for IIH. Head injuries, cerebral and intraventricular hemorrhages, and postinfectious conditions are also frequent causes of communicating hydrocephalus.

7.1.2 Previously Known Hydrocephalus

(a) *Arrested hydrocephalus*: In this group of patients, it is suspected that a congenital, perinatal, or childhood event led to the development of hydrocephalus that was either of short duration or chronic. However due to normal or borderline intracranial pressure, these patients have remained asymptomatic for many years. Some of the patients might have been previously diagnosed with "a large head," visual impairments, or mild cognitive deficits, but otherwise they were thriving and developing acceptably well. For unknown reason, a subset group of patients will lose the frail equilibrium between the intraventricular-intraparenchymal pressure and the cerebral perfusion pressure and will become symptomatic.

(b) *Previously shunted patients*: Many patients were previously operated at a younger age for hydrocephalus. While these patients were followed properly by their initial provider or pediatric neurosurgeons, some teenagers are too often lost in their follow-up either for logistical reasons (absence of transitional care between pediatric and adult hospitals) or simply because they were doing well. However, due to the rapid physical growth of teenagers, many shunts will show evidence of hardware failure. Frequently, distal catheters will be found to be extraperitoneal for ventriculoperitoneal shunt (VP shunt) or within the jugular vein or superior vena cava for ventriculoatrial shunt (VA shunt). Disconnection from the valve or connectors and catheter breakage (often associated with calcification of the catheter) are also common findings (Fig. 7.1). Proximal obstruction can also occur due to the pulling out of ventricular catheters and scarring or adhesions in the ependymal lining or the choroid plexus.

7.2 Clinical Presentation of Active Hydrocephalus in Adolescence

The cranial sutures being already interlocked in this age group, the usual signs and symptoms of active hydrocephalus are those of intracranial hypertension as seen in adults. Headache, nausea and vomiting, visual disturbance, and in severe cases decreased level of consciousness can be reported. De novo seizures or increased seizure activity in previously known epileptic patients can occur. Some patients will present with symptoms typical of Chiari type 1 malformation (neck pain, lower cranial nerves palsies, sleep apnea, loss of pain, temperature sensations in the upper extremities, etc.). However in many cases, symptoms might be subtle and can reflect

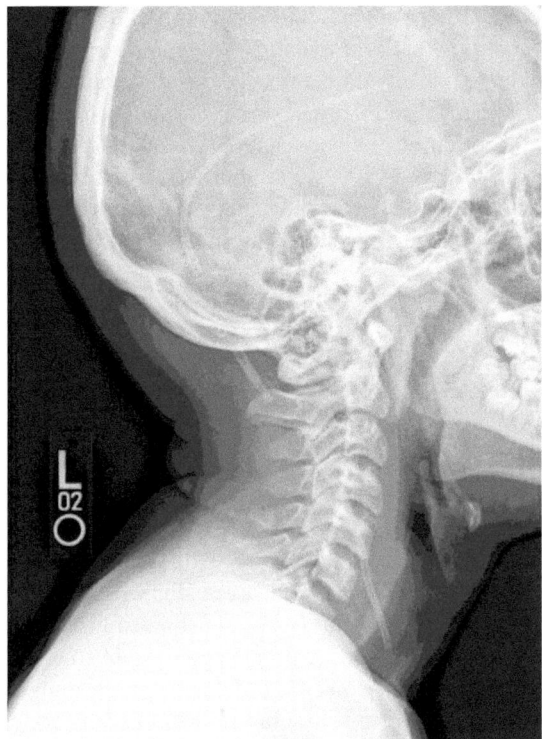

Fig. 7.1 Radiograph showing two-point breakage in the cervical segment of a VP shunt

Table 7.1 Insidious symptoms of hydrocephalus in adolescence

1.	Decreased cognitive functions (decline in academic performance)
2.	Speech problems
3.	Secondary amenorrhea
4.	Loss of interest in daily activities and/or withdrawal of social activities

early or intermittent episode of raised intracranial pressure. This situation is often faced amongst patients previously operated for shunt insertion [3]. Some of these insidious clinical manifestations are summarized in Table 7.1.

7.3 Follow-Up of Shunted Patients

7.3.1 Who Should Follow Adolescents Previously Shunted?

The answer to this question is not always clear and depends on the available medical resources and the structure of the health-care system where the patients reside. Regardless if the follow-up is provided by a pediatric or an adult neurosurgeon, previously shunted patients deserve regular assessment even if they are asymptomatic.

During those wellness assessments, physicians might detect hardware breakage or early shunt malfunctions that will eventually become symptomatic. In these situations, surgery might be considered before patients develop raised intracranial hypertension. Elective or planned shunt revisions are always preferable compared to urgent procedures that, too often, are done in suboptimal conditions. Planned procedures allow patients to be treated by the most appropriate neurosurgical team who will proceed with thorough assessment of the patient and might even consider other therapeutic alternatives (see shunt revision). In situations where a previously shunted patient is followed by a new provider, it is imperative for the neurosurgeon to have access to the previous medical and imaging records. The knowledge of previous surgeries, complications, medical past history, and allergies represents invaluable data that can decrease significantly adverse effects or suboptimal outcome after surgery.

7.3.2 Clinical Examination of Shunted Patients

(a) *Physical exam*
    1. Head circumference has limited value in this age group but can help in understanding the chronicity of the hydrocephalus when macrocrania is present.
    2. Neuro-ophthalmologic exam: Gaze palsies causing diplopia or even Parinaud syndrome can be observed. A fundoscopic exam to confirm papilledema is crucial but is often non-contributing due to preexisting optic nerve atrophy [4, 5]. Optical coherence tomography (OCT) scans are extremely valuable exams and allow neuro-ophthalmologists to objectively quantify the degree of papilledema and more importantly to follow these patients (Fig. 7.2) [6]. OCT can even confirm the success of a surgical procedure such a shunt insertion and an endoscopic third ventriculostomy [7]. Color desaturation, when present, is an early indicator of raised intracranial pressure.
    3. Neurological examination: Obtaining a neurological examination is important, mainly in 'asymptomatic' shunted patients. This baseline exam will allow the neurosurgeon to follow properly these patients and detect early changes that might indicate shunt malfunction. For patients presenting with shunt malfunction, the neurological findings are usually the ones found for adults with intracranial hypertension.

Fig. 7.2 Representation of the Codman Medos cylindrical model with the proximal reservoir or pre-chamber (**a**) and the valve chamber (**b**)

(b) *Shunt assessment*: Valves can be difficult to assess when they have been implanted for many years. They can be buried in bone and covered by scar tissue or thick skin. In some instances, the valve palpation can even be misleading. In fact, in cases of distal obstruction, it is common to find a completely filed valve chamber that can be depressed by manual palpation. Although the valve chamber will not empty itself due to the distal obstruction, it will allow some depression and rebound that will falsely lead the physician to assume an intact functioning. This finding should never replace an imaging study (CT or MRI) mainly when a patient has new symptoms suggestive of a shunt malfunction. On the other hand, there are some valve models with a proximal reservoir or pre-chamber that can provide valuable information to the physician (e.g., the Codman Medos cylindrical model with pre-chamber) (Fig. 7.3). With these systems, the physician can block the refilling of the valve by maintaining a constant pressure on the proximal reservoir. With another finger, the physician will press on the valve chamber to flush it completely. While releasing the pressure on the valve chamber, the physician will keep his finger at the same position on the proximal reservoir waiting for three possible scenarios: (1) If the valve chamber stays full, it indicates a distal obstruction. (2) While releasing the pressure on the proximal reservoir, the valve chamber might refill completely indicating a good functioning of the shunt, or (3) the valve chamber will not refill and stay collapsed, indicating a proximal obstruction.

(c) *Radiological assessment*
   1. *Shunt series*: Multiple views of the skull, chest, and abdomen are necessary to evaluate the position and the integrity of a VP shunt. For instance, extra-peritoneal located catheters (pre-peritoneal) can be missed with a single AP view of the abdomen. Careful assessment of the entire length of the catheters is mandatory since discreet changes such calcifications, kink, and breakage can be present, explaining some of the symptoms described by the patients.
   2. *CT head*: CT scan is often the only exam available in emergency situations. They usually provide most of the needed information but do not allow comprehensive assessment of the ventricular system and basal cisterns.
   3. *Brain MRI*: Brain MRI represents the gold standard for the assessment of a VP shunt. It allows precise assessment of the location of ventricular catheter(s) and the anatomy of the ventricular system, the basal and prepontine cisterns. CSF flow and navigation studies are often added if an endoscopic third ventriculostomy or navigation-guided procedures are planned (endoscopic procedures, shunt insertion, etc.). If a patient has already a programmable valve, a post-MRI valve x-ray or direct reprogramming of the valve should be planned since some of the valve settings might change during the MRI.

(d) *Ancillary tests*: Dynamic changes in the pulsatility index (PI) and the resistance index (RI), as calculated from the transcranial Doppler (TCD), are very useful data that can provide indirect but objective measurements of the intracranial pressure [8, 9].

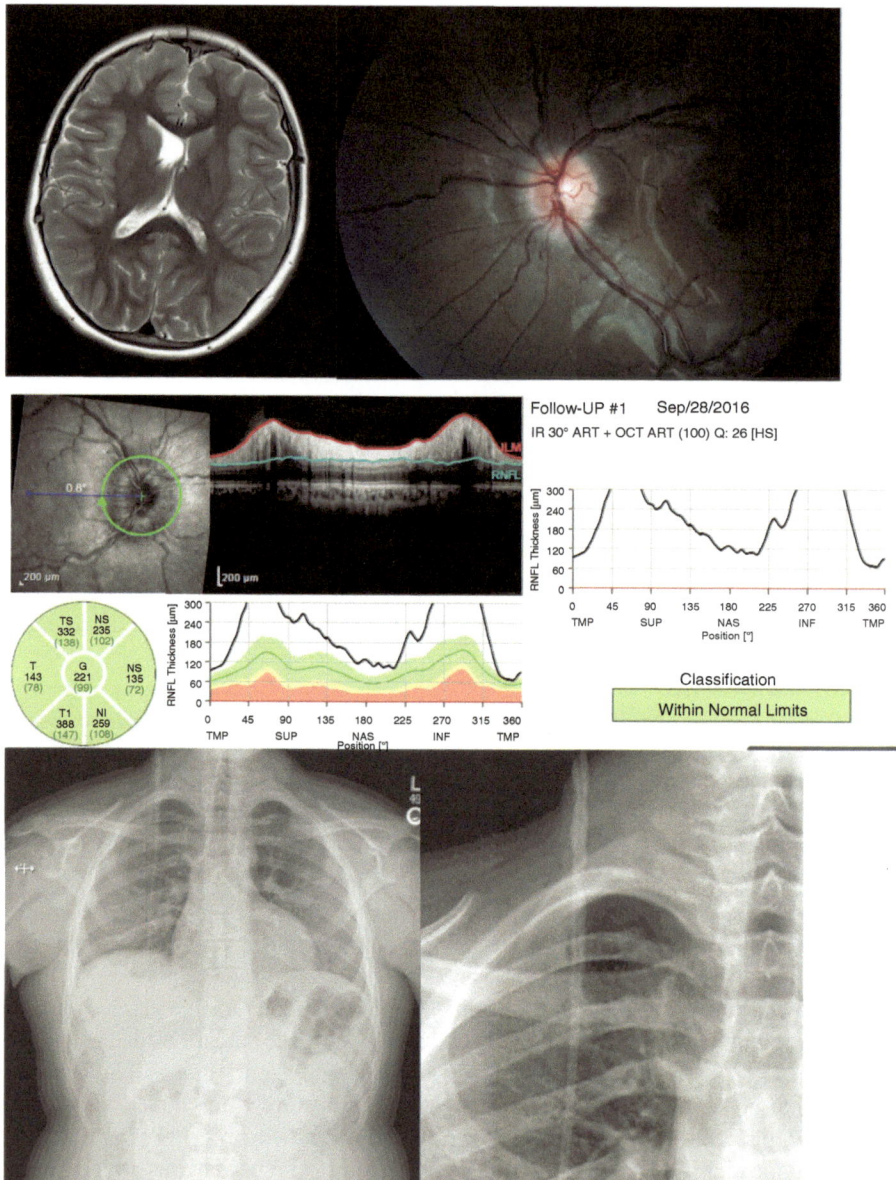

Fig. 7.3 Assessment of a 16 year-old patient shunted at a younger age with recent history of headache. The brain MRI showed mild enlargement of the right lateral ventricle. The neurological exam was normal except for papilledema. The plain x-ray showed discreet calcifications and narrowing of the cervical part of the distal catheter. The patient underwent distal revision of his VP shunt with complete resolution of the symptoms

7.4   Management of Hydrocephalus in Adolescence

7.4.1   Management of Patients with "Arrested Hydrocephalus"

This group of patients is probably the most challenging one since the clinical presentation is often unclear and the management is not standardized. One simple way to address this problem is to divide these patients in four groups. *Group 1* are patients with new onset of symptoms (often headache), positive neuro-ophthalmological findings (papilledema, Parinaud syndrome, etc.), and MRI confirming hydrocephalus. These patients should be treated with ETV (in case of aqueductal stenosis) or VP shunt (for communicating hydrocephalus). *Group 2* is composed of patients with on and off headaches and normal neuro-ophthalmological exams (absence of papilledema or optic nerve atrophy) but positive MRI. For these patients, in addition of being referred in neurology for headache and migraine consultation, a close follow-up is recommended and includes neuro-ophthalmological exams, MRI every 6 months, and transcranial Doppler including the measurement of pulsatility index and resistance index [8]. In cases where the TCD shows clear signs of raised intracranial pressure or gradual increase of the intracranial pressure, surgery is recommended. *Group 3* includes patients with similar clinical presentation to Group 2 but with fluctuating ICP (based on the TCD). These patients should be followed closely but are usually treated conservatively. Finally *Group 4* includes asymptomatic patients with pure incidental finding of "arrested hydrocephalus", normal neuro-ophthalmological exam and normal TCD follow-ups (including PI and RI). These patients have probably only constitutional or acquired ventriculomegaly and should be treated conservatively.

7.4.2   Primary Treatment Options for Active Hydrocephalus

Shunts have saved millions of lives and remain excellent treatment options for hydrocephalus. However, due to the non-negligible associated complications, other therapeutic options should always be considered. When possible, the cause of obstructive hydrocephalus should be treated first, even if a temporary external ventricular drain has to be inserted for few days. This includes tumor or lesion removal and endoscopic fenestration of arachnoid cysts or entrapped ventricles. Endoscopic third ventriculostomy represents an excellent treatment option unless the CT or MRI scans show contraindications for such procedure (stenotic foramina of Monro, virtual prepontine space, etc.) or in the presence of active central nervous system infection or hemorrhage. Finally, shunt insertion still represents a great procedure and, when done meticulously, has a low morbidity and good long-term outcome. When a decision is made to implant a shunt, a ventriculoperitoneal shunt should be the first option, keeping ventriculoatrial shunts for specific indications.

7.4.3   Shunt Revision

(a) *Conversion to an endoscopic third ventriculostomy (ETV)*: Some patients presenting with shunt malfunction can develop severe intracranial hypertension
and will need urgent intervention. In these cases, the most critical investigation
is usually obtained with a head CT scan or occasionally with a MRI.
Unfortunately, too often these urgent situations do not always allow the surgeon
to develop other therapeutic plans. However, one should keep in mind that a
shunt obstruction might represent a great opportunity to replace a shunt by an
endoscopic procedure (endoscopic third ventriculostomy, cyst fenestration,
etc.). In these cases, proper investigation should include a brain MRI to assess
the anatomy of the ventricular system, the basal cisterns, and vascular structures. At the same time, the neurosurgical team should also be aware that many
patients were previously shunted due to recurrent childhood infections, failed
ETV attempts, or even congenital anatomical variations, conditions that can be
considered as relative or absolute contraindications for endoscopic procedures.
Having said that, when feasible, performing an ETV might represent the best
treatment for these shunted patients.
(b) *Shunt replacement*
   1. *Simplifying hydrocephalus*: At the moment of a shunt revision, simplifying a
      complex hydrocephalus case into a simple one can also represent a great
      improvement for patients. During shunt revision procedures, endoscopic
      fenetration of cysts, or lysis of adhesions and septations, can be done with a
      low morbidity rate when performed by an experienced team.
   2. *Improving accuracy of proximal catheter placement*: Proximal catheters can
      be placed also more accurately during revisions by using either navigation or
      endoscopic assistance. However, one should also consider the slight
      increased in infection rate with longer procedure or the usage of
      endoscopy.

7.4.4   Removing a Shunt

Due to personal considerations, some teenagers will consult neurosurgeons with the
intention of removing their shunt (Fig. 7.4). These situations have to be dealt with
compassion and understanding. A poor encounter might bring this patient to consult
another less experienced neurosurgeon who will agree to proceed with such request.
However one should remember that removing a functioning shunt could result in a
disastrous situation. Most of these patients have poor cerebral compliance, and inter-
rupting this fragile equilibrium can lead to rapid intracranial hypertension. Shunt
removal might not be a wise choice in situations where patients had several shunt
revisions, history of symptomatic slit ventricles, or recurred episodes of "headache".
However, in cases where the primary indication was unclear, MRI findings show
normal anatomy and normal CSF flow study, or in the presence of a long-standing
history of asymptomatic hardware breakage or disconnection, then a staged removal

Fig. 7.4 Fifteen-year-old patient seen in consultation for shunt removal due to cosmetic considerations

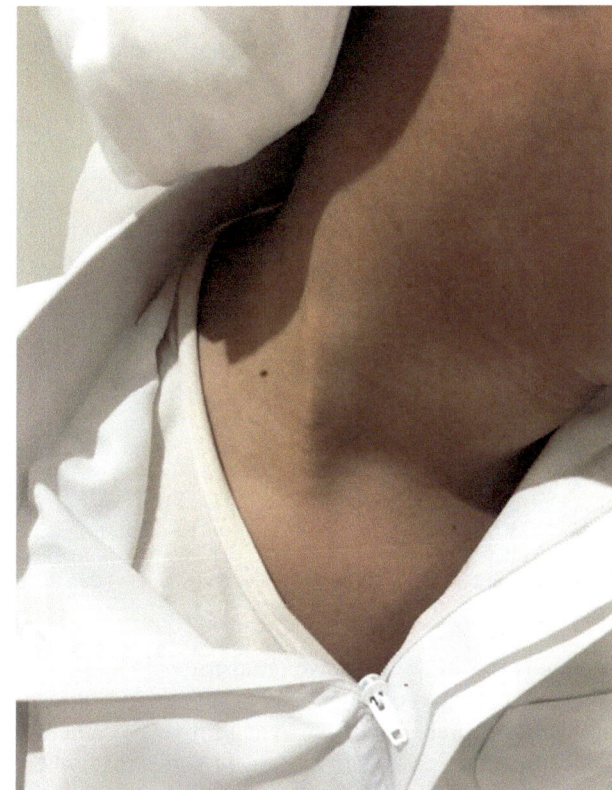

could be considered. However the patient and the parents should have a full understanding of the potential lethal implications of such procedure. If this option is chosen, patients should be admitted in a high dependency unit with continuous monitoring. Once the complete preoperative investigation is completed, ligation of the distal catheter can be performed under local anesthesia. The ligation point should be done at a site that can be rapidly access and removed in case of emergencies. MRI scanning is performed after 24–48 h, and patients can be discharged home after few days with clear instructions given to the parents to urgently bring back the patient to the emergency department in case of any new symptoms. After an arbitrary trial of 2–3 weeks, the brain MRI is repeated, and if no radiological changes are noted, the shunt can be removed under general anesthesia. Prior to the procedure, coagulation parameters should be verified, and the patient and parents should be informed of the potential complications, in particular cerebral or intraventricular hemorrhage leading to permanent neurological deficits, CSF leakage, and the possible need to reinsert another permanent shunt. During surgery, although most ventriculoperitoneal shunt can be removed through one cranial incision, complete skin preparation and draping is recommended. Distal catheters can be more friable, adherent, and anchored to the

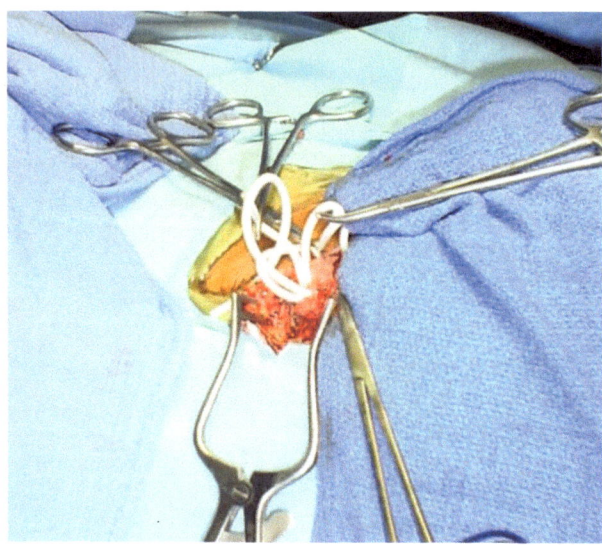

Fig. 7.5 Intraoperative photograph depicting knots within the distal catheter

abdominal wall or even have intraperitoneal loops and knots (Fig. 7.5). In the same manner, proximal catheters can be adherent to the choroid plexus, ependyma, and cerebral or cerebellar parenchyma. Proximal catheters and valve can be completed embedded and covered by bone. In some case where the catheter is found to be lounged into the brain stem or in the close vicinity of a vascular structure, a craniotomy or an endoscopic approach might be needed for direct visualization of the catheter removal. Needless to say that in these cases the shunt removal should be reconsidered. All the wounds should be closed in an anatomical fashion to prevent CSF leakage or poor cosmetic results.

7.4.5   Scar Revision

Sometimes the problem is not the shunt itself but the scar appearance. A small area of alopecia can be treated easily with a scar revision or hair transplant. The risk of infection should be considered but is relatively low.

References

1. Osborn AG, Salzman KL, Jhaveri MD. Diagnostic imaging: brain. 3rd ed. Philadelphia: Elsevier; 2015.
2. Markey KA, et al. Understanding idiopathic intracranial hypertension: mechanisms, management, and future directions. Lancet Neurol. 2016;15(1):78–91.
3. Iskandar BJ, McLaughlin C, Mapstone TB, Grabb PA, Oakes WJ. Pitfalls in the diagnosis of ventricular shunt dysfunction: radiology reports and ventricular size. Pediatrics. 1998;101:1031–6.

4. Mizrachi IBB, Trobe JD, Gebarski SS, Garton HJL. Papilledema in the assessment of ventriculomegaly. J Neuroophthalmol. 2006;26:260–3.
5. Nazir S, O'Brien M, Qureshi NH, Slape L, Green TJ, Phillips PH. Sensitivity of papilledema as a sign of shunt failure in children. J AAPOS. 2009;13:63–6.
6. Heidary G, Rizzo JF III. Use of the optical coherence tomography to evaluate papilledema and pseudopapilledema. Semin Ophthalmol. 2010;25(5–6):198–205. Review.
7. Koktekir E, et al. Resolution of papilledema after endoscopic third ventriculostomy versus cerebrospinal fluid shunting in hydrocephalus: a comparative study. J Neurosurg. 2014;120:1465–70.
8. Jindal A, Mahapatra AK. Correlation of ventricular size and transcranial Doppler findings before and after ventricular peritoneal shunt in patients with hydrocephalus: prospective study of 35 patients. J Neurol Neurosurg Psychiatry. 1998;65:269–71.
9. Rainov NG, Weise JB, Burkert W. Transcranial Doppler sonography in adult hydrocephalic patients. Neurosurg Rev. 2000;23:34–8.

Clinical Presentation of Hydrocephalus in Adults

8

Christopher Witiw, Laureen Hachem, and Mark Bernstein

8.1 Clinical Presentation of Acute Hydrocephalus

Acute hydrocephalus may be caused by a variety of primary conditions, and the majority of cases occur in close temporal association with the causative pathology. Table 8.1 provides a distribution of causes from a consecutive series of 103 patients where a bedside external ventricular drain was placed for relief of elevated ICP secondary to acute hydrocephalus. Numerous other pathologies have the potential to cause hydrocephalus through increased resistance to CSF flow. A full review of the potential etiologies of acute hydrocephalus is outside the scope of this chapter, but detailed history taking will alert the practitioner to potential causative etiologies and prompt more detailed questioning into the presence of signs and symptoms that may be associated with elevated intracranial pressure from hydrocephalus.

8.2 Signs and Symptoms of Raised Intracranial Pressure Secondary to Hydrocephalus

The cardinal feature of acute hydrocephalus in adults is an increase in intracranial pressure (ICP). Regardless of the underlying etiology, raised ICP is associated with a constellation of clinical signs and symptoms, which may be subtle in the early

C. Witiw, M.D., M.S. • L. Hachem, M.D
Division of Neurosurgery, University of Toronto, Toronto, ON, Canada
e-mail: christopher.witiw@mail.utoronto.ca; laureen.hachem@mail.utoronto.ca

M. Bernstein, M.D., M.H.Sc., F.R.C.S.C. (✉)
Faculty of Medicine, Division of Neurosurgery, Toronto Western Hospital,
University of Toronto, Toronto, ON, Canada

Joint Centre for Bioethics, University of Toronto, Toronto, ON, Canada
e-mail: mark.bernstein@uhn.ca

© Springer International Publishing AG 2017
A. Ammar (ed.), *Hydrocephalus*, DOI 10.1007/978-3-319-61304-8_8

Table 8.1 Distribution of etiologies of acute hydrocephalus from a consecutive series of 103 patients for which a bedside external ventricular drain was placed

Diagnosis	n (%)
Subarachnoid hemorrhage	56
Intracerebral hematoma	13
Tumor	12
Arteriovenous malformation	9
Intraventricular hemorrhage	8
Trauma	3
Dural arteriovenous fistula	2

Adapted from Roitberg et al. [36]

stages but worsen rapidly over time. A clear understanding of the clinical presentation of high ICP and its progression is therefore essential for early detection and management.

8.2.1 Deterioration of Consciousness

An altered level of consciousness is often one of the most concerning signs of raised ICP. In the initial stages, changes in mental status may present insidiously with subtle alterations in personality, confusion, disorientation, and lethargy. In the later stages, elevated ICP can lead to central brain herniation resulting in damage to the reticular activating system and the brainstem cardiac and respiratory centers. Severe compression can manifest in the classic Cushing's triad of systolic hypertension, bradycardia, and irregular breathing, and patients can deteriorate to stupor, coma, and eventually death [1]. Although the rate of deterioration is dependent on the severity of the resulting elevation in ICP, patients can progress to coma within hours, making early detection of raised ICP a key tenet of appropriate management.

8.2.2 Headaches

Headaches are the most common symptom of raised ICP. The classic headache from raised ICP is thought to be caused by compression of pain-sensitive coverings of the meninges, venous sinuses, and blood vessels as a result of the raised cranial pressures [2]. Not all instances of an elevated ICP lead to headache, and the rate of change of CSF pressure that is the key determinant is likely. The headaches of increased ICP are usually described as a dull, generalized, non-throbbing pain that is worse in the morning due to reduced venous drainage from lying supine at night and dilation of cerebral vessels secondary to elevated blood carbon dioxide levels during sleep. Headaches are further aggravated by other factors that transiently elevate ICP including coughing and Valsalva maneuver [3].

8.2.3 Nausea and Vomiting

Projectile vomiting, particularly worse in the morning and in conjunction with headaches, may be suggestive of elevated ICP. This effect is likely mediated by activation of meningeal mechanoreceptors due to irritation from elevated ICP which in turn stimulates histamine receptors in the brainstem vomiting center [4]. While nausea may also be present in some patients, vomiting in the setting of raised ICP is generally not preceded by symptoms of nausea.

8.2.4 Papilledema

Papilledema is considered a hallmark sign of raised intracranial pressure. True papilledema resulting from high ICP is seen on fundoscopic examination as bilateral swollen optic discs, with blurring at the disc margin. Furthermore, venous pulsations are generally absent, and splinter hemorrhages can be seen around the disc [5]. If raised ICP is left untreated, chronically elevated cerebral pressures can lead to permanent optic atrophy. Although papilledema has high specificity for elevated ICP, its sensitivity is poor and may take several days to develop after the onset of elevated ICP. Therefore, its absence on exam should not preclude a diagnosis of elevated ICP especially in the presence of other positive signs or symptoms [3].

8.2.5 Parinaud's Syndrome

Focal neurological deficits may also be associated with raised ICP secondary to acute or subacute hydrocephalus due to the compressive effects of brain herniation on specific anatomical structures or from direct enlargement of the third ventricle in the case of obstructive hydrocephalus. The most common site of damage is the dorsal midbrain where compression disrupts underlying structures including the rostral interstitial nucleus of medial longitudinal fasciculus (riMLF) and the posterior commissure leading to a constellation of ocular impairments known as Parinaud's syndrome. Patients may initially present with lid retraction (Collier's sign) and light-near dissociation of the pupils where pupils constrict to accommodation but not to light [5]. The hallmark of Parinaud's syndrome is impairment in upward gaze which can first manifest as an upbeat nystagmus and later progress to upward gaze palsy usually with spared downward gaze [1, 6, 7]. Since the damage is supranuclear, vestibuloocular and oculocephalic reflexes remain intact, and upward gaze may still be elicited by passive movement of the head (doll's maneuver) [8]. As ICP progresses, these reflexes can be lost leading to a negative doll's sign and a primary conjugate downward gaze known as the "setting-sun sign." Although the latter sign has been well described in infants with hydrocephalus, this finding may be less common in adults.

8.2.6   Abducens Palsy

Abducens palsy may also be present in the setting of an acute elevation in ICP where patients present with a clinically evident esotropia and complain of horizontal diplopia. The abducens nerve is most susceptible of the cranial nerves to damage from high ICP and brain herniation due to its long course from the brainstem to the cavernous sinus [9, 10].

8.3   Chronic Clinical Presentation of Hydrocephalus

Contrary to the immediacy with which acute hydrocephalus must be recognized and managed, the practitioner is afforded more time to fully work up hydrocephalus of the chronic nature. However, this does not diminish the importance of a clear appreciation of the clinical presentation. In fact, in chronic cases the clinical presentation typically plays a critical role in diagnosis because the symptoms are often rather subtle and may mimic other neurological conditions. In such case, treatment decision-making hinges upon an accurate clinical diagnosis. Without this, a patient may not be afforded the potential benefits of intervention or may be subjected to the unwarranted risks of a needless procedure. In this section, we will focus on the three most common categories of chronic hydrocephalus: idiopathic normal pressure hydrocephalus (NPH), hydrocephalus associated with Chiari malformations, and compensated hydrocephalus.

8.4   Idiopathic Normal Pressure Hydrocephalus

Normal pressure hydrocephalus (NPH) is the most common form of chronic hydrocephalus in adults. While the symptoms of NPH are nonspecific, it is classically associated with the triad of gait disturbance, urinary incontinence, and dementia with ventricular enlargement [11]. The presentation of NPH can closely mimic that of other neurological conditions, and careful appreciation of the clinical symptoms is crucial to effective management.

8.4.1   Gait Disturbance

Gait disturbance is considered the primary presenting symptom of NPH and the most likely to improve after shunting. The gait disturbances are due to a motor apraxia in which there is difficulty in planning motor movements. Initially, impairment in gait may be mild and present as slight imbalance and slowed steps. Over time, patients present with a "magnetic gait" characterized by a wide stance, short stride length, and reduced foot-floor clearance. Postural instability along with difficulty in initiating steps and turning are also signs that can be present. A summary of the gait impairments seen in NPH is shown in Table 8.2. The exact mechanisms

Table 8.2 Gait impairments in NPH

Feature	Clinical finding
Step frequency in 10 m	>13 steps
Step width	Distance between toes >1 foot length
Step length	<1 foot length
360° turn	Requires >4–6 steps
Bipedal gait	Correction of foot position in >25% of steps
Foot posture	Externally rotated

Adapted from Kiefer and Unterberg [14]

underlying these disturbances remain controversial and are likely the result of a combination of abnormalities in cortical and subcortical motor circuitry along with impairment in executive function from the associated dementia [12, 13].

Although the gait disturbances associated with NPH may present similarly to those seen in other movement disorders, a number of features can aid in its distinction. Unlike Parkinson's disease, patients with NPH do not have a resting tremor and generally have absent or very mild cogwheel rigidity. Asymmetrical motor symptoms, which can be seen in Parkinson's disease, are rarely present in NPH and should point toward another diagnosis [14].

Furthermore, NPH does not present with the characteristic signs of dysarthria or dysmetria seen in cerebellar ataxia. Weakness and incoordination are also uncommon features of NPH [15].

8.4.2   Urinary Incontinence

Bladder dysfunction occurs in 45–90% of patients with NPH and generally presents late in the course of the condition [16]. Early signs manifest as urinary frequency and urgency but can progress to complete incontinence. Initially, impairments in bladder function may be due to the damage caused by the enlarged ventricles on descending pathways that normally inhibit bladder contractions. This leads to detrusor muscle over activity and spastic incontinence with preserved bladder sphincter control [17, 18]. Later in the disease, incontinence can be compounded by the loss of a drive for micturition from the concurrent dementia. Fecal incontinence is not usually present in NPH but may occur in very severe cases [19].

8.4.3   Dementia

It is estimated that up to 6% of adult dementias are due to NPH [20]. The cognitive impairments seen with NPH affect frontal lobe function and are subcortical in nature. As such, patients present with reduced information processing, slowed psychomotor speed, and impairment in executive functioning along with deficits in learning, attention, and memory rather than the specific cortical deficits seen in cortical dementias like Alzheimer's disease [18]. The spectrum of these symptoms

Table 8.3 Distinguishing NPH from other dementias

	NPH	Alzheimer's	Parkinson's
Gait disturbance	✓	✓	✓
Postural instability	✓		✓
Impaired memory	✓	✓	✓
Urinary dysfunction	✓	✓	✓
Changes in behavior		✓	✓
Limb rigidity			✓
Resting tremors			✓
Bradykinesia			✓
Type of dementia	Subcortical	Cortical	Subcortical

ranges from very mild changes that may be overlooked as normal signs of aging to severe impairments that are refractory to shunting treatments [21]. NPH can closely resemble dementias associated with neurodegenerative conditions, and approximately 75% of patients with NPH also have a concurrent Alzheimer's or vascular dementia [14]. There are, however, a number of features, which may help distinguish NPH from other common causes of cognitive impairment which are summarized in Table 8.3.

The onset and severity of the dementia can provide important clues to the responsiveness of NPH to medical treatment. Severe dementia or a dementia presenting prior to the onset of gait disturbance may be an indicator of a worse prognosis and unresponsiveness to shunting [22]. Mild forms of dementia seen in NPH respond well to treatment, and thus early detection is essential. Standard psychological tests like the Mini-Mental Status Examination are often insufficient to detect subtle impairments in cognition, and thus more specific testing of executive function should be done when suspecting a patient with NPH. It is important to note however that not all cases of NPH present with cognitive impairment, and thus the constellation of presenting signs and symptoms must be considered when reaching a diagnosis.

8.5   Hydrocephalus Related to Chiari Malformation

A number of congenital conditions can present with hydrocephalus in adulthood. Most notable are the Chiari malformations. The Chiari III and IV malformation are rare and are all but unseen in adults. Conversely, Chiari I malformations are frequently diagnosed in adulthood, and Chiari II are often cared for by adult neurosurgeons once they reach adulthood. The associated hydrocephalus seen in Chiari malformations is thought to be a consequence of the underlying disruption of CSF dynamics at the level of the foramen magnum [23]. To date, there still remains significant debate surrounding the exact pathogenesis of Chiari-related hydrocephalus; however, an understanding of the clinical symptoms and manifestations of this condition is critical in accurate diagnosis and effective management.

8.5.1 Type I Chiari Malformation

Type I Chiari malformation is characterized by herniation of the cerebellar tonsils at least 5 mm below the foramen magnum. While the abnormality is present from birth, patients often remain asymptomatic until early in adulthood with the onset of symptoms generally occurring in the second or third decade of life. The clinical manifestations of Chiari I malformation can be grouped into those caused by direct compression of hindbrain and brainstem structures and those due to disturbances in CSF dynamics. Compressive related symptoms include sensory and motor deficits, hypo- or hyperreflexia, ataxia, nystagmus, and respiratory abnormalities [23]. Direct compression of cranial nerves may also manifest in visual disturbances, difficulty in swallowing, and hoarseness.

Occipital-suboccipital headaches are the most common symptom of Chiari I malformation in adults and are likely a result of the changes in CSF dynamics. The exact mechanism remains unknown but has been postulated to be a result of either a "craniospinal pressure dissociation" which exacerbates herniation, thus putting traction on pain-sensitive dural areas, or transient increases in ICP from the obstructed CSF outflow tract [24, 25]. Either theory may explain why the classic Chiari headache is worsened with Valsalva, coughing, or exertion.

Disruption of CSF flow is also thought to underlie the findings of syringomyelia and hydrocephalus in patients with Chiari I malformation, although there is currently no accepted theory to explain their pathogenesis. Approximately 65% of patients present with syringomyelia which can lead to central cord syndrome and progressive limb weakness due to expansion of the syrinx. In contrast, hydrocephalus is seen in 7% of Chiari I patients and is more common in those with concomitant syringomyelia [26]. The hydrocephalus can present with typical signs of increased ICP along with imaging evidence of enlarged ventricles [27]. Dynamic MR imaging studies among symptomatic adults have also found significant elevations in CSF peak systolic velocities at the foramen magnum [28].

It is important to note that many cases of Chiari I malformation can remain asymptomatic well into adulthood and may only be detected incidentally imaging on imaging done for unrelated symptoms [29]. Furthermore, the extent of herniation does not predict the severity of hydrocephalus or other symptoms, and thus the presenting clinical picture must guide management plans.

8.5.2 Type II Chiari Malformation

In contrast to type I, type II Chiari malformation is associated with more extensive herniation including portions of the cerebellar vermis, medulla, and fourth ventricle. Type II Chiari almost always presents in infants with myelomeningocele with symptoms similar to those seen in type I but often more severe. Due to the underlying myelomeningocele, hydrocephalus is strongly associated with type II malformation and is present in over 80% of patients. Although Chiari II malformation rarely remains asymptomatic into adulthood, patients who underwent treatment in infancy may present as adults with decompensated hydrocephalus and signs and symptoms of elevated ICP [30].

8.6 Syndrome of Compensated Hydrocephalus

Compensated hydrocephalus is often used interchangeably with the term arrested hydrocephalus. Currently there isn't an agreed upon definition for these terms, but two well-defined clinicopathologic entities fall into the category. These are late-onset idiopathic aqueductal stenosis (LIAS) and long-standing overt ventriculomegaly in adults (LOVA). It was commonly thought that patients with "compensated hydrocephalus" are asymptomatic, but it is now becoming clearer that these patients are not truly without symptoms and in fact may benefit from a CSF diversion procedure but only if the subtle associated clinical symptoms are recognized [31].

8.6.1 Late-Onset Idiopathic Aqueductal Stenosis

LIAS is a noncommunicating hydrocephalus where CSF flow is obstructed at the aqueduct of Sylvius without any extrinsic compression or history of an etiology that may predispose to aqueductal stenosis [32]. From a group of 31 patients with adult onset idiopathic aqueductal stenosis, acute onset of symptoms was uncommon [33]. Only two patients presented with symptoms that were less than 1 month in duration. Most of the patients presented with symptoms that had been present for longer than 6 months. Of these, about half presented primarily with headaches, and the others presented primarily with NPH-type symptoms (more than one of the classic triad of NPH symptoms—gait disturbance, memory disturbance, or incontinence) (Table 8.4). Patients who presented primarily with headaches were significantly younger than the patients with NPH-type symptoms. Overall the initial success rate of ETV was greater than 80% in the overall group, highlighting the importance of symptom recognition [33].

8.6.2 Long-Standing Overt Ventriculomegaly in Adults

LOVA is a term first described in 2000 [34]. This clinicopathological entity is comprised of chronic hydrocephalus starting in infancy, evolving slowly then manifesting with clinical abnormalities in adulthood. In the original series of 20 patients, aqueductal stenosis was the underlying etiology in all cases [34]. The primary distinguishing feature between LIAS and LOVA is that there is evidence of hydrocephalus at an early age in cases of LOVA (head circumference greater than two standard deviations above the mean and/or neuroimaging evidence of a significantly expanded or destroyed sella turcica). The most common symptoms found in patients diagnosed with LOVA were headaches and imbalance, but numerous presenting symptoms were noted (Table 8.5) [35]. Overall these patients have been found to benefit from CSF diversion necessitating an awareness of the subtle clinical manifestations of this condition [35].

Table 8.4 Type and duration of symptoms from a cohort of 31 patients with diagnosed late-onset idiopathic aqueductal stenosis

	Duration		
	≤1 month	1–6 months	>6 months
Headache	5	1	14
Gait disturbance	2	1	13
Memory disturbance	2	1	12
Incontinence	2	0	8
Blurred vision	0	0	4
Tremor	0	0	3
Seizure	0	0	1
Swallowing difficulty	0	0	1
Parinaud's syndrome	1	0	0

Adapted from Fukuhara and Luciano [33]

Table 8.5 Presenting symptoms in adulthood of a cohort of patients with long-standing overt ventriculomegaly in adults

Diagnosis	n (%)
Headache	19 (95%)
Imbalance	15 (75%)
Loss of consciousness	10 (50%)
Memory problems	9 (45%)
Visual obscuration	8 (40%)
Cognitive problems	5 (25%)
Vomiting	3 (15%)
Nausea	2 (10%)
Dizziness	2 (10%)
Urinary incontinence	2 (10%)

Adapted from Al-Jumaily et al. [35]

Conclusions

Regardless of whether the presentation of hydrocephalus is acute or chronic, an accurate clinical diagnosis is necessary for effective clinical management. In acute cases, clinical signs and symptoms will often be the earliest indicators of the onset of hydrocephalus allowing for expeditious workup and management, while in chronic cases, the symptoms may be subtle or overlapping with other neurological conditions. Imaging findings act as valuable diagnostic adjuncts, but they do not supplant the need for clinical acumen.

References

1. van Gijn J, Hijdra A, Wijdicks EF, Vermeulen M, van Crevel H. Acute hydrocephalus after aneurysmal subarachnoid hemorrhage. J Neurosurg. 1985;63(3):355–62.
2. Ducros A, Biousse V. Headache arising from idiopathic changes in CSF pressure. Lancet Neurol. 2015;14(6):655–68.
3. Dunn LT. Raised intracranial pressure. J Neurol Neurosurg Psychiatry. 2002;73(Suppl 1):i23–7.
4. Holland JC, Breitbart WS, Jacobsen PB, Lederberg MS, Loscalzo MJ, McCorkle R. Psycho-oncology. 3rd ed. Oxford: Oxford University Press; 2015.

124

C. Witiw et al.

5. Corbett JJ. Neuro-ophthalmologic complications of hydrocephalus and shunting procedures. Semin Neurol. 1986;6(2):111–23.
6. Koga H, Mori K, Kawano T, Tsutsumi K, Jinnouchi T. Parinaud's syndrome in hydrocephalus due to a basilar artery aneurysm. Surg Neurol. 1983;19(6):548–53.
7. Swash M. Periaqueductal dysfunction (the Sylvian aqueduct syndrome): a sign of hydrocephalus? J Neurol Neurosurg Psychiatry. 1974;37(1):21–6.
8. Pierrot-Deseilligny CH, Chain F, Gray F, Serdaru M, Escourolle R, Lhermitte F. Parinaud's syndrome: electro-oculographic and anatomical analyses of six vascular cases with deductions about vertical gaze organization in the premotor structures. Brain. 1982;105(Pt 4):667–96.
9. Azarmina M, Azarmina H. The six syndromes of the sixth cranial nerve. J Ophthalmic Vis Res. 2013;8(2):160–71.
10. Hanson RA, Ghosh S, Gonzalez-Gomez I, Levy ML, Gilles FH. Abducens length and vulnerability? Neurology. 2004;62(1):33–6.
11. Edwards RJ, Dombrowski SM, Luciano MG, Pople IK. Chronic hydrocephalus in adults. Brain Pathol. 2004;14(3):325–36.
12. Zaaroor M, Bleich N, Chistyakov A, Pratt H, Feinsod M. Motor evoked potentials in the preoperative and postoperative assessment of normal pressure hydrocephalus. J Neurol Neurosurg Psychiatry. 1997;62(5):517–21.
13. Yogev-Seligmann G, Hausdorff JM, Giladi N. The role of executive function and attention in gait. Mov Disord. 2008;23(3):329–42. Quiz 472.
14. Kiefer M, Unterberg A. The differential diagnosis and treatment of normal-pressure hydrocephalus. Dtsch Arztebl Int. 2012;109(1–2):15–25. Quiz 26.
15. Bradley WG. Normal pressure hydrocephalus: new concepts on etiology and diagnosis. AJNR Am J Neuroradiol. 2000;21(9):1586–90.
16. Meier U, Zeilinger FS, Kintzel D. Signs, symptoms and course of normal pressure hydrocephalus in comparison with cerebral atrophy. Acta Neurochir. 1999;141(10):1039–48.
17. Sakakibara R, Kanda T, Sekido T, et al. Mechanism of bladder dysfunction in idiopathic normal pressure hydrocephalus. NeurourolUrodyn. 2008;27(6):507–10.
18. Tsakanikas D, Relkin N. Normal pressure hydrocephalus. Semin Neurol. 2007;27(1):58–65.
19. Hakim CA, Hakim R, Hakim S. Normal-pressure hydrocephalus. Neurosurg Clin N Am. 2001;12(4):761–73, ix.
20. Casmiro M, D'Alessandro R, Cacciatore FM, Daidone R, Calbucci F, Lugaresi E. Risk factors for the syndrome of ventricular enlargement with gait apraxia (idiopathic normal pressure hydrocephalus): a case-control study. J Neurol Neurosurg Psychiatry. 1989;52(7):847–52.
21. Graff-Radford NR. Normal pressure hydrocephalus. Neurol Clin. 2007;25(3):809–32, vii–viii.
22. Black PM, Ojemann RG, Tzouras A. CSF shunts for dementia, incontinence, and gait disturbance. Clin Neurosurg. 1985;32:632–51.
23. Tubbs RS, Lyerly MJ, Loukas M, Shoja MM, Oakes WJ. The pediatric Chiari I malformation: a review. Childs Nerv Syst. 2007;23(11):1239–50.
24. Williams B. Cough headache due to craniospinal pressure dissociation. Arch Neurol. 1980;37(4):226–30.
25. Sansur CA, Heiss JD, DeVroom HL, Eskioglu E, Ennis R, Oldfield EH. Pathophysiology of headache associated with cough in patients with Chiari I malformation. J Neurosurg. 2003;98(3):453–8.
26. Milhorat TH, Chou MW, Trinidad EM, et al. Chiari I malformation redefined: clinical and radiographic findings for 364 symptomatic patients. Neurosurgery. 1999;44(5):1005–17.
27. Hayhurst C, Osman-Farah J, Das K, Mallucci C. Initial management of hydrocephalus associated with Chiari malformation Type I-syringomyelia complex via endoscopic third ventriculostomy: an outcome analysis. J Neurosurg. 2008;108(6):1211–4.
28. Haughton VM, Korosec FR, Medow JE, Dolar MT, Iskandar BJ. Peak systolic and diastolic CSF velocity in the foramen magnum in adult patients with Chiari I malformations and in normal control participants. AJNR Am J Neuroradiol. 2003;24(2):169–76.
29. Meadows J, Kraut M, Guarnieri M, Haroun RI, Carson BS. Asymptomatic Chiari Type I malformations identified on magnetic resonance imaging. J Neurosurg. 2000;92(6):920–6.

30. Jenkinson MD, Hayhurst C, Al-Jumaily M, Kandasamy J, Clark S, Mallucci CL. The role of endoscopic third ventriculostomy in adult patients with hydrocephalus. J Neurosurg. 2009;110(5):861–6.
31. Larsson A, Stephensen H, Wikkelso C. Adult patients with "asymptomatic" and "compensated" hydrocephalus benefit from surgery. Acta Neurol Scand. 1999;99(2):81–90.
32. Spennato P, Tazi S, Bekaert O, Cinalli G, Decq P. Endoscopic third ventriculostomy for idiopathic aqueductal stenosis. World Neurosurg. 2013;79(2 Suppl):S21.e13–20.
33. Fukuhara T, Luciano MG. Clinical features of late-onset idiopathic aqueductal stenosis. Surg Neurol. 2001;55(3):132–6. Discussion 136–7.
34. Oi S, Shimoda M, Shibata M, et al. Pathophysiology of long-standing overt ventriculomegaly in adults. J Neurosurg. 2000;92(6):933–40.
35. Al-Jumaily M, Jones B, Hayhurst C, et al. Long term neuropsychological outcome and management of 'decompensated' longstanding overt ventriculomegaly in adults. Br J Neurosurg. 2012;26(5):717–21.
36. Roitberg BZ, Khan N, Alp MS, Hersonskey T, Charbel FT, Ausman JI. Bedside external ventricular drain placement for the treatment of acute hydrocephalus. Br J Neurosurg. 2001;15(4):324–7.

Radiological Diagnosis of Hydrocephalus

9

Sari Saleh AlSuhibani, Abdulrahman
Hamad Alabdulwahhab, and Ahmed Ammar

9.1 Making Diagnosis

9.1.1 Conventional Radiograph

The conventional radiograph is an initial step in hydrocephalus evaluation. It is based on the findings of increased intracranial pressure like sutural diastasis, craniofacial disproportion, bulging of anterior fontanelle, erosion of dorsum sella, and intracranial calcification. The most common sign noticed in the case of chronic hydrocephalus is known as the copper beaten skull (Fig. 9.1) exhibited by a prominent gyral impressions on the inner table of the skull. This appearance is sensitive for chronic raised intracranial pressure and seen in other conditions such as posterior fossa mass, craniosynostosis, and metabolic disorder like hypophosphatemia [1]. Table 9.1 shows the common findings in skull radiographs due to increased intracranial pressure.

9.1.2 Ultrasonogram

The ultrasound modality is highly specific for diagnosing hydrocephalus in the prenatal period. It can be early detected by the end of first trimester of pregnancy. The fetal ventricular system can be visualized as early as 20–24 weeks of gestational period [2].

S. Saleh AlSuhibani (✉) • A. Hamad Alabdulwahhab
Department of Radiology, King Fahd University Hospital, Imam Abdulrahman Bin Faisal University, Al Khobar, Saudi Arabia
e-mail: ssuhibani@uod.edu.sa

A. Ammar
Department of Neurosurgery, King Fahd University Hospital, Imam Abdulrahman Bin Faisal University, Al Khobar, Saudi Arabia
e-mail: ahmed@ahmedammar.com

© Springer International Publishing AG 2017
A. Ammar (ed.), *Hydrocephalus*, DOI 10.1007/978-3-319-61304-8_9

Fig. 9.1 Lateral view of skull x-ray demonstrates diffuse confluent thumbprinting appearance. There is indentation of the occipital skull representing copper beaten skull. The visulaized proximal portion of the VP shunt shows partial discontinuity adjacent to reservoir, suggestive of VP shunt fracture

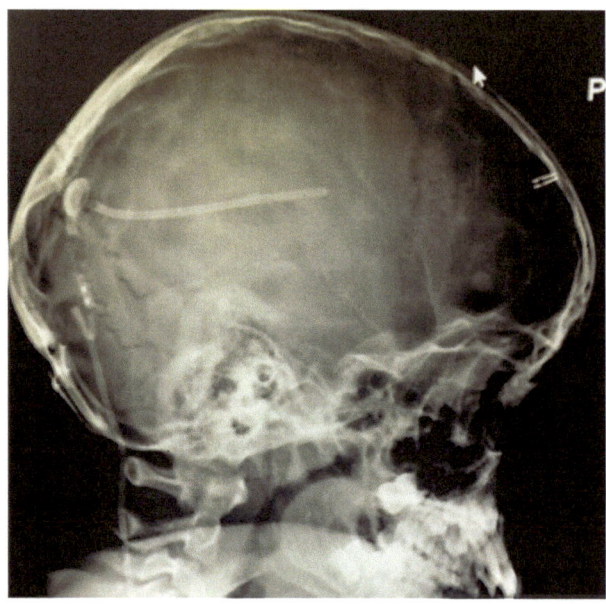

Table 9.1 Common findings in skull radiograph due to increased intracranial pressure

No.	Findings
1	Diastasis of the cranial sutures
2	Bulging of the frontal fontanelle
3	Erosion of dorsum sella
4	Craniofacial disproportion
5	Intracranial calcification
6	Erosion of the inner table of occipital bone

Cranial ultrasound is still useful in infants and young children up to 18 months when the anterior fontanelle is still patent to assess and evaluate the morphology of the lateral ventricle and can detect extracerebral fluid collection and parenchymal lesion and assess intraparenchymal hemorrhage in preterm patients (Fig. 9.2). However, there is limitation in evaluation of third and fourth ventricles as well as the subarachnoid spaces, and for this reason, the definite cause and diagnosis of hydrocephalus is difficult and sometimes impossible to be made by cranial ultrasound alone and needs further evaluation [2].

Cranial ultrasound has the following advantages: (1) easy to handle, (2) no radiation involved, (3) comes in a portable machine, (4) less expensive, and (5) no sedation requirement. A phase array probe is a high-frequency small footprint probe used specifically for pediatric cranial ultrasound.

The following images are samples of cranial sonogram with labeled anatomy (Figs. 9.3a–c and 9.4).

Fig. 9.2 Cranial ultrasound (coronal image) demonstrates both bodies of the lateral ventricles are enlarged suggestive of hydrocephalus. **A** measures the transverse diameter of the right lateral ventricle

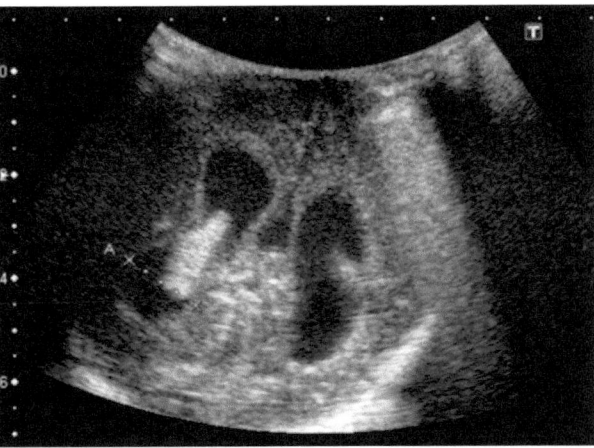

Fig. 9.3 (**a**) Anterior coronal image. Illustration: *red arrow*, frontal lobe; *yellow arrow*, superior aspect of the orbit. (**b**) More posterior coronal image. Illustration: *red arrow*, frontal horn of lateral ventricle; *yellow arrow*, corpus callosum; *white arrow*, interhemispheric fissure; *green arrow*, Sylvian fissure. (**c**) A more posterior coronal image. Illustration: *red arrow*, body of lateral ventricle; *yellow arrow*, corpus callosum; *white arrow*, thalamus; *green arrow*, Sylvian fissure

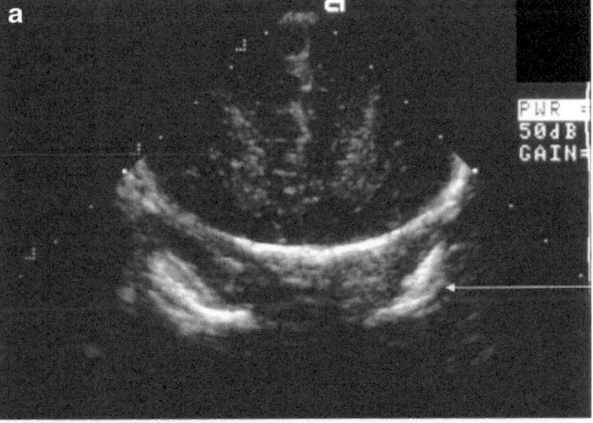

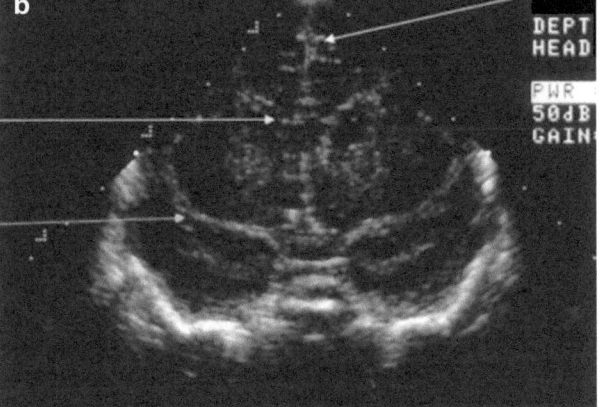

Fig. 9.3 (continued)

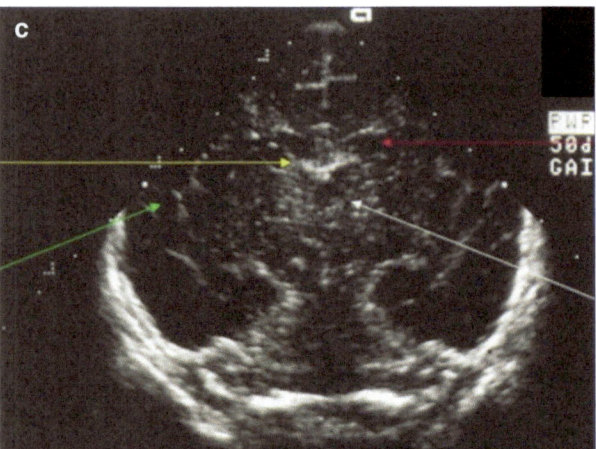

Fig. 9.4 Sagittal image. Illustration: *red arrow*, corpus callosum; *yellow arrow*, fourth ventricle; *white arrow* (*up*), thalamus; *white arrow* (*down*), pons; *green arrow*, third ventricle

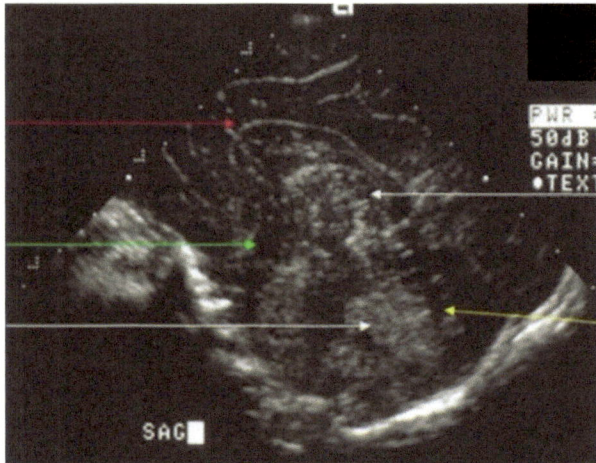

9.1.3   Computed Tomography (CT) Scans

Computed tomography (CT) scan is a diagnostic radiological tool for definitive diagnosis, follow-up, and evaluation of complications of hydrocephalus. It is apparently used to evaluate the ventricular shape, size and identify the cause of hydrocephalus and the complications of shunt failure. It is widely available, rapid, and compatible with life support devices. But few children needed to be sedated to reduce head movements during the CT scan exam. Radiation exposure from multiple CT follow-ups is a disadvantage. Some institutions prefer an alternative way like low-dose CT or T2 HASTE MR scans, which is an acceptable method for follow-up [3]. The signs of hydrocephalus detected in CT/MRI examination are shown in Table 9.2.

Figures 9.5 and 9.6a, b are unenhanced CT scans showing enlargement of ventricular system.

Table 9.2 CT/MRI appearance of hydrocephalus

No.	Features
1	Dilatation of the ventricles above the level of obstruction
2	Rounding of the frontal horns of the lateral ventricles (earliest sign)
3	Dilated temporal horns anteroposteriorly, more than 2 mm
4	Bulging of the floor and lateral walls of the third ventricle
5	Confluent periventricular hypo-attenuation representing transependymal flow (interstitial edema), seen in the case of active hydrocephalus
6	Thinning and bowing of corpus callosum
7	Depression of fornices, which is seen in chronic hydrocephalus
8	Sylvian and interhemispheric fissures are not visible
9	The ratio between the largest width of the frontal horns and the internal diameter from inner table to inner table at this level should be greater than 0.5
10	Herniation of the third ventricle into the sella turcica and rarely causes erosion in chronic cases
11	Irreversible demyelination of white matter in case of long-standing untreated hydrocephalus

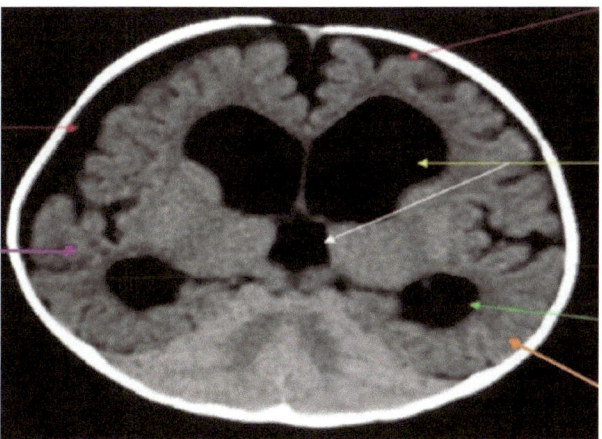

Fig. 9.5 An unenhanced CT scan (trans-axial cut) demonstrates enlargement of the ventricular systems and bowing of frontal lobe. Note presence of white matter changes. Illustration: *red arrow*, extra-axial CSF spaces; *yellow arrow*, frontal horn of lateral ventricle; *white arrow*, third ventricle; *green arrow,* temporal horn of lateral ventricle; *purple arrow*, gray and white matter junction; *orange arrow*, white matter changes

9.1.4 Magnetic Resonance Imaging (MRI)

Magnetic resonance imaging (MRI) is the best modality to delineate the ventricles and has a superior resolution to evaluate the cause of hydrocephalus. It has a major role in preoperative surgical planning such as endoscopic third ventriculostomy (ETV). The protocol for conventional MR technique differs from one institution to another, including axial and sagittal turbo spin echo T1-WI, axial fluid-attenuated

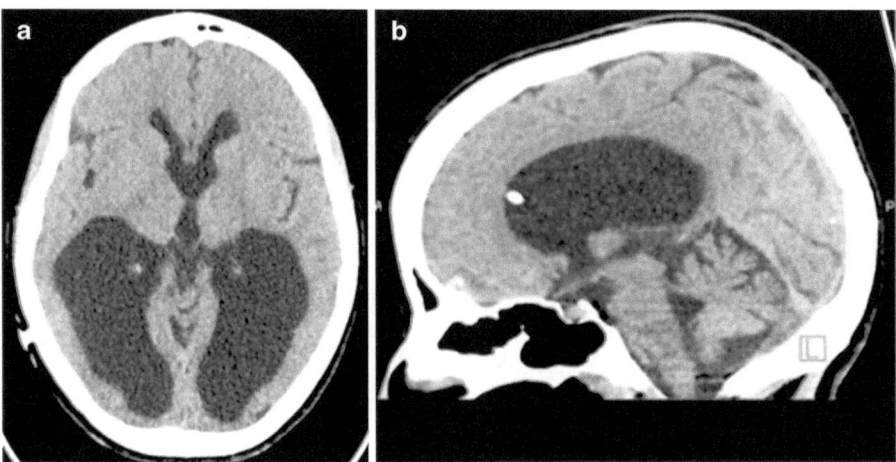

Fig. 9.6 (**a, b**) Selected images of unenhanced CT scan. 7A in axial cut demonstrate enlargement of the posterior horns of both lateral ventricles and third ventricle. 7B in sagittal cut demonstrate enlargement of lateral ventricles with normal fourth ventricle. Note the presence of shunt in the center of the lateral ventricle

Table 9.3 Radiological criteria for diagnosis of hydrocephalus [4, 5]

No.	Criteria
1	Radiological evidence of ventriculomegaly using Evans' index[a] measurement; the abnormality is more than 0.3
2	Enlargement of third ventricle recess
3	Fenestration of the corpus callosum
4	Enlargement of lateral ventricular horns
5	Decreased distance of mamillopontine
6	Reduced angle of frontal horn
7	Periventricular white matter hyperintensities
8	Aqueduct flow void phenomenon in T2W images (in case of communicating hydrocephalus)

[a]The Evans' index is a ratio of maximum width of the frontal horns of the lateral ventricle and the maximal internal diameter of the skull at the same level employed in axial CT and MRI images

inversion recovery, and axial, coronal turbo spin echo T2-WI. The diffusion imaging (DWI) can help in detecting the resolution of interstitial edema in posttreatment of acute hydrocephalus patients by using average diffusion constants. Three-dimensional (3D) constructive interface in steady state (CISS) is an effective technique for evaluation of intraventricular obstructive hydrocephalus [3]. There are nonspecific criteria for diagnosis of hydrocephalus as illustrated in Table 9.3.

There are new limited advanced MRI techniques to evaluate hydrocephalus, which include phase-contrast MRI (PC-MRI), three-dimensional (3D) heavily T2W sequences, and contrast material-enhanced MR cisternography (CE-MRC).

Phase-contrast MRI (PC-MRI) estimates CSF circulation. However, it is limited by flow artifacts [6, 7]. 3D heavily T2W sequences support to provide accurate

Fig. 9.7 MRI of the brain; trans-axial cut demonstrates widening of the right lateral ventricle with periventricular edema suggestive of active hydrocephalus. Illustration: *yellow arrow*, dilated body of right lateral ventricle; *red arrow*, posterior interhemispheric fissure; *purple arrow*, decompensated left lateral ventricle; *green arrow*, periventricular edema

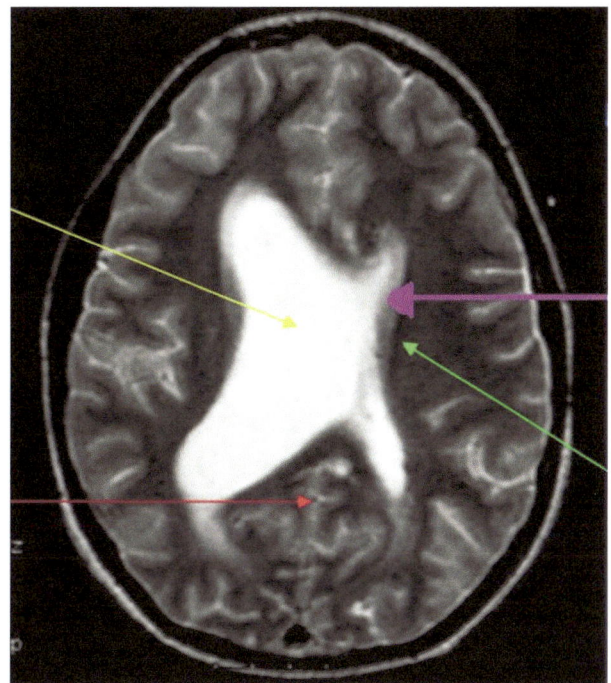

anatomical data. CE-MRC is an invasive technique and can be used to assess the patency of endoscopic third ventriculostomy or explore spontaneous third ventriculostomy [8].

The 3D space (three-dimensional sampling perfection with application-optimized contrast using the variable flip angle evolution) is the most sensitive technique for hydrocephalus evaluation, commonly in obstructive hydrocephalus. This technique uses high-resolution multiplanar reformatted images to scan the whole cranium using small isotrope voxels of limited specific absorption rate. The technique is noninvasive and sensitive to artifacts with the ability to use routine sequences [9, 10].

Samples of MRI scan anatomy (Figs. 9.7 and 9.8)

9.2 Differentiation Between Large Ventricle in Cerebral Atrophy and Hydrocephalus

The detected cause of ventriculomegaly either by pathological cause like *hydrocephalus or degenerative cerebral atrophic changes* is difficult and overlaps in radiological image interpretation. The most common sensitive sign to differentiate hydrocephalus from cerebral atrophy is dilatation of the temporal horns in relation to the body of the lateral ventricle. Table 9.4 presents the criteria that favor the diagnosis of hydrocephalus [11, 12].

Fig. 9.8 Image of a sagittal T1-WI demonstrates enlargement of the lateral and third ventricle associated with superior funneling of aqueduct. Normal fourth ventricle. Note the presence of shunt centered in lateral ventricle (dot of low signal intensity)

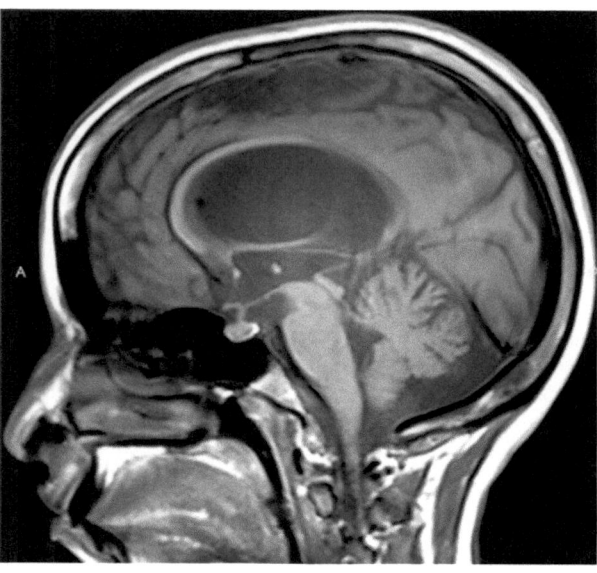

Table 9.4 Radiological findings that favor diagnosis of hydrocephalus from brain atrophy

No.	Criteria
1	Dilatation of the temporal horns of lateral ventricle in relation to the body of lateral ventricle (most sensitive sign)
2	Periventricular interstitial edema, in case of acute hydrocephalus
3	Increased frontal horn radius
4	Large or rapidly enlarging head
5	Intraventricular flow void from CSF movement on MR
6	Upward displacement of corpus callosum
7	Depression of the posterior fornix
8	Narrow callosal angle
9	Unremarkable parahippocampal fissures

Similarly, Fig. 9.9a, b are images showing the differentiation between hydrocephalus from brain atrophy.

9.3   Shunt Assessment and Complications

The main objective of applying shunts in hydrocephalus patients is to establish communication between the CSF and drainage source like the peritoneum. The shunt is a tube of one-way flow of CSF between two different gradient pressures.

Clinical assessment of patients who are suspected to have shunt obstruction present with symptoms of increased intracranial pressure (e.g., vomiting, headache, and papilledema). Some obstructive shunt cases are complaining of unusual presentation like gait abnormality and neck pain, or personality changes maybe noted. In examination of shunt obstruction, changes in the head like circumferential widening of the head size, bulging of the fontanelle, edematous veins of the scalp, sixth cranial nerve

Fig. 9.9 (**a**) Ventricular angle tends to be smaller in hydrocephalus than in atrophy. (**b**) Frontal horn radius tends to be larger in hydrocephalus than in atrophy

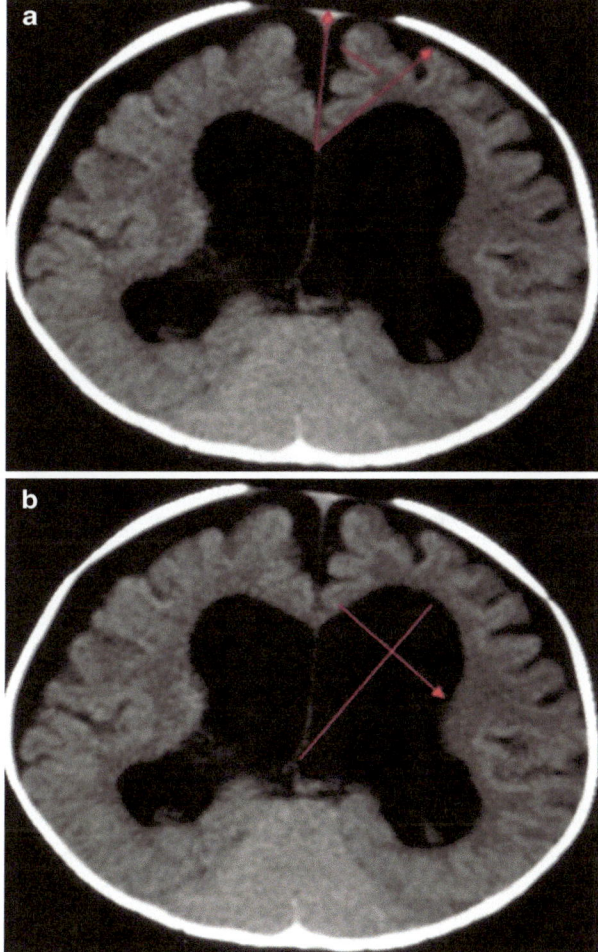

palsy, seizure, and upward gaze palsy are commonly found [13]. The malfunction of the shunt can be partial or variable based on flow of cerebral blood and production of CSF or secondary to increase intracranial pressure [14]. Evaluation of shunt for infection and gastrointestinal disease should be included in the diagnostic workup.

The radiological evaluation of the shunt includes a radiograph series of the shunt to evaluate its tract and its apparatus. The radiograph series includes lateral radiograph of the skull, neck, chest, and abdomen. An unenhanced head CT scan can provide quick evaluation of ventricular size in case of follow-up and complications [15].

The complications of VP shunt are more common in pediatric age group. A recent study shows a postoperative rate of shunt failure which is approximately 40% in the first year and 50% in the second year [16]. Infections are commonly seen in first 6 month after shunt insertion, which are likely secondary to contamination of the shunt during surgery [17, 18]. The tissue debris plugging in the lumen of the shunt is the most common cause of mechanical shunt failure. Other causes such as breakage of the shunt and migration of shunt tip are also seen (Fig. 9.10). The

Fig. 9.10 Frontal cranial
radiograph showed
discontinuity of the VP
shunt with sutural diastasis

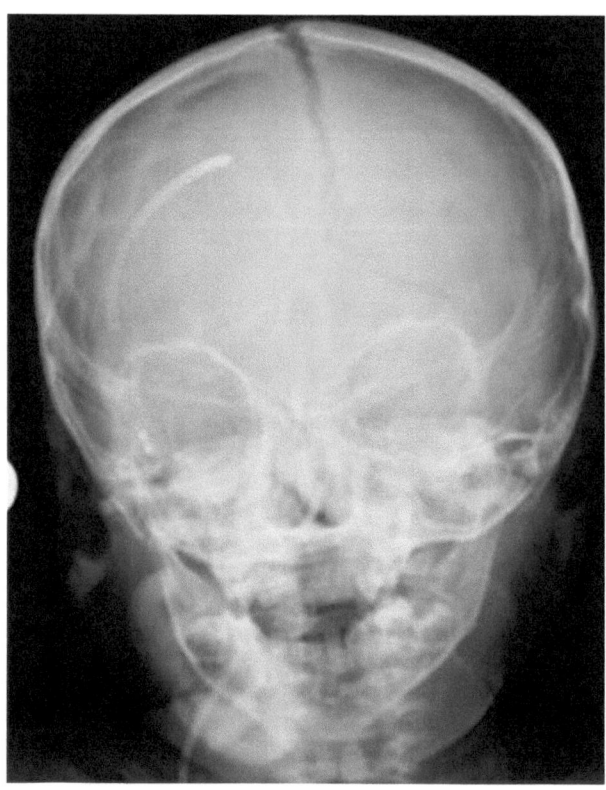

Table 9.5 Common shunt complications by location

Ventricular end	Peritoneal end	Atrial end
Blockage	Infection causing peritonitis	Thrombosis
Disconnection	CSFoma "pseudotumor"	Infection
Migration		
Hemorrhage		
Infection		
Isolated or "trapped" fourth ventricle		
Secondary craniosynostosis (see Fig. 9.11)		
Slit ventricle syndrome		

factors associated with high shunt failure include age at initial shunt insertion, indication of the shunt, symptoms like headache and fever, purulent drainage, skin erosion, abdominal pain, and signs of peritonitis [19].

The mortality rate from initial shunt insertion is estimated to be around 0.1% and from shunt failure is approximately up to 4% [20]. The nuclear medicine, Tc99m with DTPA, has a role for evaluation of shunt patency. Common shunt complications depend on the location, either in the ventricle, peritoneum, or atrium as illustrated in Table 9.5.

9.4   Pathophysiology of Transependymal CSF Permeation

Transependymal CSF permeation through the ventricle is under the term of interstitial edema (hydrocephalic edema), which is explained by increased intraventricular pressure causing breakdown of ependymal surface of the ventricle and CSF migration into extracellular space. Thus, the fluid in the brain concentrates around the ventricle with increase hydrostatic pressure within the white matter region with subsequent reduction in the volume. This type of edema is seen mainly in active hydrocephalus (Fig. 9.11) and in normal pressure hydrocephalus [21, 22].

9.5   Endoscopic Third Ventriculostomy (ETV)

The mechanism of hydrocephalus is complex. One of the several theories of hydrocephalus pathophsiology is hyperdynamic flow theory which is explained by CSF obstruction in the acute form and by changes in intracranial compliance in the chronic form [37, 38].

Endoscopic third ventriculostomy (ETV) is considered as a treatment option for hydrocephalus. It is commonly used in case of noncommunicating hydrocephalus whatever the underlying etiology. Recent study shows that the successful rate of

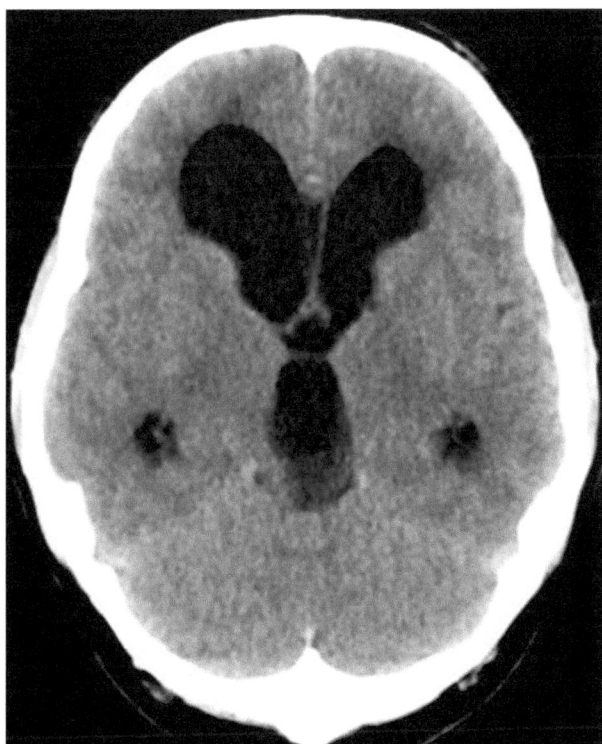

Fig. 9.11 Trans-axial CT scan of the brain showing dilated lateral and third ventricles. Presence of periventricular hypodensity around the ballooning frontal horns suggestive of transependymal CSF permeation that indicates acute hydrocephalus

ETV in obstructive hydrocephalus is higher than in communicating hydrocephalus [23].

In 1923, the first trial endoscopic third ventriculostomy was performed by William Mixter who tried to use urethroscope to perform third ventriculostomy in a child with obstructive hydrocephalus [24]. The aim of this procedure is to make communication between third ventricle and interpeduncular cistern to create bypasses of CSF obstruction.

The preoperative assessment of ETV is important to know the exact location of posterior communicating artery and distance from the midline to generate prepontine interval to identify anatomical data for the surgeon. Locating Liliequist's membrane located in the prepontine cistern using three-dimensional sequences with multiplanar reformat images will also help [25, 26]. The measurement ratio of CSF stroke volume of the interpeduncular and prepontine cisterns is important in preoperative assessment [27]. MRI cine phase-contrast imaging can be used as a method of distinguishing between types of hydrocephalus and any other abnormality located in basal cisterns [28].

The endoscopic third ventriculostomy outcome depends on the size of the ventricle and should regress postoperatively to obtain perfect clinical outcome [29]. Magnetic resonance ventriculography is an effective method for subarachnoid assessment and patency of stoma after ETV [30].

In postoperative assessment, MRI of CSF flow can locate the area of narrowing by detecting flow void appearance, which is inconsistent in other routine techniques. The best sequence to detect flow void sign is sagittal T2-weighted turbo inversion recovery MR images, which is important to assess stoma patency. Cine phase-contrast MR imaging is used for evaluating CSF flow, which is a sensitive and noninvasive quantification method. Cine PC MRI may be used to determine the patency of the stoma and estimated stroke volume in ventriculostomy [31].

9.6   Normal Pressure Hydrocephalus (NPH)

Normal pressure hydrocephalus (NPH) is defined as ventriculomegaly in a background of absent CSF raising pressure, supported by a clinical triad of dementia, gait abnormality, and urinary incontinence. The first to describe NPH was Hakim and Adams. They noticed that many patients have ventriculomegaly with normal opening of CSF pressure after lumbar puncture [32].

In 50% of the cases, there is no obvious identified cause, and they are classified as idiopathic normal pressure hydrocephalus. In the remaining 50%, it may be secondary to other conditions like meningitis and subarachnoid hemorrhage, but this condition can be reversible after treatment. However, the pathophysiology, which leads to normal pressure hydrocephalus, is still not clear [33].

Neuroimaging scan such as CT and MRI plays an important role in the diagnosis of normal pressure hydrocephalus in addition to the clinical triad and

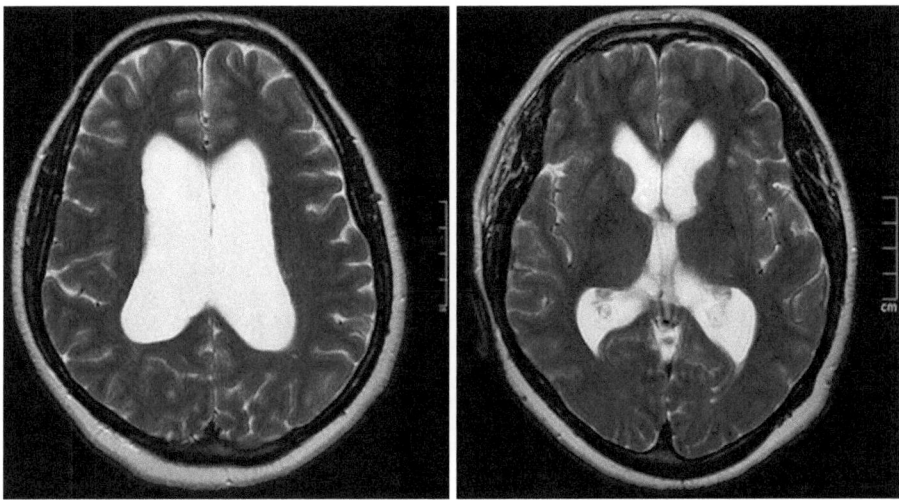

Fig. 9.12 Selected images of axial cuts of T2-WI of the brain demonstrates enlarge lateral ventricles out of proportion to sulcal atrophy, suggestive of normal pressure hydrocephalus

lumbar puncture. MRI has superior resolution to evaluate the ventricle and extra-axial CSF spaces. Ventricular enlargement out of proportion to sulcal atrophy is typically seen (Fig. 9.12). The isotopic cisternography (In111 DTPA cisternography) has a major role in evaluating the dynamic CSF outcome in postoperative condition [34].

9.7   CSF Flow Technique in MRI

Cerebrospinal fluid (CSF) flow technique using a variety of MRI sequences that measures quantification of pulsatile CSF flow through synchronized cardiac gating and applying a high-resolution axial phase contrast with axis should be perpendicular to the proximal cerebral aqueduct and applying phase encoding in a caudocranial direction. The phase-contrast images are displayed on a gray scale, the low signal intensity indicates caudal flow, and the bright signal intensity represents cranial flow [35].

9.8   Pre- and Post-shunting Assessment of Tract in Hydrocephalus

Long-standing hydrocephalus affects the white matter secondary to mechanical changes and compression due to ventricular dilatation and metabolic derangement. It affects the pyramidal tract, which results to irreversible damage [36].

References

1. Mahomed N, Sewchuran T, Mahomed Z. The copper beaten skull. SA J Radiol. 2012;16:25–6.
2. Fudge RA, editor. About hydrocephalus—a book for families. (Brochure). San Francisco: University of California; 2000.
3. Nielsen N, Breedt A. Hydrocephalus. Nursing care of the pediatric Neurosurgery patient. Published by A Springer. 2013; p. 52–3.
4. Dincer A, Ozek MM. Radiologic evaluation of pediatric hydrocephalus. Childs Nerv Syst. 2011;27(10):1543–62. doi:10.1007/s00381-011-1559-x.
5. Pople IK. Hydrocephalus and shunts: what the neurologist should know. J Neurol Neurosurg Psychiatry. 2002;73(Suppl 1):i17–22.
6. Algin O, Hakyemez B, Parlak M. Phase-contrast MRI and 3D-CISS versus contrast-enhanced MR cisternography on the evaluation of the aqueductal stenosis. Neuroradiology. 2010;52:99–108. doi:10.1007/s00234-009-0592-x.
7. Algin O, Hakyemez B, Parlak M. Phase-contrast MRI and 3D-CISS versus contrast-enhanced MR cisternography on the evaluation of spontaneous third ventriculostomy existence. J Neuroradiol. 2011;38(2):98–104. doi:10.1016/j.neurad.2010.03.006.
8. Algin O, Turkbey B. Intrathecal gadolinium-enhanced MR cisternography: a comprehensive review. AJNR Am J Neuroradiol. 2013;34(1):14–22. doi:10.3174/ajnr.A2899.
9. Algin O, Turkbey B. Evaluation of aqueductal stenosis by 3D sampling perfection with application-optimized contrasts using different flip angle evolutions sequence: preliminary results with 3 T MR imaging. AJNR Am J Neuroradiol. 2012;33(4):740–6. doi:10.3174/ajnr. A2833.
10. Algin O, Turkbey B, Ozmen E, Ocakoglu G, Karaoglanoglu M, Arslan H. Evaluation of spontaneous third ventriculostomy by three-dimensional sampling perfection with application-optimized contrasts using different flip-angle evolutions (3D-SPACE) sequence by 3 T MR imaging: preliminary results with variant flip-angle mode. J Neuroradiol. 2013;40(1):11–8. doi:10.1016/j.neurad.2011.12.003.
11. Segev Y, Metser U, Beni-adani L, et al. Morphometric study of the midsagittal MR imaging plane in cases of hydrocephalus and atrophy and in normal brains. AJNR Am J Neuroradiol. 2001;22(9):1674.
12. Maytal J, Alvarez LA, Elkin CM, et al. External hydrocephalus: radiologic spectrum and differentiation from cerebral atrophy. AJR Am J Roentgenol. 1987;148(6):1223–30.
13. Stellman-Ward GR, Bannister CM, Lewis MA, et al. The incidence of chronic headache in children with shunted hydrocephalus. Eur J Pediatr Surg. 1997;7:12–4.
14. Ditmyer S. Hydrocephalus. In: Allen PJ, Vessey JA, editors. Primary care of the child with a chronic condition. St Louis: Mosby; 2004. p. 543–60.
15. Iskandar BJ, McLaughlin C, Mapstone TB, et al. Pitfalls in the diagnosis of ventricular shunt dysfunction: radiology reports and ventricular size. Pediatrics. 1998;101(6):1031.
16. Browd SR, Ragel BT, Gottfried ON, Kestle JR. Failure of cerebrospinal fluid shunts art I: obstruction and mechanical failure. Pediatr Neurol. 2006;34(2):83–92.
17. Duhaime AC, Bonner K, McGowan KL, et al. Distribution of bacteria in the operating room environment and its relation to ventricular shunt infections. Childs Nerv Syst. 1991;7:211–4.
18. Kulkarni AV, Drake JM, Lamberti-Pasculli M. Cerebrospinal fluid shunt infection: a prospective study of risk factors. J Neurosurg. 2001;94:195–201.
19. Piatt JH Jr, Garton HJ. Clinical diagnosis of ventriculoperitoneal shunt failure among children with hydrocephalus. Pediatr Emerg Care. 2008;24(4):201–10.
20. McGirt MJ, Leveque JL, Wellons JC III, et al. Cerebrospinal fluid shunt survival and etiology of failures: a seven-year institutional experience. Pediatr Neurosurg. 2002;36:248–55.
21. Milhorat TH, Clark RG, Hammock MK. Experimental hydrocephalus. 2. Gross pathological findings in acute and subacute obstructive hydrocephalus in the dog and monkey. J Neurosurg. 1970;32:390–9.

22. Milhorat TH, Clark RG, Hammock MK, McGrath PP. Structural, ultrastructural, and permeability changes in the ependyma and surrounding brain favoring equilibration in progressive hydrocephalus. Arch Neurol. 1970;22:397–407.
23. Roopesh Kumar SV, Mohanty A, Santosh V, Satish S, Devi BI, Praharaj SS, et al. Endoscopic options in management of posterior third ventricular tumors. Childs Nerv Syst. 2007;23:1135–45.
24. Yadav YR, Parihar V, Pande S, Namdev H, Agarwal M. Endoscopic third ventriculostomy. J Neurosci Rural Pract. 2012;3(2):163–73.
25. Nishikawa T, Takehira N, Matsumoto A, Kanemoto M, Kang Y, Waga S. Delayed endoscopic intraventricular hemorrhage (IVH) removal and endoscopic third ventriculostomy may not prevent consecutive communicating hydrocephalus if IVH removal was insufficient. Minim Invasive Neurosurg. 2007;50:209–11.
26. Souweidane MM, Morgenstern PF, Kang S, Tsiouris AJ, Roth J. Endoscopic third ventriculostomy in patients with a diminished prepontine interval. J Neurosurg Pediatr. 2010;5:250–4.
27. Anik I, Etus V, Anik Y, Ceylan S. Role of interpeduncular and prepontine cistern cerebrospinal fluid flow measurements in prediction of endoscopic third ventriculostomy success in pediatric triventricular hydrocephalus. Pediatr Neurosurg. 2010;46:344–50.
28. Di X, Ragab M, Luciano MG. Cine phase-contrast MR images failed to predict clinical outcome following ETV. Can J Neurol Sci. 2009;36(5):643–7.
29. Santamarta D, Martin-Vallejo J, Díaz-Alvarez A, Maillo A. Changes in ventricular size after endoscopic third ventriculostomy. Acta Neurochir. 2008;150:119–27.
30. Singh I, Haris M, Husain M, Husain N, Rastogi M, Gupta RK. Role of endoscopic third ventriculostomy in patients with communicating hydrocephalus: an evaluation by MR ventriculography. Neurosurg Rev. 2008;31:319–25.
31. Faggin R, Calderone M, Denaro L, Meneghini L, d'Avella D. Long-term operative failure of endoscopic third ventriculostomy in pediatric patients: the role of cine phase-contrast MR imaging. Neurosurg Focus. 2011;30:E1.
32. Pujari S, Kharkar S, Metellus P, et al. Normal pressure hydrocephalus: long-term outcome after shunt surgery. J Neurol Neurosurg Psychiatry. 2008;79(11):1282–6.
33. Hamlat A, Adn M, Sid-ahmed S, et al. Theoretical considerations on the pathophysiology of normal pressure hydrocephalus (NPH) and NPH-related dementia. Med Hypotheses. 2006;67(1):115–23.
34. Shprecher D, Schwalb J, Kurlan R. Normal pressure hydrocephalus: diagnosis and treatment. Curr Neurol Neurosci Rep. 2008;8(5):371–6.
35. Yousef MI, Abd El Mageed AE, et al. Use of cerebrospinal fluid flow rates measured by phase-contrast MR to differentiate normal pressure hydrocephalus from involutional brain changes. Egypt J Radiol Nucl Med. 2016;47(3):999–1008.
36. Hattori T, Yuasa T, et al. Altered microstructure in corticospinal tract in idiopathic normal pressure hydrocephalus: comparison with Alzheimer disease and Parkinson disease with dementia. AJNR Am J Neuroradiol. 2011;32:1681–7.
37. Greitz D. Radiological assessment of hydrocephalus: new theories and implications for therapy. Neurosurg Rev. 2004;27(3):145–65. doi:10.1007/s10143-004-0326-9.
38. Men S. BOS akım hastalıkları ve hidrosefali. In: Erden I, editor. Nöroradyoloji manyetik rezonans uygulamaları, 1st ed. Ankara: Türk manyetik rezonans derneği; 2006. p. 80–95 (in Turkish).

Hydrocephalus and Its Neuro-ophthalmic Complications

10

Wafa Al Bluwi, Mohanna Al Jindan, and Ahmed Ammar

10.1 Papilledema

The term papilledema is reserved for those patients with bilateral optic disc edema (symmetrical or asymmetrical) caused by increased intracranial pressure. Knowing that the optic nerve is surrounded by the subarachnoid space, any increase in intracranial pressure will be transmitted to this and lead to an increase in tissue pressure, stasis of the axoplasmic flow, nerve fiber swelling in the optic nerve head, and eventually optic disc edema. Those swollen nerve fibers will compress the small vessels in the pre-laminar region and surface nerve fiber layer of the optic nerve head, which leads to venous stasis and dilation, microaneurysm development, and disc and peripapillary radial hemorrhages. [6] Persistent increases in intracranial pressure lasting for 1–5 days will lead to papilledema [7, 8].

However, sudden increase in the intracranial pressure caused by acute subarachnoid hemorrhages may result in papilledema which rapidly develops within hours [9].

Disc edema typically begins to appear at the lower part of the disc and is followed by the upper part, nasal part, and finally by the temporal part of the disc [7, 8].

There are four stages of papilledema (2):

1. In the early form, there is an obscuration of the underlying vessels and blurring of the disc margins (Fig. 10.1).
2. In the fully developed form, blurred disc margins, hemorrhages, and cotton wool spots occur.

W. Al Bluwi (✉) • M. Al Jindan
Department of Ophthalmology, King Fahd Hospital of University,
Imam Abdulrahman Bin Faisal University, Al Khobar, Saudi Arabia
e-mail: dr.wafab@gmail.com

A. Ammar
Department of Neurosurgery, King Fahd Hospital of University,
Imam Abdulrahman Bin Faisal University, Al Khobar, Saudi Arabia
e-mail: ahmed@ahmedammar.com

© Springer International Publishing AG 2017
A. Ammar (ed.), *Hydrocephalus*, DOI 10.1007/978-3-319-61304-8_10

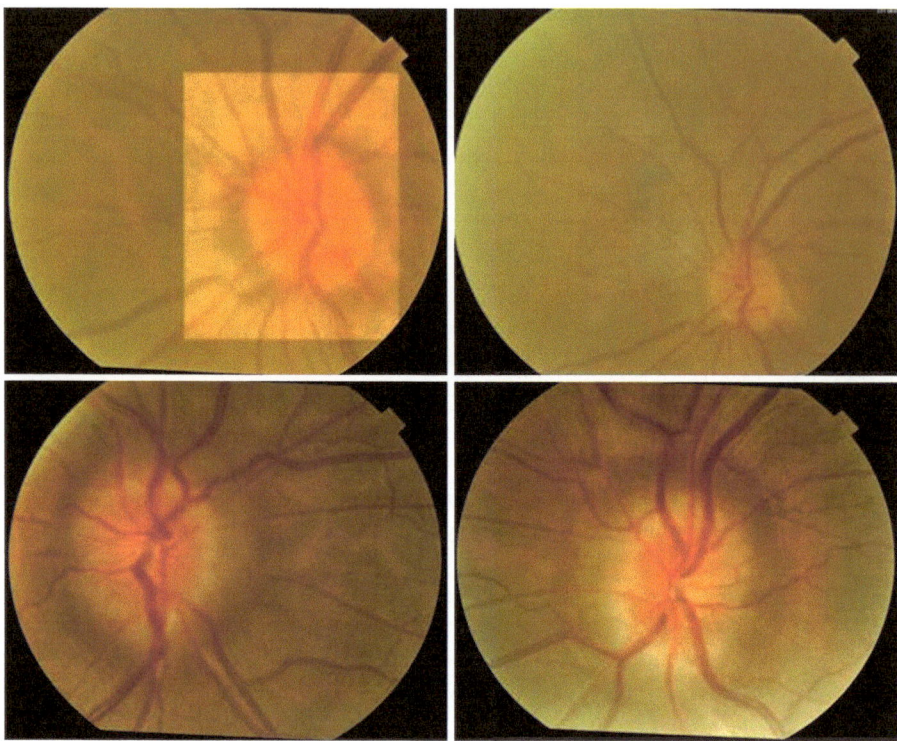

Fig. 10.1 Early papilledema

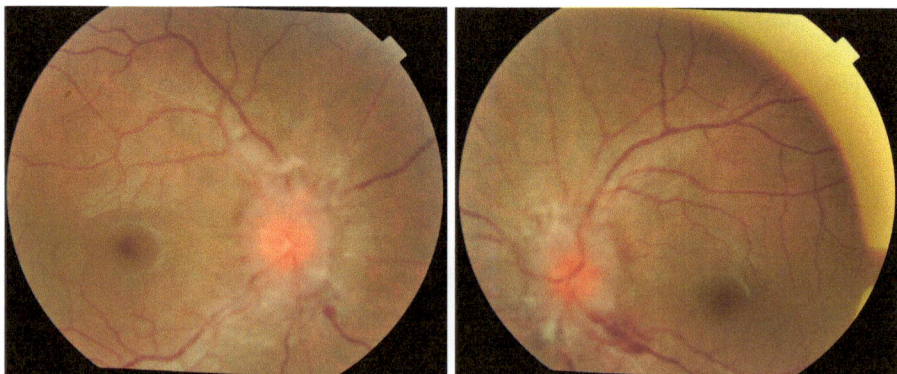

Fig. 10.2 Fully developed papilledema

3. In chronic papilledema, the disc takes on a "champagne-cork" appearance after papilledema continues for weeks or months (Fig. 10.2).
4. In atrophic papilledema, the swollen nerve fibers die, the disc atrophies, and the swelling becomes pale and less prominent (Fig. 10.3).

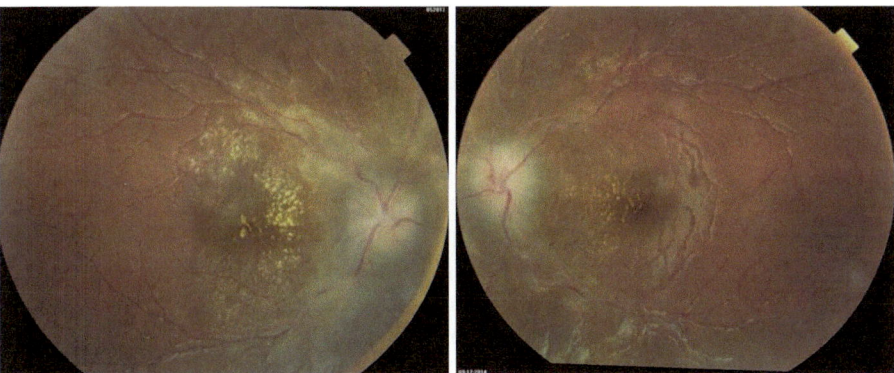

Fig. 10.3 Chronic papilledema

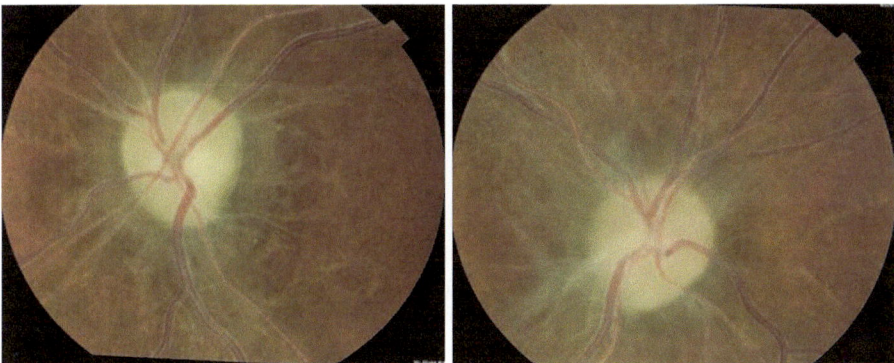

Fig. 10.4 Atrophic papilledema

This form of papilledema is the form that we are attempting to prevent [8, 10].

Although papilledema is considered to be the first ophthalmic manifestation of increased ICP [1, 7], some reports have demonstrated that it is not always observed in infants with hydrocephalus; only 12% of 200 infants exhibited papilledema in one series [11]. This is because open sutures in infants will provide a space for cranial enlargement and a decrease in degree of the elevation in intracranial pressure. However, it should be kept in mind that the effect of open sutures might be outweighed by an acute rapid rise in intracranial pressure and lead to the development of papilledema.

Monitoring of papilledema is not only helpful in the follow-up of patients; its resolution has also been found to be of clinical value in monitoring intracranial pressure reduction after shunts [12]. Furthermore, the presence of this sign during follow-up has been found to be of more predictive value for shunt success assessment than the radiological monitoring signs (i.e., changes in ventricular size and shape, flattening of the subarachnoid space, and altered periependymal signals) (Fig. 10.4).

Iskandar et al. found that depending on the radiological findings alone as indicators of shunt failure is associated with a 33% rate of false-negative results [13].

Buxton et al. and Mizrachi et al. stressed the importance of clinical ophthalmic fundus examinations during follow-up and concluded that the existence of papill-edema may be an important factor in the management of hydrocephalus [14, 15].

10.2 Optic Atrophy

Hydrocephalus is a well-known cause of optic atrophy in children [4, 16–19]. Optic atrophy is usually bilateral and often asymmetric but can also present unilaterally. It is mainly a consequence of long-term papilledema but can also be due to stretching of the chiasm by a displaced brainstem due to high cerebral volume, pressure on the chiasm by dilated third ventricle, or optic nerve stretching due to an expanding skull [20].

Unilateral optic nerve atrophy usually results from unequal expansion of an enlarged third ventricle [5]. Dowu and Balogun compared congenital hydrocephalus with or without myelomeningocele in a study that aimed to demonstrate incidence of optic nerve deficits in those children. The results revealed that optic atrophy was highly associated with hydrocephalus without myelomeningocele, which was explained by the early presentation in cases of hydrocephalus with myelomeningocele [21].

10.3 Ocular Motility Disorder

Strabismus is a common ophthalmic complication of hydrocephalus and occurs in 30–40% of patients [17, 22]. Patients can present with different forms of strabismus including the following: exotropia, esotropia, hypotropia, hypertropia, and hetero-phoria [23]. Strabismus usually exhibits a good response to medical and surgical treatment [24]. The rates of strabismus and significant refractive error are higher in children with shunt revisions [24]. Additionally, functional amblyopia is frequent and mainly related to strabismus [25] with or without significant refractive error. Biglan concluded that with a proper ophthalmic evaluation during the follow-up, 94% of patients will have 6/12 visual acuity or better [24].

Hydrocephalus can also cause paretic squint by causing unilateral or bilateral sixth nerve palsy, which usually results from an increase in intracranial pressure that causes traction at Dorello's canal or shunt placement. Fourth nerve palsy (unilateral or bilateral) has been found to be less frequent than sixth nerve palsy and found to be a localizing sign of superior medullary velum involvement due to pressure from an enlarged third ventricle [26]. Hydrocephalus rarely causes third nerve palsy [27].

10.4 Visual Loss

Vision in humans requires intact anterior visual pathways (i.e., the eyes, optic nerves, and chiasm), intact geniculo-striate visual pathways (i.e., the LGB, optic radiation, primary visual cortex, and "area 17"), and an intact extra-geniculate visual pathway (areas 18 and 19) [28]. Visual loss and cortical visual impairment (CVI) with or without recovery are major morbidities of hydrocephalus [3, 29–33].

In hydrocephalus, an enlarged third ventricle will compress the optic nerve, optic chiasm, optic tract, and the adjacent blood vessels, but the major mechanism responsible for the decrease in vision has been found to be post-papilledema optic atrophy. Any change in visual acuity is considered to be a warning sign of shunt failure [34], but fortunately, with early diagnosis and intervention, some visual recovery may occur [33, 34].

Patients with aqueductal stenosis, long-term shunting without revision and small non-compliant ventricles are at a risk of visual loss, which can be partially or completely recovered if shunt revision is performed rapidly [34]. Additionally, acute visual loss might occur following a rapid reduction in the ICP post-shunting or following decompressive craniotomy due to hypoperfusion to the laminar part of the optic nerve [31]. Damage to the posterior visual pathway is more strongly related to shunt failure than a primary result of hydrocephalus [16].

Cortical visual impairment is an important cause of low vision in children in developed countries [28]. CVI is defined as a loss in visual function caused by any disease that affects the geniculate and extra-geniculate pathways in the presence of intact anterior visual pathways and ocular structures. The most common etiology found is perinatal hypoxia, but other etiologies such as hydrocephalus, cerebral malformations, premature birth, and meningitis have been found to be associated [28, 32]. Indeed, hydrocephalus was found to be the commonest cause of CVI in a study by Chen et al. [35].

Hydrocephalus can lead to acute and chronic CVI [35, 36] through the dilation of the ventricles, which causes posterior cortex distension and occlusion of the posterior cerebral arteries and results in infarction in the occipital cortex [37]. However, many cases have been found to be due to dilated ventricles without infarction. Additionally, CVI has been found in cases with shunt malfunctions [30] or following successful shunt procedures after rapid reductions in the intracranial pressure [16]. Children with CVI may have experience some recovery of vision or dramatic improvements in vision within hours, the latter situation has been observed in patients with chronic hydrocephalus who underwent shunt revision [37].

Those who present with better initial visual acuity are those who exhibit the greatest improvement [32]. Therefore, we can state that a prolonged period of blindness due to hydrocephalus does not indicate a hopeless prognosis [36]. The occurrence of this recovery of vision has been explained by several reasons. This recovery could be because during the growth of an infant, a more normal maturation of the visual system occurs and allows more residual visual potential to develop with time [38]. Another reason is that the damage to the visual cortex or optic radiations might be partial, which would allow for residual visual function [39]. However, many investigators believe that the brain plasticity of an infant is the reason for this recovery [28, 40]. For a better explanation of this elasticity, we need to detail both the existing understanding of visual cortex injury recovery mechanisms in adults and the recent animal models that have specifically addressed infant visual cortex ablation.

Different visual field defects observed in patients with hydrocephalus that could be reversible include inferior altitudinal hemianopia, inferior binasal quadrantanopia, bilateral central scotoma, incongruous homonymous hemianopia, and concentric visual field constriction [41–44]. These defects could result from unpredicted movements of the shunt catheter that might affect the visual pathway or compress the vessels; mechanical forces involving distention of the third ventricle with compression of the optic nerve, chiasma, or optic tract; increased intracranial pressure; or shunt malfunction [5, 41–43, 45].

10.5   Sunset Sign

This is a valuable early sign of hydrocephalus that requires urgent neuroimaging and surgical intervention. The sunset sign usually results from the continuous effect of an increase in intracranial pressure on the thin orbital roof such that the eyes are displaced downward and the superior sclera is visible. This sign is observed in 40% of children with obstructive hydrocephalus. Initially, it is intermittent but becomes continuous with time, and the child will lose upward conjugate gaze [19].

10.6   Dorsal Midbrain Syndrome

The presence of dorsal midbrain syndrome during the course of hydrocephalus is a sign of disease deterioration. This sign may herald a significant ventriculomegaly, and an urgent assessment is therefore needed before the condition becomes irreversible [46]. Dorsal midbrain syndrome results from pressure on the posterior commissure (i.e., the upgaze center) due to dilation of the third ventricle [47]. The syndrome initially presents with light-near dissociation and mild upgaze limitation, which are followed by upgaze paralysis, pathologic lid retraction, and convergence-retraction nystagmus [16].

10.7   Arrested Hydrocephalus

"Compensated" or "arrested" hydrocephalus affects a subpopulation of hydrocephalus patients and involves the intracranial pressure returning to normal and the ventricles remaining slightly dilated [48]. This subgroup of patients is challenging when they present with increased intracranial pressure. The modalities for assessment include CT scan, which is of limited value because it may not illustrate significant changes from previous studies, and there is the risk of radiation exposure with repeated tests. Another method is the measurement of the optic nerve sheath diameter (ONSD) with ocular ultrasonography, which has been found to be a safe and easy method of assessment in these cases [49].

10.8   Role of Periodic Ophthalmic Examination
During Follow-Up

Signs of malfunctioning shunts usually include nausea, vomiting, headache, lethargy, changes in consciousness, papilledema, disturbances in eye movements and alignment, diplopia, and less commonly loss of vision [4, 11, 30, 50]. Neuroimaging is an important indicator of functioning shunts. However, in some children,

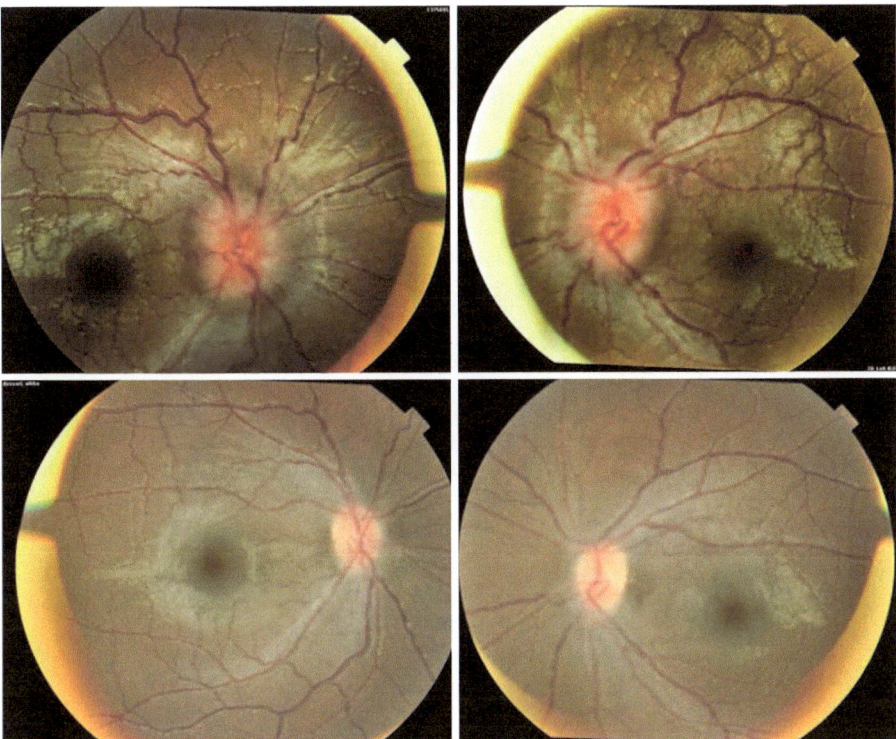

Fig. 10.5   Resolution of papilledema after a shunt procedure

ophthalmologic signs are the only or the earliest indicators of failure, which aids the rapid diagnosis of shunt failure [4, 51] (Fig. 10.5).

Because ophthalmic manifestations require specialized techniques, shunt-dependent patients should undergo periodic neuro-ophthalmic clinic visits that involve examinations of the visual acuity/field, ocular motility, and fundus after the patient is diagnosed with hydrocephalus, and the patients should then be monitored according to their needs. With such teamwork, we can facilitate the detection of the neuro-ophthalmic warning signs of unrecognized shunt malfunction to ensure proper and urgent interventions to achieve the best possible standard of vision.

References

1. Chou SY, Digre KB. Neuro-ophthalmic complications of raised intracranial pressure, hydro-cephalus, and shunt malfunction. Neurosurg Clin N Am. 1999;10:587–608.
2. Kanski JJ. Clinical ophthalmology. A systematic approach. 4th ed. Oxford: Butterworth-Heinemann; 1999.
3. Persson EK, Anderson S, Wiklund LM, Uvebrant P. Hydrocephalus in children born in 1999-2002: epidemiology, outcome and ophthalmological findings. Childs Nerv Syst. 2007;23(10):1111–8.

4. Tzekov C, Cherninkova S, Gudeva T. Neuroophthalmological symptoms in children treated for internal hydrocephalus. Pediatr Neurosurg. 1991–1992;17(6):317–20.
5. Osher RH, Corbett JJ, Schatz NJ, Savino PJ, Orr LS. Neuro-ophthalmological complications of enlargement of the third ventricle. Br J Ophthalmol. 1978;62:536–42.
6. Hayreh SS. Optic disc edema in raised intracranial pressure: pathogenesis. Arch Ophthalmol. 1977;95:1553–65.
7. Hayreh MS, Hayreh SS. Optic disc edema in raised intracranial pressure: 1. Evolution and resolution. Arch Ophthalmol. 1977;95:1237–44.
8. Liu, Grant T ; Volpe, Nicholas J, Galetta, Steven. Neuro-Ophthalmology: Diagnosis and Management. Saunders Elsevier. ClinicalKey Flex. 2010; Chapter 6, p. 199–236.
9. Pagani LF. The rapid appearance of papilledema. J Neurosurg. 1969;30:247–9.
10. Hedges TR. Papilledema: its recognition and relation to increased intracranial pressure. Surv Ophthalmol. 1975;19:201–23.
11. Ghose S. Optic nerve changes in hydrocephalus. Trans Ophthalmol Soc UK. 1983;103(pt 2):217–20.
12. Singhal A, Yang MMH, Sargent MA, Cochrane DD. Does optic nerve sheath diameter on MRI decrease with clinically improved pediatric hydrocephalus? Childs Nerv Syst. 2013;29:269–74.
13. Iskandar BJ, Mclaughlin C, Mapstone TB, Grabb PA, Oakes WJ. Pitfalls in the diagnosis of ventricular shunt dysfunction: radiology reports and ventricular size. Pediatrics. 1998;101:1031–6.
14. Buxton N, Turner B, Ramli N, Vloeberghs M. Changes in third ventricular size with neuroendoscopic third ventriculostomy: a blinded study. J Neurol Neurosurg Psychiatry. 2002;72:385–7.
15. Mizrachi IBB, Trobe JD, Gebarski SS, Garton HJL. Papilledema in the assessment of ventriculomegaly. J Neuroophthalmol. 2006;26:260–3.
16. Corbett JJ. Neuro-ophthalmologic complications of hydrocephalus and shunting procedures. Semin Neurol. 1986;6:111–23.
17. Gaston H. Ophthalmic complications of spina bifida and hydro-cephalus. Eye. 1991;5(pt 3):279–90.
18. Chinta S, Wallang BS, Sachdeva V, Gupta A, Patil-Chhablani P, Kekunnaya R. Etiology and clinical profile of childhood optic nerve atrophy at a tertiary eye care center in South India. Indian J Ophthalmol. 2014;62(10):1003–7. doi:10.4103/0301-4738.145996.
19. Rizvi R, Anjum Q. Hydrocephalus in children. J Pak Med Assoc. 2005;55(11):502–7.
20. Harcourt B, Jay B. Bilateral optic atrophy in childhood. Br J Ophthalmol. 1968;52(11):860–1.
21. Idowu OE, Balogun MM. Visual function in infants with congenital hydrocephalus with and without myelomeningocele. Childs Nerv Syst. 2014;30(2):327–30.
22. Mankinen-Heikkinen A, Mustonen E. Ophthalmic changes in hydrocephalus. A follow-up examination of 50 patients treated with shunts. Acta Ophthalmol (Copenh). 1987;65(1):81–6.
23. Altintas O, Etus V, Etus H, Ceylan S, Caglar Y. Risk of strabismus and amblyopia in children with hydrocephalus. Graefes Arch Clin Exp Ophthalmol. 2005;243:1213–7.
24. Biglan AW. Ophthalmologic complications of meningomyelocele: a longitudinal study. Trans Am Ophthalmol Soc. 1990;88:389–462.
25. Billard C, Santini JJ, Nargeot MC, Gillet P, Adrien J, Dudin A. What future is there for hydrocephalus children? Intellectual and visual neurological prognosis in series of 77 cases of nontumor hydrocephalus. Arch Fr Pediatr. 1987;44(10):849–54.
26. Guy JR, Friedman WF, Mickle JP. Bilateral trochlear nerve paresis in hydrocephalus. J Clin Neuroophthalmol. 1989;9(2):105–11.
27. Cultrera F, D'Andrea M, Battaglia R, Chieregato A. Unilateral oculomotor nerve palsy: unusual sign of hydrocephalus. J Neurosurg Sci. 2009;53(2):67–70.
28. Good W, Jan J, Desa L, Barrovich J, Groenveld M. Visual impairment in children. Surv Ophthalmol. 1994;38(4):351–64.
29. Calogero JA, Alexander E. Unilateral amaurosis in a hydrocephalic child with an obstructed shunt. Case report. J Neurosurg. 1971;34:236–40.
30. Arroyo HA, Jan JE, Mccormick AQ, Farrell K. Permanent visual loss after shunt malfunction. Neurology. 1985;35:25–30.

31. Cedzich C, Schramm J, Wenzel D. Reversible visual loss after shunt malfunction. Acta Neurochir. 1990;105:121–3.
32. Khetpal V, Donahue SP. Cortical visual impairment: etiology, associated findings, and prognosis in a tertiary care setting. J AAPOS. 2007;11(3):235–9. Epub 2007 Apr 24.
33. Oyama H, Hattori K, Kito A, Maki H, Noda T, Wada K. Visual disturbance following shunt malfunction in a patient with congenital hydrocephalus. Neurol Med Chir (Tokyo). 2012;52(11):835–8.
34. Kraus R, Hanigan WC, Kattah J, Olivero WC. Changes in visual acuity associated with shunt failure. Childs Nerv Syst. 2003;19:226–31.
35. Chen TC, Weinberg MH, Catalano RA, et al. Development of object vision in infants with permanent cortical visual impairment. Am J Ophthalmol. 1992;114:575–8.
36. Lorber J. Recovery of vision following prolonged blindness in children with hydrocephalus or following pyogenic meningitis. Clin Pediatr. 1967;6:699–703.
37. Connolly MB, Jan JE, Cochrane DD. Rapid recovery from cortical visual impairment following correction of prolonged shunt malfunction in congenital hydrocephalus. Arch Neurol. 1991;48:956–7.
38. Hoyt CS, Nickel BL, Billson FA. Ophthalmological examination of the infant: developmental aspects. Surv Ophthalmol. 1982;26:177–89.
39. Whiting S, Jan JE, Wong PK. Permanent cortical visual impairment in children. Dev Med Child Neurol. 1985;27:730–9.
40. Lindberg R, Walsh FB, Sacks JG. Neuropathology of vision: an atlas. Philadelphia: Lea and Febiger; 1973. p. 446–66.
41. Rudolph D, Sterker I, Graefe G, Till H, Ulrich A, Geyer C. Visual field constriction in children with shunt-treated hydrocephalus. Clinical article. J Neuro-surg Pediatr. 2010;6:481–5.
42. Kojima N, Kuwamura K, Tamaki N, Matsumoto S. Reversible congruous homonymous hemianopsia as a symptom of shunt malfunction. Surg Neurol. 1984;22:253–6.
43. Kojima N, Tamaki N, Hosoda K, et al. Visual field defects in hydrocephalus. No To Shinkei. 1985;37:229–36.
44. Holsgrove D, Leach P, Herwadkar A, Gnanalingham KK. Visual field deficit due to downward displacement of optic chiasm. Acta Neurochir (Wien). 2009;151(8):995–7.
45. Molia L, Winterkorn JM, Schneider SJ. Hemianopic visual field defects in children with intracranial shunts: report of two cases. Neurosurgery. 1996;39(3):599–603.
46. Lerner MA, Kosary IZ, Cohen BE. Parinaud's syndrome in aqueduct stenosis: its mechanism and ventriculographic features. Br J Radiol. 1969;42:310–2.
47. Cobbs WH, Schatz NJ, Savino PJ. Midbrain eye signs in hydro-cephalus. Ann Neurol. 1978;4:172.
48. Mclone DG, Aronyk KE. An approach to the management of arrested and compensated hydrocephalus. Pediatr Neurosurg. 1993;19:101–3.
49. Newman WD, Holman AS, Dutton GN, Carachi R. Measurement of optic nerve sheath diameter by ultrasound: a mean of detecting acute increase in hydrocephalus. Br J Ophthalmol. 2002;86(10):1109.
50. Sekhar LN, Moossy J, Guthkelch AN. Malfunctioning ventriculoperitoneal shunts. Clinical and pathological features. J Neurosurg. 1982;56:411–6.
51. Phillips PH, Ku B, Qureshi N, Adada B. Ophthalmologic signs of shunt failure in children with hydrocephalus. J AAPOS. 2006;10(1):91.

Post-traumatic Hydrocephalus in Adults and Paediatrics

11

Tomasz Klepinowski and Nabeel S. Alshafai

11.1 Incidence and Timing

The total incidence of adult PTH varies based on the cohort observed in contemporary studies between 2.6 and 14.0% [1, 2, 6, 7]. In paediatrics, the global incidence of PTH is also low: 2–7% of children suffering TBI will eventually require a permanent shunt to treat PTH [8, 9]. Fourteen percent of *middle-age* patients (mean age = 49.6) who survived an acute stage of severe TBI and were admitted to the rehabilitation centre with Glasgow coma scale (GCS) 3–9 or 10–12 with concomitant significant focal deficits developed hydrocephalus at some point of the 1-year observation [2]. On the other hand, amongst *young* adults (median age = 25), it has been reported that the incidence of PTH is a bit lower as it revolves around 8% [1]. Nevertheless, mere radiological proof of ventricular enlargement after having experienced TBI is considered to happen way more often than the real active hydrocephalus. That radiologically proven ventriculomegaly resulting from severe to moderate head injury will occur in approximately 70% [10].

Regarding timing, although it was reported that PTH may take place even within hours after head trauma [11], the vast majority does not occur that quickly. Up to 27% of all PTH tend to appear within the acute stage[1] of trauma, which is defined as a time from injury to the beginning of rehabilitation (median 20 days) [2, 12].

---

[1] More comprehensive definition of the acute stage: The number of days necessary to sufficiently stabilize the patient in terms of requirement for artificial ventilation and surgery (e.g. neurosurgery, orthopaedic surgery, etc.)

T. Klepinowski
Department of Neurosurgery, Collegium Medicum, Jagiellonian University, Cracow, Poland

Alshafai Neurosurgical Academy A.N.A., Toronto, ON, Canada

N.S. Alshafai (✉)
Alshafai Neurosurgical Academy A.N.A., Toronto, ON, Canada
e-mail: nabeel.alshafai@ymail.com

© Springer International Publishing AG 2017
A. Ammar (ed.), *Hydrocephalus*, DOI 10.1007/978-3-319-61304-8_11

Licata et al. [12] and Kammersgaard et al. [2] found out that timing of the post-traumatic hydrocephalus after traumatic brain injury within a 1-year observation.

Interpreting Timing of PTH Onset in Everyday Practice
In clinical practice, patients should be carefully followed up for at least 5 months after moderate to severe brain injury because that is the time range when nearly all of post-traumatic hydrocephalus is revealed [2, 12, 13].

- Acute stage—14–27% of all PTH
- Rehabilitation stage—more than 73%

In children, low admission GCS score correlated with early PTH presentation (within <3 days). On the other hand, high GCS score correlated with PTH presentation after 1 week [9].

11.2   Risk Factors

There are a few risk factors which have been found to have strong association with developing PTH. In our opinion most important factors are decompressive craniectomy (DC), its superior margin being less than 25 mm from the midline, reoperation, delayed cranioplasty, traumatic subarachnoid haemorrhage (tSAH), traumatic intracranial haematoma, and interhemispheric hygroma (IHH). No correlation was found between sex or location of SAH.

Choi et al. believe that higher risk of developing PTH had the patients who underwent an extended craniectomy or who needed reoperation [14]. Waziri et al. concluded that DC has an influence on the physiology of ICP and leads to disruption of pressure-dependent arachnoid granulations that drain the CSF, thus causing hydrocephalus [15]. Also, they observed a spontaneous resolution of hydrocephalus after restoration of normal intracranial pressure by performing cranioplasty. Delay in cranioplasty may cause a permanent dysfunction of the arachnoid granulations, similarly as in hydrocephalus induced by long-term CSF drainage. De Bonis et al. [16], however, suggest that the factor that really is the cause of trouble is not the DC itself but the position of its superior margin. Craniectomy which is too close to the midline may disrupt the venous outflow and result in hydrocephalus. On the other hand, multivariate analysis performed by Heng-Li Tian et al. did not show any correlation between DC and PTH [17]. In severe TBI in *children*, decompressive craniectomy (DC) or craniotomy is also controversial but often constitutes the treatment option to reduce the elevated intracranial pressure. Noteworthy, in contrast to adults, there has been evidence of substantial benefit for the paediatric population [18].

Extravasated blood from tSAH or traumatic intracranial haematoma may cause a mechanical blockage or inflammation of arachnoid granules leading to the same effect—decreased CSF resorption and finally, PTH [14]. As a matter of fact, subdural fluid collection or SAH in one study was found in 93% of PTH patients [1]. However, performing a multivariate analysis on factors such as SAH or IVH did not prove any significant association with post-traumatic hydrocephalus [19, 20]. There is however independent factor of PTH— inter-hemispheric hygroma. More than 80% of patients with IHH develop hydro-cephalus within first 6 months [20]. It is agreed that epidural haematomas have no effect on PTH [8].

11.3 Clinical Presentation

A traumatic patient developing hydrocephalus differs enormously from other hydro-cephalus sufferers. This is ascribed to the aftermath of the primary injury itself. Therefore, in order not to overlook this impairing condition, one need to seek such subtle symptoms as slow progression of the recuperation or lack thereof. It has been reported [13] that only about 23% of PTH patients manifest typical hydrocephalus symptoms such as headache, nausea with or without vomiting, seizures, increased spasticity, or Adams-Hakim triad (incontinence, gait abnormality, progressive dementia). It means that in over three-fourths of patients, the only clue to their insidious hydrocephalus is prolonged poor consciousness level [1, 13]. Vigilant atti-tude is especially important in the acute phase of trauma when a severely injured patient will be comatose and intubated.

11.4 Investigations

Diagnosis is based upon the history of cerebral trauma, radiological proof of enlarged ventricles (or enlarged subarachnoid spaces in case of external hydroceph-alus), and clinical status. Probably the most important difficulty a physician may face when it comes to establishing a diagnosis of PTH is to differentiate ventriculo-megaly due to atrophy of the cerebral tissue (*hydrocephalus* ex vacuo) from true hydrocephalus. To date, numerous ways to ensure proper diagnosis of PTH have been developed, and they are as in Table 11.1 [1, 21–24].

Another investigation study that is more of a predictive factor is cerebrospinal fluid (CSF) dynamics assessed on grounds of intraventricular.

Table 11.1 Details of clinical course and investigations that are found helpful in distinguishing active hydrocephalus from hydrocephalus ex vacuo in traumatic patients

Investigation	Finding
Clinical course	Deterioration, recovery slower than expected, or total lack thereof
Lumbar puncture Opening pressure	Increased >15 mmHg[a]
CT: Periventricular area	Hypodense
MRI: Periventricular area	Hyperintense on T2-weighted images
MRI: Mamillopontine distance	Decreased
MRI: Mamillocommissural distance	Increased
MRI: Third ventricular floor	Concave
DTI: Periventricular fractional anisotropy	Pathological

Usual results of active hydrocephalus are illustrated
[a]Note that normal pressure still might be associated with clinically important hydrocephalus both in adults and children [1, 21–24]; *DTI* diffusion tensor imaging, *CT* computed tomography, *MRI* magnetic resonance imaging

11.5 Management

Shunt placement remains the mainstay of long-term care for post-traumatic hydrocephalus patients. Other methods developed over the years but still not so popular in case of PTH which includes external lumbar drainage (ELD) and endoscopic third ventriculostomy (ETV) (Fig. 11.1). Ventriculo-peritoneal shunting is the most common surgical procedure implemented in PTH. Although it is surely possible to place other shunt types, such as ventriculo-atrial, reports of these in PTH are extremely scarce [25]. A good question appears: when is the best time to insert a shunt? So far, however, it is still controversial and has been shown that a time span between the initial injury and the shunt operation may or may not play any major role in adults—Kim et al. [8] did not indicate any statistically significant result between the group treated in the first month and the groups managed later on ($p > 0.05$). But recently, in 2016, Weintraub et al. got results suggesting that the most severely injured patients who had not followed commands at the admission benefited from shunts inserted prior to 69 days from injury as compared to those who were given similar treatment but after this period. Another surprising finding is that shunt success is not affected by the preoperative intracranial pressure. Those treated with preoperative CSF pressure below 16 cmH$_2$O shared similar postoperative recovery to their counter subgroup. Recovery was defined as Glasgow outcome score at 14 days, 3 months, and 6 months after surgery (Table 11.2) [8]. Hence, for the time being ventriculo-peritoneal shunts might be recommended to both types of patients: with and without concomitant elevated intracranial pressure. In fact, shunting is so far the only established surgical treatment for post-traumatic *normal* pressure hydrocephalus in adults and paediatrics [9, 26].

A principal issue with shunting is the lack of criteria that would determine which subgroup of traumatic patients is going to benefit the most from shunt insertion. Recently, it was proposed that evaluating CSF dynamics by the means of

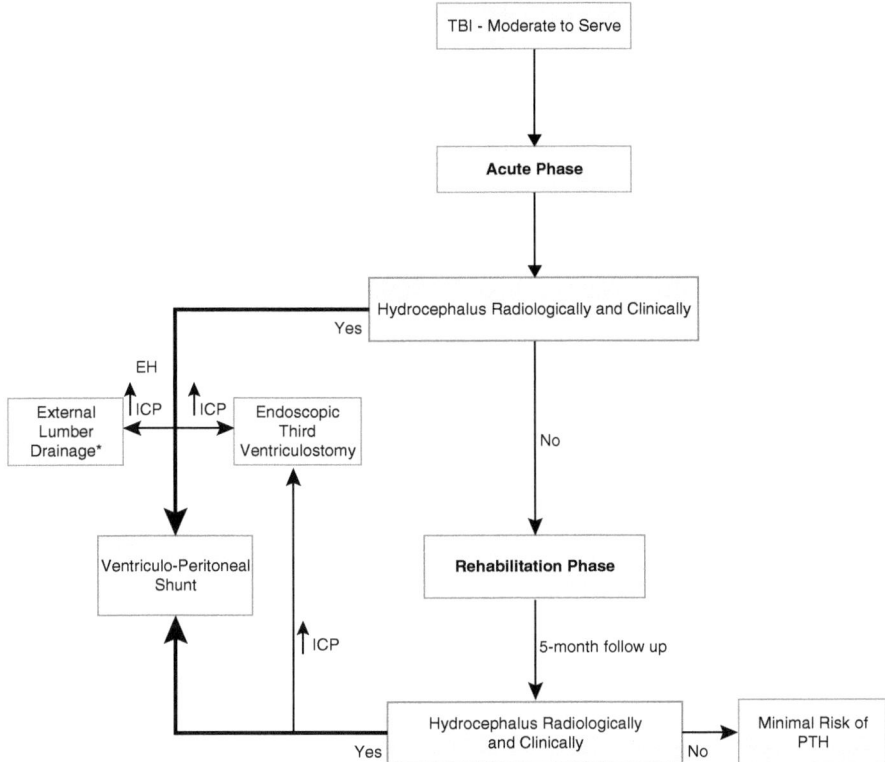

Fig. 11.1 Proposed algorithm for the management of PTH. *Arrows* leading to shunting are larger illustrating that nowadays it is the most pervasive surgical treatment worldwide employed in PTH patients. *ICP* intracranial pressure, *EH* external hydrocephalus, *TBI* traumatic brain injury, *ELD* external lumbar drain. *ELD after having ruled out intracranial masses and realized the ICP is refractory to the first-line conservative and invasive therapy including external ventricular drainage. External lumbar drainage has been proven to successfully lower ICP with uneventful following period [12, 28, 29]

Table 11.2 Glasgow outcome scale

Glasgow outcome scale		
Score	Grade	Description
5	Death	–
4	Vegetative state	–
3	Severe disability	Disabled. Needs others for daily support
2	Moderate disability	Disabled but stays independent in terms of self-care and daily activities
1	Good recovery	Back to normal activity and employment

intraventricular infusion test could be a great way to predict good response to this therapy [23]. The parameters that are critical and of the highest value in this assessment are outflow resistance (OR) and intracranial elastance index (EI). OR greater than 10 mmHg/mL/min and EI greater than 0.3 indicate those individuals that are most likely to substantially improve after shunting. Sensitivity of OR was 100%, and specificity of EI was 100%. However, limitations of this study were a small number of patients included ($n = 15$) and the fact that they consisted solely of men. Therefore, a larger prospective study is in demand so as to corroborate these results.

Another method of treatment of PTH, which is somehow underrated and, thus, rarely reported in the literature for PTH, is ETV. In case of obstructive hydrocephalus, which is a rare mechanism of PTH [27], ETV seems like an obvious internal shunt which allows CSF to bypass the blocked passage. Nonetheless, in most cases of PTH reported so far, there is usually a communicating mechanism of hydrocephalus with a drop in the CSF reabsorption [16, 23, 28]. It is why ETV for PTH was believed to be a relative contraindication. The reviewed data, however, clearly suggests that this procedure might be actually very beneficial for PTH sufferers (93% patients improved), especially if there is concomitant increase in intracranial pressure (>15 mmHg) [28].

Lately, external lumbar drainage (ELD) was introduced to the neurosurgeon's armamentarium for PTH [29]. This, however, to date was tested *only* during the acute phase of TBI and for clearly external hydrocephalus. Prior to carrying out ELD, brain CT has to be conducted so as to rule out the presence of intracranial haematomas and to measure subarachnoid spaces (SAS). ELD might be considered if initial emergent interventions failed to lower ICP.

In *paediatric population* with severe TBI and hydrocephalus diagnosed at admission, the need of EVD is often a temporary treatment option. However, children receiving an EVD for PTH were significantly less likely to be converted to a shunt than for hydrocephalus derived from other causes such as neoplasms, obstructive hydrocephalus, or vascular IVH [30]. Beside EVD and classic shunting, there have been reports about the application of endoscopic third ventriculostomy not only in adults but also in paediatrics, as described above [28].

11.6   Outcome

Overall outcome depends on the variety of factors: severity of head trauma, coexisting pathologies like subarachnoid haemorrhage or diffuse axonal injury, decompressive craniectomy, cranioplasty, and many more. Due to the complexity of this problem and many controversies around treatment, in order to explain the outcome of patients who suffered from PTH, we will distinguish two simple categories: favourable and unfavourable outcomes, mainly based on a score in Glasgow outcome scale (Table 11.2). Patients with score of 4 and 5 were classified as favourable outcome, whereas 1, 2, and 3 meant unfavourable outcome [14]. Favourable outcome was achieved in 18.2–54.5% depending on the study [14, 31]. The general trend showed that patients with less extensive craniectomy and smaller level of hydrocephalus had better outcome, and patients with extended craniectomy usually had worse outcome.

The level of hydrocephalus (based on the radiological size of ventricles compared to expected) correlated significantly with neurobehavioural outcome, post-traumatic epilepsy, and irritative electroencephalographic abnormalities as suggested by Mazzini et al. [31]. This subject, however, needs stronger data because it still remains controversial. In the study by De Bonis et al., no association with worse outcome after developing PTH has been found [16]. Neither did Linnemann et al. find any independent influence of PTH on outcome in their cohort [32].

Acknowledgments We would like to appreciate Radosław Trzciński medical student from University of Gdansk, Poland and Magda Garzon consultant neurosurgeon from University of Barcelona, Spain for their assistance in preparing the manuscript.

References

1. Weintraub AH, Gerber DJ, Kowalski RG. Post-traumatic hydrocephalus as a confounding influence on brain injury rehabilitation: incidence, clinical characteristics and outcomes. Arch Phys Med Rehabil. 2016;98(2):312–9. doi:10.1016/j.apmr.2016.08.478.
2. Kammersgaard LP, Linnemann M, Tibæk M. Hydrocephalus following severe traumatic brain injury in adults. Incidence, timing, and clinical predictors during rehabilitation. NeuroRehabilitation. 2013;33:473–80. doi:10.3233/NRE-130980.
3. Kravchuk AD, Likhterman LB, Shurkhai VA. [Monosymptomatic clinical course of posttraumatic normal pressure hydrocephalus]. Zh Vopr Neirokhir Im N N Burdenko. 2011;75:42–6; discussion 46.
4. Kato T, Iseki C, Takahashi Y, et al. [iNPH (Idiopathic normal pressure hydrocephalus) and AVIM (asymptomatic ventriculomegaly with features of iNPH on MRI)]. Rinsho Shinkeigaku. 2010;50:963–5.
5. Iseki C, Kawanami T, Nagasawa H, et al. Asymptomatic ventriculomegaly with features of idiopathic normal pressure hydrocephalus on MRI (AVIM) in the elderly: a prospective study in a Japanese population. J Neurol Sci. 2009;277:54–7. doi:10.1016/j.jns.2008.10.004.
6. Low CYD, Low YYS, Lee KK, et al. Post-traumatic hydrocephalus after ventricular shunt placement in a Singaporean neurosurgical unit. J Clin Neurosci. 2013;20:867–72. doi:10.1016/j.jocn.2012.06.007.
7. Fotakopoulos G, Tsianaka E, Siasios G, et al. Posttraumatic hydrocephalus after decompressive craniectomy in 126 patients with severe traumatic brain injury. J Neurol Surg A Cent Eur Neurosurg. 2015;77:88–92. doi:10.1055/s-0035-1558411.
8. Kim HS, Lee SU, Cha JH, et al. Clinical analysis of results of shunt operation for hydrocephalus following traumatic brain injury. Korean J Neurotrauma. 2015;11:58–62. doi:10.13004/kjnt.2015.11.2.58.
9. Vadivelu S, Rekate HL, Esernio-Jenssen D, et al. Hydrocephalus associated with childhood nonaccidental head trauma. Neurosurg Focus. 2016;41:E8. doi:10.3171/2016.8.FOCUS16266.
10. Poca MA, Sahuquillo J, Mataró M, et al. Ventricular enlargement after moderate or severe head injury: a frequent and neglected problem. J Neurotrauma. 2005;22:1303–10. doi:10.1089/neu.2005.22.1303.
11. Takagi H, Tamaki Y, Morii S, Ohwada T. Rapid enlargement of ventricles within seven hours after head injury. Surg Neurol. 1981;16:103–5.
12. Licata C, Cristofori L, Gambin R, et al. Post-traumatic hydrocephalus. J Neurosurg Sci. 2001;45:141–9.
13. Matsushita H, Takahashi K, Maeda Y, et al. [A clinical study of posttraumatic hydrocephalus]. No Shinkei Geka. 2000;28:773–9.
14. Choi I, Park H-K, Chang J-C, et al. Clinical factors for the development of posttraumatic hydrocephalus after decompressive craniectomy. J Korean Neurosurg Soc. 2008;43:227–31. doi:10.3340/jkns.2008.43.5.227.

15. Waziri A, Fusco D, Mayer SA, et al. Postoperative hydrocephalus in patients undergoing decompressive hemicraniectomy for ischemic or hemorrhagic stroke. Neurosurgery. 2007;61:489–93. doi:10.1227/01.NEU.0000290894.85072.37.

16. De Bonis P, Pompucci A, Mangiola A, et al. Post-traumatic hydrocephalus after decompressive craniectomy: an underestimated risk factor. J Neurotrauma. 2010;27:1965–70. doi:10.1089/neu.2010.1425.

17. Tian HL, Xu T, Hu J, et al. Risk factors related to hydrocephalus after traumatic subarachnoid hemorrhage. Surg Neurol. 2008;69:241–6. doi:10.1016/j.surneu.2007.02.032.

18. Pechmann A, Anastasopoulos C, Korinthenberg R, et al. Decompressive craniectomy after severe traumatic brain injury in children: complications and outcome. Neuropediatrics. 2014;46:5–12. doi:10.1055/s-0034-1393707.

19. Nor MAM, Abdul Rahman NA, Adnan JS. Post-traumatic hydrocephalus. Malays J Med Sci. 2013;20:95–6.

20. Kaen A, Jimenez-Roldan L, Alday R, et al. Interhemispheric hygroma after decompressive craniectomy: does it predict posttraumatic hydrocephalus? J Neurosurg. 2010;113:1287–93. doi:10.3171/2010.4.JNS10132.

21. Segev Y, Metser U, Beni-Adani L, et al. Morphometric study of the midsagittal MR imaging plane in cases of hydrocephalus and atrophy and in normal brains. Am J Neuroradiol. 2001;22:1674–9.

22. Greenberg M. Posttraumatic hydrocephalus. In: Handbook of neurosurgery. 7th ed. New York: Thieme; 2010. p. 906.

23. De Bonis P, Mangiola A, Pompucci A, et al. CSF dynamics analysis in patients with posttraumatic ventriculomegaly. Clin Neurol Neurosurg. 2013;115:49–53. doi:10.1016/j.clineuro.2012.04.012.

24. Marmarou A, Abd-Elfattah Foda MA, Bandoh K, et al. Posttraumatic ventriculomegaly: hydrocephalus or atrophy? A new approach for diagnosis using CSF dynamics. J Neurosurg. 1996;85:1026–35. doi:10.3171/jns.1996.85.6.1026.

25. Soto-Hernández JL, Ramírez-Crescencio MA, Moreno Estrada VM, del Valle RR. Candida albicans cerebral granulomas associated with a nonfunctional cerebrospinal fluid shunt: case report. Neurosurgery. 2000;47:973–6.

26. Wen L, Wan S, Zhan RY, et al. Shunt implantation in a special sub-group of post-traumatic hydrocephalus—patients have normal intracranial pressure without clinical representations of hydrocephalus. Brain Inj. 2009;23:61–4. doi:10.1080/02699050802635265.

27. McNamee S, Cifu D, Damiano T. Syringomyelia from complicated posttraumatic hydrocephalus. Am J Phys Med Rehabil. 2008;87:967–8. doi:10.1097/PHM.0b013e31818a6b84.

28. De Bonis P, Tamburrini G, Mangiola A, et al. Post-traumatic hydrocephalus is a contraindication for endoscopic third-ventriculostomy: isn't it? Clin Neurol Neurosurg. 2013;115:9–12. doi:10.1016/j.clineuro.2012.08.021.

29. Manet R, Schmidt EA, Vassal F, et al. CSF lumbar drainage: a safe surgical option in refractory intracranial hypertension associated with acute posttraumatic external hydrocephalus. Acta Neurochir Suppl. 2016;122:55–9. doi:10.1007/978-3-319-22533-3_11.

30. Walker CT, Stone JJ, Jacobson M, et al. Indications for pediatric external ventricular drain placement and risk factors for conversion to a ventriculoperitoneal shunt. Pediatr Neurosurg. 2012;48:342–7. doi:10.1159/000353608.

31. Mazzini L, Campini R, Angelino E, et al. Posttraumatic hydrocephalus: a clinical, neuroradiologic, and neuropsychologic assessment of long-term outcome. Arch Phys Med Rehabil. 2003;84:1637–41. doi:10.1053/S0003-9993(03)00314-9.

32. Linnemann M, Tibæk M, Kammersgaard LP. Hydrocephalus during rehabilitation following severe TBI. Relation to recovery, outcome, and length of stay. NeuroRehabilitation. 2014;35:755–61. doi:10.3233/NRE-141160.

Anaesthesia for Hydrocephalic Patients and Large Head Patients

12

Mohamed El Tahan

12.1 Anaesthesia and Prematurity

A healthy term baby born between 37 and 42 postconception weeks usually weighs 2.5–3 kg. A premature baby is born before 37 completed postconception weeks. Babies may be small for dates, premature or both. Very low birth weight babies are defined as <1500 g and extremely low birth weight <1000 g.

Nowadays, anaesthesia for the premature is no longer just the art of adapting older children techniques and equipment to smaller infants but a sophisticated concern of its own.

12.1.1 Abnormalities Associated with Prematurity [1]

1. Postoperative apnoea, defined as a cessation of respiration for 15–20 s during the first 24 h following general anaesthesia or sedation in the more premature babies, could potentially be associated with oxygen desaturation and bradycardia. The

Electronic supplementary material The online version of this chapter (doi:10.1007/978-3-319-61304-8_12) contains supplementary material, which is available to authorized users.

M. El Tahan, M.D.
Anesthesiology Department, Imam Abdulrahman Bin Faisal University, King Fahd Hospital of the University, Dammam, Saudi Arabia
e-mail: mohamedrefaateltahan@yahoo.com; mohamedrefaateltahan@hotmail.com

apnoeic episodes can be central, obstructive or mixed in aetiology and are usually self-limiting or require mild stimulation of the baby to encourage respiration:

- **Risk factors**:
  - Age younger than 56–60 post-gestational weeks
  - Neurological disease
  - Anaemia
  - General anaesthesia
  - Sedation
  - Opiate use
  - Known respiratory disease, e.g. bronchopulmonary dysplasia
  - Preoperative oxygen requirement
- **Management**:
  - Stimulation to the baby.
  - Bag-mask ventilation.
  - Nasal continuous positive airway pressure (CPAP) ventilation.
  - Optimize preoperative conditions (e.g. treat anaemia).
  - Pulse oximeter and apnoea monitoring for at least 24 h postoperatively.
  - Consider pre- and postoperative use of caffeine or aminophylline.
  - Postpone elective surgery, and do not treat expremature babies as day cases until after 60 post-gestational weeks.

2. Retinopathy of prematurity: can happen after using high-inspired oxygen fraction (FiO_2) for premature babies younger than 8 months of age.
3. Anaemia.
4. Respiratory distress syndrome.
5. Chronic lung disease.
6. Bronchopulmonary dysplasia.
7. Intra-ventricular haemorrhage.
8. Visual or hearing loss.
9. Increased incidence of congenital abnormalities.
10. Necrotizing enterocolitis.
11. Acquired subglottic stenosis.
12. Persistent patent ductus arteriosus.
13. Failure to thrive.
14. Developmental delay.
15. Cerebral palsy.

12.1.2 Anaesthetic Considerations for Premature Infants [2]
(Table 12.1)

Table 12.1 Anaesthetic considerations for premature infants (*Reproduced with a permission obtained from Professor Jenny Thomas*)

No.	General golden rules
1	Neonatal anaesthesia is best provided by two anaesthetists working together
2	Be fully prepared with all consumables and equipment that may be needed
3	Be patient with obtaining of the vascular access
4	Minimal handling using fingertips rather than the whole hand
5	Meticulous attention should be paid to pressure complications from the blood pressure cuff, peripheral oxygen saturation (SpO_2) probes, the back of the head, and heels, and nasogastric tube traction on the nose
6	Meticulous drug administration includes choice of medication, dilutions and the volumes of fluid flushes
7	Balance risk of invasive monitoring with the risk of a compromised limb that could be lessening with using a local anaesthesia block
8	Good pain control because neonates in pain and distress will not grow
9	Intraoperative ventilation: • Minimize FiO_2, aiming for SpO_2 of 88–95% • Consider the use of positive end-expiratory pressure (PEEP) of 3–5 cmH$_2$O, I:E ratio of 1:1, 40–60 breaths per minute and permissive hypercapnia
10	Postoperative ventilation: for very ill infant and those that received high-dose opioid anaesthetic and muscle weakness

12.2 Neonatal Anaesthesia

Neonates, defined as infants younger than 1 month of age, have the highest incidence of potential perioperative risks for the patient and much stress for the anaesthesiologist. Proper communication with the parents, surgeons and neonatologists is crucial to safe and successful anaesthetic care for neonates.

12.2.1 Anaesthetic Considerations for Neonates [2]

Anaesthetic considerations for neonates include:

(a) *Anatomical airway differences* such as:
- Infants are obligate nasal breather until 5 months of age.
- The head is larger relative to body size and has a prominent occiput.

- The hypopharynx is relatively shorter in height and narrower in width.
- The larynx is relatively higher in the neck. The cricoid ring is located approximately at the level of the C_4 vertebrae at birth, C_5 at age 6 years, and C_6 as adult.
- The epiglottis in children is more "U" shaped (compared to flat in adults), and it is less in line with the trachea and may lie across the glottic opening.
- Children have smaller size of the airway at the glottic opening than at the cricoid level, despite the cricoid is functionally the narrowest part of the airway due to the relatively non-distensibility of its cartilage [3].

(b) *Respiratory concerns*:
- Higher oxygen consumption
- Higher closing volume
- Higher minute ventilation to maintain elevated respiratory rate
- Compliant ribs and less type 1 muscle in the diaphragm
- Significant incidence of postoperative apnoea
- Risks of retinopathy of prematurity

(c) *Cardiovascular concerns*:
- A return to transitional circulation.
- Persistent cardiac defects.
- Immature myocardium.
- A small fixed stroke volume (1–2 mL kg^{-1}) and a rate-dependent cardiac output (180–240 mL kg^{-1} min^{-1}).
- Normal heart rate is 140 ± 20 beats min^{-1}, blood pressure is $\pm70/52$ mmHg and mean blood pressure (MAP) equals gestational age in weeks.
- Parasympathetic dominance until about 6 months of age.
- Reactive pulmonary vasculature and pulmonary hypertension can be a significant cause of morbidity and mortality.
- Impact of anaesthetic agents on cardiovascular stability.

(d) *CNS concerns*:
- Neurological immaturity
- Focal neurological deficits
- Peri- or intra-ventricular haemorrhage

(e) *Haematology concerns*:
- Blood volume is 90–100 mL kg^{-1}.
- Foetal haemoglobin: 18–20 g dl^{-1} that decreases at 2–3 months of age.
- Erythropoietin disappears shortly after birth and reappears at 2–3 months.
- Anaemia.
- Coagulation abnormalities.

(f) *Drugs*:
- Pharmacology changes in neonates.
- Off-label use of medications may include using the medication for an unregistered indication, route of administration, dose and/or age group of the patient.
- Dilution errors.

(g) *Metabolic/electrolytes/fluids*:
  – Hypovolaemia, however, neonates could not tolerate rapid infusion of fluids.
  – Hypos and hypers: glucose, calcium, potassium and sodium (Na^+).
(h) *Environment*:
  – Ambient temperature (risk of hypothermia)
  – Transport
  – Minimal handling
(i) *Pain*:
  – In similar to adults, neonates demonstrate metabolic, hormonal and neurophysiologic responses to the pain and surgery.
  – Pain management during minor surgery is often provided by the combination of local anaesthesia (either as a block or as infiltration of the wound) and intravenous or rectal paracetamol 15 mg kg^{-1}.
  – Following major surgery, pain management with ultrashort-acting medication such as remifentanil, alfentanil or fentanyl, as well as the traditional longer-acting morphine, can be used in neonates provided they can be monitored adequately in a high-dependency area, particularly in babies <5 kg.
(j) *Type of surgery*:
  – The urgent or emergency procedure may have additional attendant risks (full stomach, abnormal electrolytes, hypovolaemia, precipitously high intracranial pressure (ICP)).
(k) *Vascular access*:
  – Difficult arterial and venous accesses

12.3 Anaesthesia for Hydrocephalic Surgery

12.3.1 General Considerations

Anaesthetic consideration for hydrocephalus surgery requires optimization of cerebral perfusion pressure (CPP), defined as the difference between the MAP and the sum of ICP and the central venous pressure (CVP):

$$CPP = MAP - (ICP + CVP)$$

Normal ICP in children may be as low as 2–4 mmHg, compared to 8–15 mmHg in adults. The infants are less able to compensate for the changes in MAP because the cerebral autoregulation limit is significantly lower with a MAP of 20–60 mmHg. Children have a higher global cerebral blood flow (CBF) than adults (100 vs. 50 mL 100 g^{-1} min^{-1}). The infants and premature babies have much lower CBF values.

Latex precautions should be considered in patients with myelomeningocele undergoing shunt placement.

12.3.2 Preoperative Evaluation and Preparation

Preoperative evaluation is focused on the airway, current cardiorespiratory and neurological status (e.g. vomiting, status of hydration), the underlying primary disease (e.g. intra-ventricular haemorrhage), concomitant medical illnesses (e.g. cervical spine abnormalities, congenital syndromes) and current medication regimen.

All neonates should have haemoglobin level and clotting function estimations before surgery to predict the need for blood transfusion. Whereas other preoperative tests should be selectively requested on the basis of clinical indication [4].

Sickle cell disease (SCD) testing should be done in neonates with positive family history or for those living in known regions with prevalence of SCD. However, SCD screening may provide false negatives in a very young baby with SCD, as HbS levels are so low, where electrophoresis can be accurate in this age group.

Patients who are on anticonvulsant therapy may have altered drug metabolism levels that may be rechecked only if there have been recent dosing changes or if seizures worsened. Preoperative medication may not be required because of associated high ICP and/or altered mental status.

12.3.3 Fasting

Infants should be fed before elective surgery. Breast milk is safe up to 4 h and non-human milks up to 6 h. Solid food should be prohibited for 6 h before elective surgery in children. Thereafter, children should be encouraged to drink clear fluids including water, pulp-free juice and tea or coffee without milk, up to 2 h before elective surgery [5].

12.3.4 Fluid Therapy [6, 7]

Children need balanced electrolyte solutions to maintain a perioperative stable water balance that corresponds to the composition of the extracellular space. The historic Holliday-Segar 4-2-1-rule should be replaced by a simpler approach.

Intraoperative maintenance requirement can be provided with a balanced electrolyte solution such as NaCl 0.45% (Na^+ 77 mmol L^{-1}) or higher strength, with and without 1% glucose, which is safe with respect to hyponatremia, hypo- and hyperglycaemia and accidental over-infusion. Fluid deficit/ongoing losses could be also replaced with isotonic saline solutions as NaCl 0.9% (Na^+ 154 mmol L^{-1}) or Ringer's acetate, according to clinical criteria. There are no existent third space losses in the brain to be compensated.

Hypotension is a threat during neonatal surgery; thus, bolus syringes containing 10 mL kg^{-1} of NaCl 0.9% should be prepared ahead of time for a quick resuscitation to a fall in MAP. However, rapid infusion of more than 60 mL kg^{-1} of NaCl 0.9% may cause hyperchloraemic acidosis. If fluid replacement reaches 100 mL kg^{-1} and hypotension persists, vasopressors may be required to restore MAP.

Allowable blood loss is generally considered to be 20–30% of the patient's blood volume. Blood loss is replaced either 3:1 with crystalloid or 1:1 with blood products. Transfusion of 10 mL kg^{-1} of packed red blood cells (PRBCs) increases haemoglobin concentration by 2 g dL^{-1}:

$$Blood\ volume\ needed = \left(Hct_1 - Hct_2 \right) / Hct_3 \times EBV$$

Hct_1 = the measured haematocrit before transfusion.
Hct_2 = the desired haematocrit required after transfusion.
Hct_3 = haematocrit of the blood to be given (60% if PRBCs).
EBV = estimated blood volume.

Paediatric patients are susceptible to dilutional thrombocytopaenia after massive blood loss and multiple PRBCs transfusions. Administration of 5–10 mL kg^{-1} of platelets increases the platelet count by 50,000–100,000 mm^{-3}.

12.3.5 Intraoperative Monitoring and Vascular Access

Routine specifically designed monitors for babies including electrocardiogram, non-invasive blood pressure, SpO_2, capnography, anaesthetic agent, temperature, FiO_2 and neuromuscular function should be used. As many neonates may have a patent ductus arteriosus, the SpO_2 probe should be placed on the right hand, if possible, to measure preductal SpO_2. The considerable alveolar-arterial CO_2 gradient leads to poor representation of arterial CO_2 by end-tidal carbon dioxide ($EtCO_2$).

One or two intravenous catheters should be placed. Arterial cannulation of the radial, femoral or axillary artery is usually not required except in the presence of serious co-morbidities. CVP is measured via the femoral or internal jugular vein.

12.3.5.1 Monitoring of the Depth of Anaesthesia

In the first 3 years of life, the monitoring of processed electroencephalogram (EEG) like as bispectral index (BIS) during anaesthesia is unlikely to reduce consumption of anaesthetics or the time to extubation [8]. The EEG narcotrend index monitoring, a measure of the hypnotic component of general anaesthesia, has the potential to reduce propofol consumption in children [9]. Spectral entropy may be a useful tool for measuring the level of hypnosis in anaesthetized children and seems to perform as well as BIS [10]. Whereas, the aepEX Plus monitor, utilizing a mid-latency auditory evoked potential-derived index of depth of hypnosis, is inferior to the BIS in terms of distinguishing different levels of sedation and hypnosis in children [11].

12.3.6 Temperature Maintenance [12]

Perioperative hypothermia could be induced with exposing large surface areas of the body for heat loss and administering of cold intravenous fluids and cold, dry oxygen and anaesthetic gases. Thus, perioperative temperature monitoring is a pre-requisite for prevention of inadvertent hypothermia. Normothermia could be maintained using different measures such as transfer/operate in an incubator and use an overhead radiant heater, forced-air warming, fluid warming, warm-inspired gases or a heat-moisture exchanger and heat conservation wrap of exposed areas. Cautiousness should be exerted to avoid accidental iatrogenic hyperthermia. Postoperative shivering can be treated using meperidine or clonidine.

12.3.7 General Anaesthesia

Induction: Intravenous or inhalational induction of general anaesthesia represents equally good options for the majority of hydrocephalus procedures. Endotracheal intubation is usually required during the delicate intra-ventricular surgical manoeuvres.

Eye Protection: Ophthalmological abnormalities, including visual impairment, are common in 80% of children with hydrocephalus. Additionally, patients with long-term ventriculoperitoneal (VP) shunts could have secondary enophthalmos due to the change in pressure gradient between cranial cavity and orbit [13]. Thus, extreme cautiousness should be exerted for eye protection during general anaesthesia to avoid ocular complications which could happen because of the loss of protective ocular reflexes. Simple taping of the eyelids closed, the instillation of ointments into the conjunctival sac, and the use of protective goggles have been recommended for eye protection [14].

Maintenance: The combination of an inhalation anaesthesia (without nitrous oxide) and wound infiltration with local anaesthetics (at the site of the burr hole at the beginning of the procedure) is usually sufficient to minimize the administration of opioids. Neonates require the use of a minimum alveolar concentration of 1.5–2% to prevent movements during surgery. However, preterm infants do not tolerate the most potent anaesthetics without unacceptable haemodynamic effects; thus, anaesthetic agents must be titrated carefully in neonates.

Controlled ventilation should be considered as soon as possible to achieve mild hyperventilation (EtCO$_2$ 30–35 mmHg) to offset any anaesthetic increase in CBF. PEEP should be carefully titrated to avoid venous congestion in the head; despite it will be useful if there are difficulties in maintaining oxygenation.

Postoperative care: The infants are usually extubated on table when they meet the following criteria:

1. Visualized crying (cannot be heard when endotracheal tube (ETT) is in place)
2. Opening eyes
3. Grasping for the ETT

After emergence from anaesthesia, the surgical team will perform the first neurological examination in the operating room. Thereafter, the patient is moved to the recovery room and ward with frequent follow-up neurological examinations.

12.4 Special Anaesthetic Concerns for Hydrocephalic Procedures

12.4.1 Patients with Klippel-Trenaunay Syndrome

It is a rare disorder, with an incidence of 1 out of 27,500 live births, that is associated with the triad of (1) capillary vascular malformation, (2) varicose veins and/or venous malformation (3) and soft tissue and/or bony hypertrophy. Anaesthetic management of those patients requires preparation for blood transfusion against massive haemorrhage and hypovolaemic shock. Furthermore, these patients have potential airway difficulty due to the soft tissue hypertrophy and upper and airway haemangiomas [15].

12.4.2 VP Shunts

VP shunt is widely conducted as the treatment for hydrocephalus. Shunt obstruction, fracture, disconnection, migration, infection, and over-drainage are known as the complications require revision procedures.

Preoperative preparation: Preoperative type and screen rather than crossmatch of blood is required because perioperative haemorrhagic shock has been reported after accidental puncture of trunc vessels during emergency shunt procedures in children presented with severe intracranial hypertension [16]. Antiepileptic drugs should be continued during the perioperative period. Antibiotic prophylaxis is mandatory against *Staphylococcus* infections of internal CSF drainage devices.

Intraoperative management: The main problems encountered in these patients are difficult intubation and full stomach associated with increased ICP. The supine position with the turned head to the contralateral side of the site of insertion of the VP shunt is usually indicated with occasional placement of a roll of towel under the shoulders to facilitate a straight line from the ear/neck to the abdomen for tunnelling of the shunt. Neck flexion may result in migration of the ETT to the main stem bronchus or may occlude the jugular vein impeding venous drainage and increasing ICP. The use of laryngeal mask airways (LMAs) like Supreme and Air-Q could be reserved for the management of children with difficult airway during VP shunting, because the frequent needs for repositioning the LMA or even proceed to orotracheal intubation secondary to the lateral neck position for surgery [17].

With the exception of ketamine and enflurane, the majority of anaesthetic drugs can be used. Tunnelling significantly increases ICP and blood pressure that could be attenuated with a prior administration of remifentanil 1 μg/kg [18].

The transversus abdominis plane block using volumes of 0.2 mL kg^{-1}, with a maximum volume of 20 mL, injected on either side has the potential to improve the quality of analgesia after VP shunting. The dosages of bupivacaine have to be limited to 2 mg kg^{-1} in neonates, 3 mg kg^{-1} in children and 4 mg kg^{-1} in adolescents to avoid local anaesthetic toxicity [19].

Postoperative management: Rapid emergence from anaesthesia and extubation should be encouraged.

12.4.3 Laparoscopic Placement of VP Shunts [20]

Laparoscopy-assisted VP shunting becomes an alternative safe, effective and minimally invasive procedure that reduces surgical trauma, complications and postoperative pain. CO_2 pneumoperitoneum is created up to a pressure of 12 mmHg that could influence the cardiorespiratory system, acid-base blood balance and the $EtCO_2$. CO_2 insufflation forces the diaphragm cephalad with potential inadvertent bronchial intubation in children with a fixed ETT at the lip. Nasogastric or orogastric tube placement and bladder catheterization are required to prevent injury to their respective organs. Nitrous oxide is usually avoided to allow for higher FiO_2, to prevent bowel distention and to avoid its diffusion into the pneumoperitoneum. Children with compromised cardiorespiratory function and volume depletion will have an exaggerated adverse response to CO_2 insufflation. In contrast to adults, volume loading and treatment with anticholinergic agents have not been universally utilized to prevent hypotension and bradycardia in children during laparoscopic procedures. CO_2 insufflation-induced hypercarbia could worsen the high ICP that requires positive pressure ventilation (PPV). The use of PEEP can prevent atelectasis and subsequent hypoxaemia. Subcutaneous emphysema secondary to fascial dissection of CO_2 during laparoscopic surgery can also occur and is associated with rapid increases in $EtCO_2$ because of an increased surface area for absorption.

12.4.4 Endoscopic Paediatric Neurosurgery [21]

Endoscopic surgery is increasingly utilized for hydrocephalus interventions such as endoscopic fenestrating or ventriculostomy of the third ventricle that has high reported success rate. Much of the preoperative considerations are not dictated by the specific endoscopic procedure, but by the patients' conditions (e.g. age, weight and current health status). A standard operating room layout requests more vigilance. The anaesthesia team and equipment are usually situated on the left side of the patient. The surgical team is positioned directly around the head of the patient, observing the video monitors, which are located at the foot of the patient while navigating the endoscope. Muscle relaxation needed to prevent patient movement during the delicate intra-ventricular surgical procedures. Cautiousness should be exercised, to administer irrigation solutions at body temperature, as cold fluids may cause bradycardia. The most common postoperative complications include CSF leak, meningitis, haemorrhage, hypothalamic injury, cranial nerve injury and seizure.

12.5 Airway Management for Large Head Patients

Neonates usually require tracheal intubation and ventilation during all neurosurgical interventions, because the risk of ventilation/perfusion mismatch. Airway management in hydrocephalus patients is a challenge to anaesthesia practitioners due to the particular larger occiput combined with a shorter neck. This makes laryngoscopy relatively more difficult by providing obstacles to the alignment of the oral, laryngeal and tracheal axes.

12.5.1 Airway Assessment

It usually starts with *good history* about any complications of birth or delivery; prior trauma or surgery to the airway or adjacent structures; prior anaesthetics; current or recent symptoms suggesting upper respiratory infection; difficulty in speaking, breathing or feeding; hoarseness; noisy breathing; snoring; or stopped breathing during sleep.

An *airway physical examination* should be conducted to detect physical characteristics of a difficult airway including (1) mandibular hypoplasia, (2) limited mouth opening, (3) conditions of the head and neck, (4) facial asymmetry or (5) ear anomalies.

Predictors of *difficult bag-mask ventilation* include (1) abnormal-shaped head (e.g. large head, encephaolcele), (2) limited neck mobility, (3) facial asymmetry (e.g. congenital facial anomaly), (4) obese children and (5) significant lung disease (e.g. severe asthma).

12.5.2 Congenital Craniofacial Abnormalities

Congenital craniofacial abnormalities have many airway implications as shown in Table 12.2 *[Reproduced with permission obtained from the Editor-in-Chief]*.

Table 12.2 Congenital craniofacial abnormalities and airway implications

Syndrome	Airway implications
Pierre Robin sequence	Micrognathia, glossoptosis, cleft palate
Goldenhar syndrome	Micrognathia (unilateral), cervical dysfunction
Treacher Collins syndrome	Micrognathia, small oral opening, zygomatic hypoplasia
Apert syndrome	Limited cervical motion, macroglossia, micrognathia, midface hypoplasia
Hunter and Hurler syndrome	Cervical dysfunction, macroglosia
Beckwith-Wiedemann syndrome	Macroglossia
Freeman-Sheldon syndrome	Cicumoral fibrosis, microstomia, limit cervical motion
Down syndrome	Atlantooccipital abnormalities, small oral cavity, macroglosia
Klippel-Feil syndrome	Cervical fusion
Hallermann-Streiff syndrome	Microstomia
Arthrogryposis	Cervical dysfunction
Cri-du-chat syndrome	Micrognathia, laryngomalacia
Edwards syndrome	Micrognathia
Fibrodysplasia ossificans progressiva	Limited cervical motion

12.5.3 Preparations for Airway Management

Preparation of age-appropriate equipment like facemasks, oropharyngeal airways, laryngoscopes, stylets, LMAs and ETTs is cornerstone for successful tracheal intubation (Tables 12.3 and 12.4). The modern low-pressure high-volume cuffed ETT becomes an attractive option that does not cause mucosal ischaemia and minimise air leak and level of environmental pollutants with anaesthetics [22]. The cuff pressure should be adjusted so that there is a small leak around the cuff at a peak inflation pressure of 20–30 cm H_2O.

Patient's parents/guardian should to be informed about a known or suspected difficult airway.

12.5.4 Positions for Intubation

- Folded towels are often required beneath the neonate until the height matched that of the large head to achieve a neutral position of the neck and open up the airway.
- A large donut or C-shaped headrests can be used to include the large head inside (Fig. 12.1).
- A lateral position can be reserved for children with a concomitant meningomyelocele or encephalocele [Video Clip 12.1].

Table 12.3 Age-appropriate endotracheal tube size

	Endotracheal tube size (mm ID)	Endotracheal tube depth (cm)
Preterm		
<1000 g	2.5	6–7
1000–2500 g	3.0	Weight (in kg) + 6
Term to 6 months	3.5	8–10
6 months–1 year	4.0	10–11
1–2 years	4.0–5.0	11–12
Beyond 2 years	Age (in years) + 16/4	Age (in years)/2 + 12

An inner diameter (ID) 0.5 mm smaller reserved for cuffed ETT

Table 12.4 Weight-appropriate intubating LMA classic, ETT, AEC and FOB

Body weight	LMA classic	ETT	Cook AEC	FOB
<5	1	3.0 mm uncuffed	7 F	2.2 mm
5–10	1.5	3.5 mm uncuffed	8 F	2.5 mm
10–20	2	4.5 mm uncuffed	11 F	3.5 mm
15–30	2.5	5.0 mm uncuffed	11 F	3.5 mm
30–50	3	6.0 mm cuffed	14 F	5.0 mm
50–70	4	7.0 mm cuffed	14 F	5.0 mm

Fig. 12.1 Position of large head on headrest

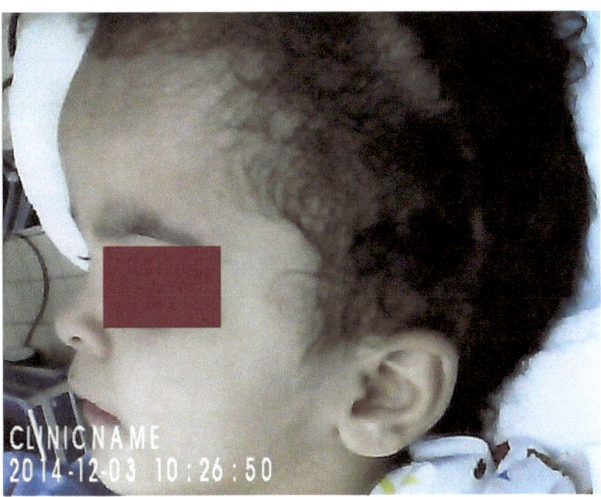

12.5.5 Suggested Algorithm for Difficult Airway Management in Children (Fig. 12.2)

The author has adopted an algorithm for management of difficult airway management in paediatrics from the released American Society of Anesthesiologists Task Force on Management of the Difficult Airway in Adults on February 2013 [23].

When difficult airway management is predicted, awake laryngoscopy and endotracheal intubation are useful using either different blades of conventional laryngoscopes (e.g. MacIntosh, Miller), a video laryngoscope (VL) [Video Clip 12.1] or a flexible fibre-optic bronchoscopy (FOB) through a LMA in an awake, unsedated patient.

When unpredicted difficult airway management has been encountered after induction of general anaesthesia, provided that mask ventilation is adequate, the number of tracheal intubation attempts should be limited to two unsuccessful attempts, using either different blades of conventional laryngoscopes, VL or intubation through the LMA.

If tracheal intubation fails, other options could be considered including, but are not limited to, surgery utilizing LMA, regional anaesthesia/nerve blockade or awakening the patient for re-preparation for awake intubation or cancelling surgery. Otherwise, invasive airway access including surgical cricothyroidotomy or retrograde intubation with or without using jet ventilation, rigid bronchoscopy or surgical tracheostomy could be considered. Needle cricothyroidotomy is not recommended under 5–8 years.

If there is difficulty with mask ventilation and the airway is considered a difficult airway, the LMA may also be placed for ventilation until awakening of patient or to proceed directly to invasive airway access. It is essential that oxygenation and an adequate depth of anaesthesia are maintained at all times during these attempts [24].

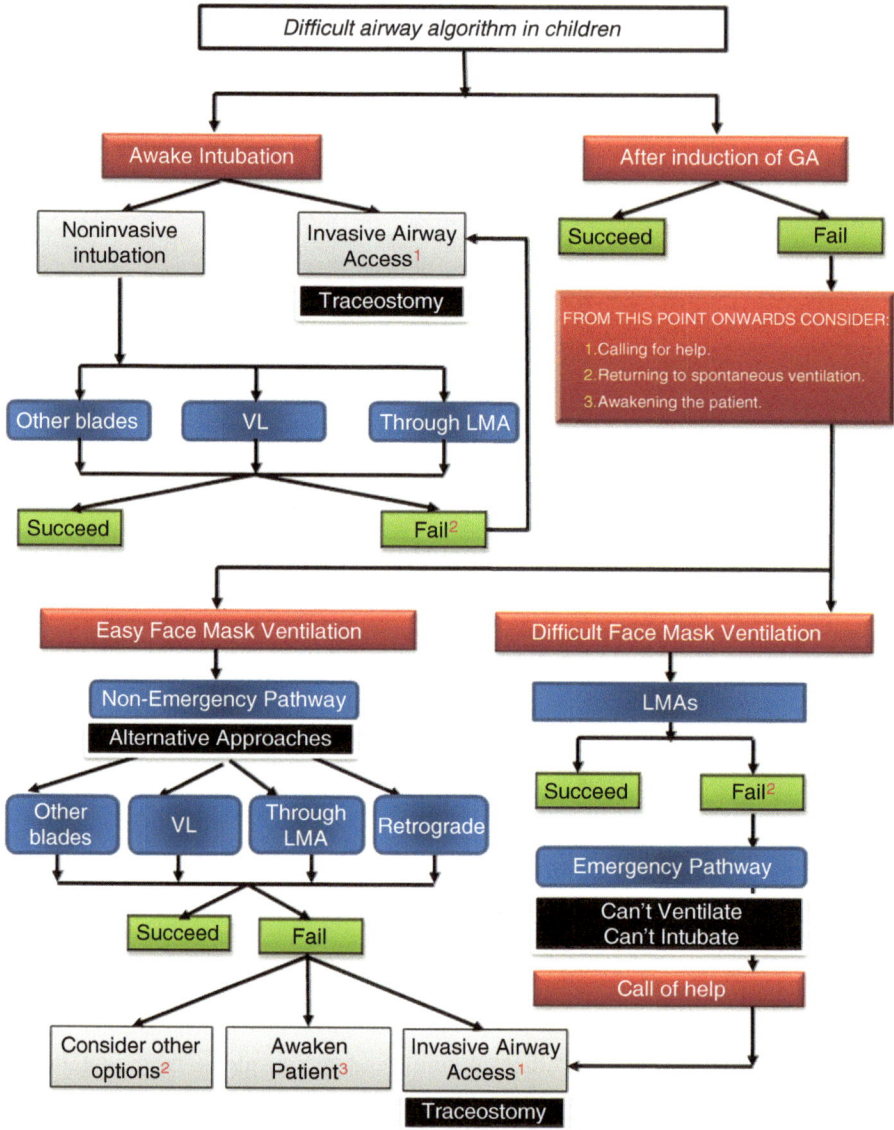

Fig. 12.2 Algorithm of difficult airway management in children

12.6   Different Available Devices for Airway Management

Securing a patent airway in children is routinely performed using direct laryngoscopy with a Macintosh or Miller laryngoscope blade.

12.6.1 Awake Fibre-Optic Tracheal Intubation

It is rarely an option in children leaving the anaesthesiologist no choice but to manage the airway after induction of anaesthesia. The ultrathin FOB (2.2 and 2.8 mm outer diameter) allows the use of ETT size 3.0 (internal diameter) or larger (Table 12.4) [25].

12.6.2 The Classic LMAs

It can be recommended as a safe airway device for those with an upper respiratory airway infection and difficult airways. However, this could not be feasible for hydrocephalus surgery. An LMA with a modified cuff (e.g. Proseal™ LMA, Supreme LMA) appears to provide a better seal over the larynx for children, allowing for more effective delivery of PPV (Table 12.4) [26].

12.6.3 Intubating LMAs

The use of LMAs (e.g. LMA Classic, Air-Q, Cobra Perilaryngeal Airway) can be used as conduits for the passage of the FOB in pediatric patients with suboptimal laryngeal exposure. Then the LMA may then stay in its position or be removed [27, 28]. However, the removal of the LMA over the ETT is impaired by its short length that could easily result in ETT dislocation from the trachea. This could be prevented with the use of bougie, the Cook Airway Exchange Catheter (AEC) [Table 12.4] or the Trachlight [29].

12.6.4 Video Laryngoscopes

Video laryngoscopes are rapidly growing to become a part of the paediatric difficult airway armamentarium that offers an expanded and high-resolution view of the airway. The available VLs for use in paediatric practice are the GlideScope®, the Storz C-MAC, the McGRATH®, the Truview, the Airtraq®, the Pentax Airway Scope (AWS) and the Bonfils intubation endoscope.

– The **Cobalt Glidescope®**, a reusable video baton and single-use laryngoscopy blades, comes in two paediatric sizes with 70° angulation of the blade. It is inserted into the mouth in the midline without displacing the tongue, and

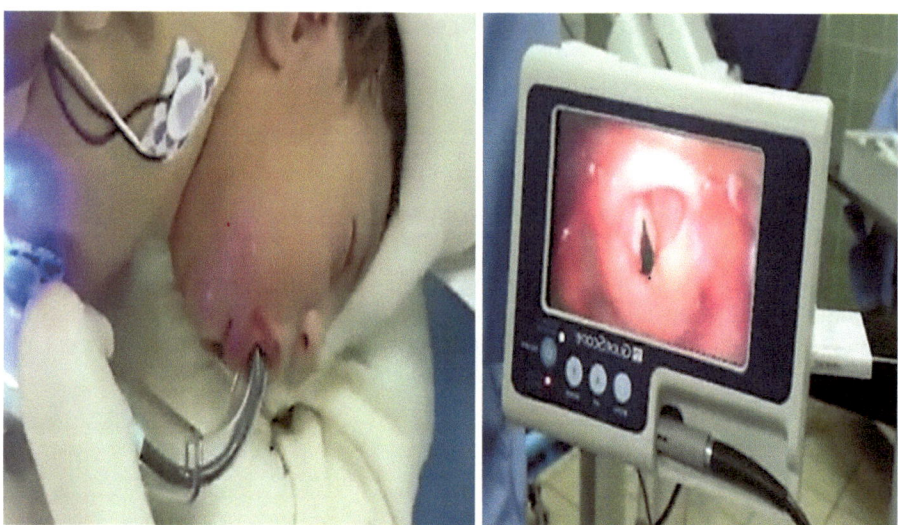

Fig. 12.3 Glidescope GVL used for tracheal intubation of hydrocephalus

the ETT is directed towards the glottis using a stylet (Fig. 12.3) [Video Clips 12.1 and 12.2]. It improves glottis visualization in children with normal airways and with potentially difficult intubation, but it could lengthen the time to intubation [30].

- The **C-MAC** integrates a fibre-optic bundle into the light source of a standard MacIntosh blade shape (size 2) or the Miller shape (sizes 0 and 1) with a camera in the handle of the device that displays and magnifies the image on a screen. It offers higher intubation success rate on first attempt in children with difficult intubation [31].
- The **McGRATH®**, a fully portable VL with a single-use blade, can be used for difficult airways in children older than 5 years.
- The **Truview PCD-Infant**, a small handy VL with an eyepiece and optics providing a wide and magnified laryngeal view at a 46° anterior refracted angle, has the potential to provide a slightly good laryngeal view in neonates and infants with normal airways [32].
- The **Airtraq®**, a channelled VL, has available sizes for tracheal intubation in infants (size 0 accommodates ETT size of 2.5–3.5 mm) and children (size 1 accommodates ETT size 3.5–5.5 mm) allowing visualization of the glottis especially in those with restricted mouth opening and micrognathia [33]. The use of a gum-elastic bougie could facilitate the introduction of the ETT tip to the trachea. The use of Airtraq® has shorter time to tracheal intubation than the GlideScope®, the C-MAC and the McGrath®, especially with growing up of the skills of the practitioners [34].
- The **Pentax AWS**, an anatomically shaped channelled VL, provides good-quality view of the glottis without the need for an intubating style in infants with hydrocephalus [35].

– The **Bonfils intubation endoscope** is a rigid metal stylet VL with a fixed 40° anterior tip curvature containing a fibre-optic bundle that facilitates intubation of the difficult airway in small babies. It is available in three sizes with outer diameters of 2, 3.5 or 5 mm and can accommodate 2.5–6-mm ETT that are railroaded under direct vision into the trachea [36].

12.7 Spontaneous Ventilation Versus Muscle Relaxation in the Difficult Airway

Classical teaching of the difficult airway management dictates inhalational induction of anaesthesia and maintenance of spontaneous ventilation until the airway has been secured by tracheal intubation. However, others advocate that paediatric patients who can be ventilated via facemask or LMA can safely be paralysed; this has the potential for airway collapse due to loss of pharyngeal tone [24].

12.8 Rapid Sequence Induction

Rapid sequence induction (RSI) is generally used in patients who are at risk for pulmonary aspiration. A 'controlled' RSI approach with deep anesthesia, profound paralysis and gentle intermittent facemask ventilation prior to intubation provides stable cardiorespiratory conditions for securing the airway in children with a suspected full stomach with no added risks for pulmonary aspiration [37]. Ketamine and etomidate potentially provide effective sedation for RSI with limited effects on haemodynamic function [38].

Conclusion

The presence of skilled vigilant anaesthesiologists, an understanding of neurophysiology, proper preparations to provide anaesthesia for the different paediatric age range and careful planning for difficult airway management allow for the safe conduct of anaesthesia for hydrocephalic interventions in the large head patients.

Conflict of Interest Statement The author declares that he has no conflicts of interest and received no financial support for the research, authorship and/or publication of this article. Dr. El Tahan received free airway device samples from Ambu in April 2014 for use in another study, and he has no direct financial or other interest in Ambu.

References

1. Bayley G, Walker I. Special considerations in the premature and ex-premature infant. Anaesth Intensive Care. 2007;9:89–92.
2. Thomas J. Reducing the risk in neonatal anesthesia. Paediatr Anaesth. 2014;24:106–13. [Permission obtained from Professor Jenny Thomas].

3. Harless J, Ramaiah R, Bhananker SM. Pediatric airway management. Int J Crit Illn Inj Sci. 2014;4:65–70. [Reproduced with a permission obtained from the Editor-in-Chief].
4. Almesbah F, Mandiwanza T, Kaliaperumal C, Caird J, Crimmins D. Routine preoperative blood testing in pediatric neurosurgery. J Neurosurg Pediatr. 2013;12:615–21.
5. Smith I, Kranke P, Murat I, et al. Perioperative fasting in adults and children: guidelines from the European Society of Anaesthesiology. Eur J Anaesthesiol. 2011;28:556–69.
6. Strauß JM, Sümpelmann R. Perioperative fluid management in infants and toddlers. Anasthesiol Intensivmed Notfallmed Schmerzther. 2013;48:264–71.
7. Drage IM, Ingvaldsen B, Dorph E, Bentsen G. New guidelines for intravenous fluid therapy for children. Tidsskr Nor Laegeforen. 2013;133:2235–6.
8. Bresil P, Nielsson MS, Malver LP, et al. Impact of bispectral index for monitoring propofol remifentanil anaesthesia. A randomised clinical trial. Acta Anaesthesiol Scand. 2013;57:978–87.
9. Weber F, Pohl F, Hollnberger H, Taeger K. Impact of the Narcotrend Index on propofol consumption and emergence times during total intravenous anaesthesia with propofol and remifentanil in children: a clinical utility study. Eur J Anaesthesiol. 2005;22:741–7.
10. Klockars JG, Hiller A, Ranta S, Talja P, van Gils MJ, Taivainen T. Spectral entropy as a measure of hypnosis in children. Anesthesiology. 2006;104:708–17.
11. Cheung YM, Scoones GP, Hoeks SE, Stolker RJ, Weber F. Evaluation of the aepEX™ monitor of hypnotic depth in pediatric patients receiving propofol-remifentanil anesthesia. Paediatr Anaesth. 2013;23:891–7.
12. Torossian A. Perioperative thermal management in children. Anasthesiol Intensivmed Notfallmed Schmerzther. 2013;48:278–80.
13. Kim JM, Chang MH, Kyung SE. The orbital volume measurement in patients with ventriculoperitoneal shunt. J Craniofac Surg. 2015;26(1):255–8.
14. Park SJ, Kim IS. Severe edema of the eyes and lips as rare side effects of eye ointment for protection of eyes under general anesthesia-A case report. Korean J Anesthesiol. 2012;63:454–6.
15. Hoshijima H, Takeuchi R, Tsukamoto M, Ogawa S, Iwase Y, Matsumoto N. Anesthetic management for a pediatric patient of Klippel-Trenaunay syndrome with giant head by hydrocephalus. Masui. 2012;61:1356–8.
16. Combettes E, Blanot S, Cuttaree H, Zérah M, Orliaguet G. Haemorrhagic shock during cerebrospinal fluid shunt procedure. Reassessment of the anaesthetic or surgical practice? Ann Fr Anesth Reanim. 2006;25:206–9.
17. Hurtado P, Valero R, Tercero J, et al. Experience with the proseal laryngeal mask in ventriculoperitoneal shunting. Rev Esp Anestesiol Reanim. 2011;58:362–4.
18. Chambers N, Lopez T, Thomas J, James MF. Remifentanil and the tunnelling phase of paediatric ventriculoperitoneal shunt insertion. A double-blind, randomised, prospective study. Anaesthesia. 2002;57:133–9.
19. Mai CL, Young MJ, Quraishi SA. Clinical implications of the transversus abdominis plane block in pediatric anesthesia. Paediatr Anaesth. 2012;22:831–40.
20. Means LJ, Green MC, Bilal R. Anesthesia for minimally invasive surgery. Semin Pediatr Surg. 2004;13:181–7.
21. Meier PM, Guzman R, Erb TO. Endoscopic pediatric neurosurgery: implications for anesthesia. Paediatr Anaesth. 2014;24:668–77.
22. Brinsmead TL, Inglis GD, Ware RS. Leak around endotracheal tubes in ventilated newborns: an observational study. J Paediatr Child Health. 2013;49:E52–6.
23. Apfelbaum JL, Hagberg CA, Caplan RA, et al. Practice guidelines for management of the difficult airway: an updated report by the American Society of Anesthesiologists Task Force on Management of the Difficult Airway. Anesthesiology. 2013;118:251–70.
24. Engelhardt T, Weiss M. A child with a difficult airway: what do I do next? Curr Opin Anaesthesiol. 2012;25:326–32.
25. Kohelet D, Arbel E, Shinwell ES. Flexible fiberoptic bronchoscopy--a bedside technique for neonatologists. J Matern Fetal Neonatal Med. 2011;24:531–5.

26. Jagannathan N, Sequera-Ramos L, Sohn L, Wallis B, Shertzer A, Schaldenbrand K. Elective use of supraglottic airway devices for primary airway management in children with difficult airways. Br J Anaesth. 2014;112:742–648.
27. Bhoi D, Dehran M, Raghavan S, Baidya DK. CobraPLA-guided tracheal intubation for airway rescue in child with large orofacial arteriovenous malformation. Acta Anaesthesiol Taiwan. 2013;51:99–100.
28. Barch B, Rastatter J, Jagannathan N. Difficult pediatric airway management using the intubating laryngeal airway. Int J Pediatr Otorhinolaryngol. 2012;76:1579–82.
29. Weiss M, Mauch J, Becke K, Schmidt J, Jöhr M. Fibre optic-assisted endotracheal intubation through the laryngeal mask in children. Anaesthesist. 2009;58:716–21.
30. Sun Y, Lu Y, Huang Y, Jiang H. Pediatric video laryngoscope versus direct laryngoscope: a meta-analysis of randomized controlled trials. Paediatr Anaesth. 2014;24:1056–65.
31. Aziz MF, Dillman D, Fu R, Brambrink AM. Comparative effectiveness of the C-MAC video laryngoscope versus direct laryngoscopy in the setting of the predicted difficult airway. Anesthesiology. 2012;116:629–36.
32. Singh R, Singh H, Vajifdar H. A comparison of Truview infant EVO2 laryngoscope with the Miller blade in neonates and infants. Pediatr Anesth. 2009;19:338–42.
33. Iwai H, Kanai R, Takaku Y, Hirabayashi Y, Seo N. Successful tracheal intubation using the pediatric Airtraq optical laryngoscope in a pediatric patient with Robin sequence. Masui. 2011;60:189–91.
34. Sørensen MK, Holm-Knudsen R. Endotracheal intubation with Airtraq® versus Storz® video-laryngoscope in children younger than two years - a randomized pilot-study. BMC Anesthesiol. 2012;12:7.
35. Matsunami S, Komasawa N, Nakao K, Nakano S, Tatsumi S, Minami T. Intubation using the Pentax-AWS Airwayscope with an infant-size Intlock in a patient with increased intracranial pressure due to acute hydrocephalus. Masui. 2014;63:412–4.
36. Krishnan PL, Thiessen BH. Use of the Bonfils intubating fibrescope in a baby with a severely compromised airway. Paediatr Anaesth. 2013;23:670–2.
37. Neuhaus D, Schmitz A, Gerber A, Weiss M. Controlled rapid sequence induction and intubation - an analysis of 1001 children. Paediatr Anaesth. 2013;23:734–40.
38. Scherzer D, Leder M, Tobias JD. Pro-con debate: etomidate or ketamine for rapid sequence intubation in pediatric patients. J Pediatr Pharmacol Ther. 2012;17:142–9.

Perioperative Management of Hydrocephalus in Preterm

13

Sherif Al Mekawi and Nermeen Galal

13.1 Introduction

Ex-preterm babies are often challenged with multiple health problems affecting different body systems. They are often at risk of developing hydrocephalic changes which may be secondary to congenital malformations, post-meningitic or posthemorrhagic. The risk for severe intraventricular hemorrhage (IVH) varies and is inversely proportionate with gestational age (GA) with an overall incidence of 7–23% [1]. Fragile blood vessels in the germinal matrix below the ventricular lining and instability of blood flow to the highly vascular area are the main mechanism behind IVH [2].

Increased fibrinolytic activity in preterms and coagulation defects presents another difficulty as coagulation abnormalities are common among preterm babies and could be reflective of underlying disease processes but are fraught with pitfalls regarding interpretation [3].

A study of the development of epidermal innate immune function from the skin surface of premature and term infants showed that the biomarker profile in premature infants was unique with difference in the structural proteins, albumen, and cytokines IL-6, IL-1beta, and IL-8 favoring more infections [4]. Another factor favoring infections is that the maternal transport of immunoglobulin to the fetus mainly occurs after 32 weeks gestation and endogenous synthesis does not begin until several months after birth [5]. All those risk factors call for caution when a decision of ventriculoperitoneal (VP) shunt insertion is considered in an ex-preterm. Many methods for treating the hydrocephalus have been used, none of which are ideal [6].

Massone ML et al. in 1994 reported a low complication risk following the outcome of early treatment with long-term external ventricular drainage (EVD) of progressive posthemorrhagic ventricular dilatation (PPHVD), following peri-intraventricular hemorrhage (PIVH) in a population of preterm newborns [7].

S. Al Mekawi, M.D. (✉) • N. Galal, M.D., M.R.C.P.C.H
Faculty of Medicine, Cairo University, Giza, Egypt
e-mail: sm@sherifalmekawi.com

© Springer International Publishing AG 2017
A. Ammar (ed.), *Hydrocephalus*, DOI 10.1007/978-3-319-61304-8_13

Brian K et al. in 2005 described ventriculosubgaleal shunts as a way of tempo-rary cerebrospinal fluid (CSF) diversion as those infants need a procedure until they gain adequate weight, and the blood and protein levels in CSF are reasonably low before permanent shunt can be placed [8]. Controlled removal of cerebrospinal fluid (CSF) by serial tapping of ventricular reservoir, such as the McComb reservoir, was used by Kormanik, Praca, Garton, and Sarkar in 2009 to decompress the ventricular system in preterm infants with rapidly progressing posthemorrhagic ventricular dilatation (PHVD) while awaiting optimal conditions for permanent CSF drainage through a ventriculoperitoneal shunt. However, the data regarding the risk of infec-tion from repeated invasive tapping of a ventricular reservoir over a prolonged period are scarce [9].

Factors in favor of early intervention include severity of IVH, rapidly progres-sive hydrocephalic changes, and/or the presence of manifestations suggestive of brainstem dysfunction [10]. Should surgery be decided upon other factors like postoperative apnea in preterm, vasomotor instability necessitates continuous monitoring and vigilant postoperative observation [11]. The risk for postoperative apnea was found to be related to bronchopulmonary dysplasia, chronic hypoxemia and hypercapnia, tracheomalacia, bronchomalacia, and pulmonary hypertension. Other risk factors include Hct < 30%, GA < 30 weeks, LBW < 1.5 kg, apnea at home, complicated neonatal intensive care unit (NICU) course, and lower weight at time of surgery [12].

Long-term developmental outcome is often unpredictable and is related to pre-ceding hypoxic ischemic injury that predisposed to IVH, cerebral perfusion follow-ing IVH, destruction of the germinal matrix, and the extent of white matter injury. Severe IVH that requires shunt insertion was found to represent an additional dif-ferential risk factor that had an additional adverse impact on neurodevelopmental outcome above the baseline risk [13].

13.2   Aim of the Study

The study aims at perioperative evaluation of VP shunt insertion in preterm infant with hydrocephalus with identification of risk factors related to intra-/postoperative incidents.

13.3   Methodology

A case series study was designed after informed consent and review board approv-als. Cases diagnosed with hydrocephalus during their neonatal intensive care unit (NICU) stay were followed up after discharge and assessed repeatedly for indica-tions of VP shunt insertion and postoperative progress if operated upon. Cases with complex congenital anomalies were excluded.

A total of 15 infants were enrolled. Three cases had arrested progress of their hydrocephalic changes on follow-up, and two cases were complicated by nosocomial infections and passed away. The remaining ten cases were subjected to detailed history taking (maternal, obstetrical, and neonatal data) and comprehensive examination with emphasis on the following.

13.3.1 Preoperative Assessment

- Resuscitation data, APGAR scores
- Gestational age
- Anthropometric measurements, including birth weight, length, head circumference (HC) using UK preterm charts with gestational correction
- Gender
- Etiology of hydrocephalus
- Grade of IVH if present
- Onset of hydrocephalic changes
- Rate of increase of HC (serial HC, ventricular diameter by ultrasound/CT scan) (Fig.13.1)
- Temporary measures (previous tapping)
- Stridor
- Respiratory distress (cause and grade if present)
- Need for surfactant, ventilation, steroids, development of chronic lung disease
- Apnea, bradycardia, or desaturation (ABD)

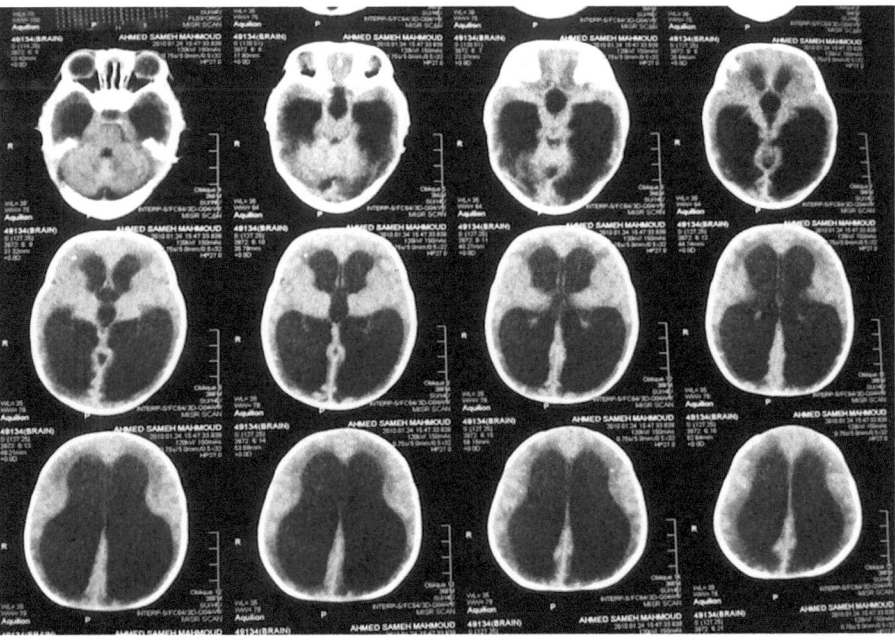

Fig. 13.1 Preoperative CT brain

- Sepsis
- Need for inotropic support
- Hemoglobin and hematocrit values
- Coagulation profile
- Vomiting, tolerance to feeds
- Length of hospital stay
- Age at surgery
- Weight at surgery

In cases with intraventricular hemorrhage and ventriculitis, an external ventricular drain was inserted and was kept until the CSF cleared clinically and upon analysis and culture and sensitivity.

13.3.2 Intraoperative Details

Upon arrival to the previously warmed operating room (OR) (25 °C), precordial stethoscope, noninvasive blood pressure, electrocardiogram, and pulse oximetry were applied. Before induction of anesthesia, a 24 G intravenous catheter was inserted, and 0.02 mg/kg atropine was administered. An infusion of 5% dextrose in 0.45% saline was then started at a rate of 5 ml/kg/h. Anesthesia was induced with fentanyl 4 $\mu g/kg^{-1}$ and vecuronium 0.1 mg/kg to facilitate tracheal intubation using size 0 straight blade laryngoscopy. Ventilation was controlled with oxygen and air (FIO_2 less than 40%). After induction of anesthesia capnography and rectal temperature, probe was applied. Third-generation cephalosporin to WT was given before skin incision.

The surgical technique was similar to previous descriptions, the skin flap used fashioned with a wide base to ensure adequate vascularity and to reduce being under tension during closure of the wound. The posterior parietal burr hole was used in all procedures; medium-pressure burr hole-type small-sized shunt device was used in all procedures. Utmost care should be provided during the introduction of the ventricular catheter regarding the direction of tapping and the length of the tube to avoid repeated tapping that may increase the risk of intraventricular hemorrhage. The length of the peritoneal catheter was measured individually, reaching the symphysis pubis of the patient, and intra-abdominal insertion was though a midline incision. Only the surgeon and the scrubbed nurse were involved in the surgical procedure with the minimum number of personnel in the operating room, trying to decrease the incidence of infection. The surgical time was about 45 min in all procedures.

At the end of the surgery, infants were transferred intubated to the pediatric intensive care unit (PICU) to be monitored for the occurrence of apnea, desaturation, and bradycardia up to 48 h.

13.3.3 Immediate Postoperative Monitoring

- Occurrence of ABD
- Hypotension, hypothermia, need for inotropic support

- Seizures, jitteriness
- Resumption of oral intake, tolerance
- Subdural/epidural hematomas, need for transfusion

13.3.4 Late Postoperative Outcome

- Function (revisions)
- Infections, fever (ventriculitis, abdominal or systemic)
- Attainment of developmental milestones (gross motor, fine motor, social behavior and play, speech, language, and hearing development) in respect to age
- Visual affection
- Motor dysfunction/sequelae
- Oral motor dysfunctions

13.4 Results

13.4.1 Initial Presentation

There were five males and five females. The gestational age ranged between 24 and 34 weeks with a mean of 29.5 weeks, standard deviation (SD) of 3.2, and interquartile range of 5. Body weight and length measurements ranged between 25th and 90th for age. The skull circumference measurements were all exceeding 99th centile for age, and values ranged between 32 and 39 cm with a mean of 35.62 and SD of 2.2 with an interquartile range of 3. The APGAR scores ranged between 5–9 and 8–10 at 1 and 5 min, respectively.

Regarding etiology six cases were due to congenital hydrocephalus (congenital aqueduct stenosis). Three were posthemorrhagic following IVH Grades III and IV [14], and a single case was post-meningitic following late-onset sepsis with *Escherichia coli*. The onset of hydrocephalic changes ranged between day 0 and day 42 (Table 13.1). The cases diagnosed with posthemorrhagic and post-meningitic hydrocephalus all necessitated external ventricular drains (that was kept for an average of 6.25 days: the longest number of days was 10 in the post-meningitic case, and the shortest number of days was 5 days), till patients were cleared for surgery. Eight of the cases had history of respiratory distress during their NICU stay with history of maternal steroid administration due to anticipated preterm labor. Mild forms of respiratory distress syndrome were treated with oxygen via continuous positive airway pressure and improved spontaneously. Three cases were ventilated and received surfactant with no sequelae. Three cases had history of apnea of prematurity during admission. Three cases were treated for neonatal sepsis during follow-up (one with *E. coli*, one with Group B *Streptococcus*, and one with *Klebsiella pneumoniae*). Two of the cases were on low-dose dopamine infusion to promote renal flow briefly during their hospital stay. Cases were all receiving expressed breast milk with preterm formula via NGT which was later changed to oral feeds except for one infant who was on prolonged NGT feeding for oral motor dysfunction. Two of the four

Table 13.1 Causes of hydrocephalus

Congenital aqueduct stenosis	Intraventricular hemorrhage	Ventriculitis
6 cases	3 cases	1 case

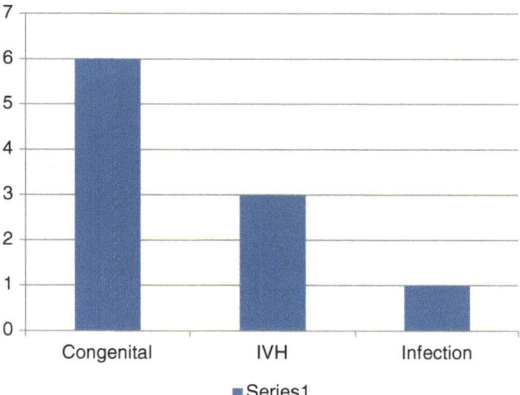

cases of congenital hydrocephalus had associated stridor. The length of hospital stay ranged between 2 and 8 weeks prior to surgery with an average of 4 weeks. The hemoglobin values ranged between 8.5 and 11 g/dl with a mean of 9.7 SD of 0.8. The coagulation profiles revealed physiological derangements in all cases and were on vitamin K and/or plasma therapy till normalized.

13.4.2 Immediate Postoperative Course

Three cases had brief seizures and were controlled by phenytoin on the day of surgery. Two cases were jittery (with normal electrolytes and temperatures) with jitteriness resolving on the second day postoperative. Oral intake was resumed by trophic feeds 12 h following surgery with gradual increase of increments according to tolerance with uneventful course. One of the cases developed transient hypotension and drop of hemoglobin (initial 8.5 dropped to 7.5 g/dl) and necessitated packed cell transfusion with no hematoma on follow-up CT. There was no correlation between gestational age/body weight and occurrence of postoperative incidents (apnea, jitteriness, seizures). No correlations were made between length of hospital stay, APGAR scores, previous ventilation, respiratory distress, and postoperative morbidity.

13.4.3 Late Postoperative Course

One case developed shunt infections on postoperative (10%). The organism retrieved was *Staphylococcus aureus*. The patient underwent shunt removal with introduction of an external ventricular drain. Infection was not controlled and the patient died. One case of distal shunt obstruction (10%), 2 months after discharge, necessitates distal revision. A single infant with oral motor dysfunction remained on NGT feeding for several months after discharge.

Cases were screened for hearing and/or visual loss by auditory and visual evoked potentials when possible and revealed no abnormalities in those screened (7/10).

As for developmental outcome, most patients showed gross motor, fine motor, speech, and social skills consistent with their normal peers. However, the oldest infant is 3 years old, and further follow-up is required to test cognitive skills and to identify associated learning difficulties.

13.5 Discussion

Watchful waiting along with serial monitoring of ventricular and head circumference size is usually adopted in preterms with hydrocephalus to determine whether shunting is required. Red flags for intervention include progressive increase of head circumference, unexplained intolerance to fed, and/or full-tense fontanels. Although ex-preterms may have active hydrocephalus, shunting is often difficult to consider because the cerebrospinal fluid may be heavily bloodstained with high protein content along with age, low weight, metabolic derangements, respiratory complications, and unstable general condition. The biggest fears from a neurosurgical prospective are bleeding tendency and susceptibility to infection along with skin thinness. From a pediatric standpoint, early intervention salvages cortical tissue and offers the infant a chance to reach his/her full development potentials. On the other hand, early surgery is often blocked with many obstacles especially for infants with stormy NICU course.

Early (few hours after birth) and late periventricular hemorrhage (PVH)/intraventricular hemorrhage (IVH) (within the first 4 days) were found to have distinct and different risk factors with early PVH being related to APGAR scores at birth and later PVH related to fluctuating blood flow, hypotension and reperfusion, hypoxia, acidosis, and hypercarbia [15]. Such evidence suggests that the whole course during NICU stay along with very early events may relate to the outcome of shunt procedure.

This study reports the incidence of shunt infection to be 10%, whereas incidence of shunt obstruction is 10%. Lan et al. reported the incidence of infection among ex-preterms with hydrocephalus is 13% [16]. Another study stated that preterm birth, postoperative shunt leakage, and intraoperative breached gloves were independent risk factors for infection [17]. A study by McGrit MJ et al. documented that insertion of a VP shunt in a premature neonate was nearly associated with a fivefold increase in shunt infection. Their study also demonstrated that staphylococcal infection was related to hospital stay >3 days before shunt insertion and predicted that *Staphylococcus aureus* would be the subsequent pathogen with a positive predictive value of 70% [18].

The developmental milestone achieved by the studied group was excellent (relevant to age norms) in the majority of cases except for the deceased case, yet still needs further follow-up as complications may arise up to years following the preliminary shunt insertion.

Conclusion

Despite being fraught with risks, early intervention for progressive hydrocephalus under meticulous perioperative care and vigilant observation is advised.

References

1. Brouwer A, Groenendaal F, Van Haastert I, Rademaker K, Hanlo P, De Vries L. Neurodevelopmental outcome of preterm infants with severe intraventricular hemorrhage and therapy for post-hemorrhagic ventricular dilatation. J Pediatr. 2008;152(5):648–54.
2. Whitelaw A. Intraventricular hemorrhage and post-hemorrhagic hydrocephalus: pathogenesis, prevention and future interventions. Semin Neonatol. 2001;6:135–46.
3. Vasudevan C, Ibhanesebhor S, Manjunatha CM, Das K, Ardyll R. Need for consensus in interpreting coagulation profile in preterm neonates. Arch Dis Child Fetal Neonatal Ed. 2010;95:F77.
4. Narendran V, Visscher MO, Abril I, Hendrix SW, Hoath SB. Biomarkers of epidermal innate immunity in premature and full term infants. Pediatr Res. 2010;67(4):382–6.
5. Ohlsson A, Lacy JB. Intravenous immunoglobulin for preventing infection in preterm and/or low-birth-weight infants. Cochrane Database Syst Rev. 2001;2:CD000361.
6. Hudgins RJ, Boydston WR, Gilreach CL. Treatment of post-hemorrhagic hydrocephalus in preterm infant with a ventricular access device. Pediatr Neurosurg. 1998;29:309–13.
7. Massone ML, Cama A, Leone D, Pellas E, Vallarino R, Carini S, Andreussi L. Results of early external ventricular diversion in post-hemorrhagic ventricular dilatation in the newborn. Minerva Anestesiol. 1994;60(11):663–8.
8. Willis BK, Kumar CR, Wylen EL, Nanda A. Ventriculo-subgaleal shunts for post-hemorrhagic hydrocephalus in premature infants. Pediatr Neurosurg. 2005;41:178–85. doi:10.1159/000086558.
9. Kormanik K, Praca J, Garton HJ, Sarkar S. Repeated tapping of ventricular reservoir in preterm infants with post-hemorrhagic ventricular dilatation does not increase the risk of reservoir infection. J Perinatol. 2010;30(3):218–21. doi:10.1038/jp.2009.
10. Kazan S, Güra A, Ucar T, Korkmaz E, Ongun Akyuz M, Di Rocco C. Hydrocephalus after intraventricular haemorrhage in preterm and low-birth weight infants: analysis of associated risk factors for ventriculo-peritoneal shunting. Surg Neurol. 2005;64:S2.77–81.
11. Maxwell LG. Peri-operative evaluation and treatment in paediatrics. Anaesthiol Clin N Am. 2004;43:22–7.
12. Murphy JJ. Peri-operative evaluation and treatment in paediatrics. J Pediatr Surg. 2008;43(5):865–8.
13. Adams-Chapman I, Hansen N, Stoll B, Higgins R. Neuro-developmental outcome of extremely low birth weight infants with post-hemorrhagic hydrocephalus requiring shunt insertion. Pediatrics. 2008;121(5):e1167–77.
14. Papile LA, Burnstein J, Burnstein R, et al. Incidence and evolution of sub-ependymal and intraventricular haemorrhage: a study of infants with birth weights less than 1,500 gm. J Pediatr. 1978;92:529–34.
15. Osborn D, Evans N, Kluckow M. Hemodynamic and antecedent risk factors of early and late periventricular/intraventricular haemorrhage in premature infants. Pediatrics. 2003;112:33–9.
16. Lan CC, Wong TT, Chen SJ, Liang ML, Tang RB. Early diagnosis of ventriculoperitoneal shunt infections and malfunctions in children with hydrocephalus. J Microbiol Immunol Infect. 2003;36:47–50.
17. Kulkarni AV, Drake JM, Lamberti-Pasculli M. Cerebrospinal fluid shunt infection: a prospective study of risk factors. J Neurosurg. 2001;94:195–201.
18. McGirt MJ, Zaas A, Fuchs H, George T, Kaye K, Sexton D. Risk factors for paediatric ventriculo-peritoneal shunt infection and predictors of infectious pathogens. Clin Infect Dis. 2003;36(7):858–62.

Management of Hydrocephalus Following SAH and ICH

14

Takahiro Murata, Tetsuyoshi Horiuchi, and Kazuhiro Hongo

14.1   Introduction

The development of hydrocephalus following aneurysmal subarachnoid hemorrhage (SAH), a well-known sequela of the SAH, was firstly documented by Bagley in 1928 [1]. Many investigators have well established the relationship between SAH and hydrocephalus, and two stages of the hydrocephalus have been recognized: acute one, observed immediately after onset, and chronic one [2]. The incidence of post-SAH hydrocephalus has been reported to be 9–67% [3–5], and this wide range may be caused by the variation of the defining criteria of hydrocephalus as well as the effect of the different treatments for ruptured aneurysm. Several mechanisms have been proposed to explain the development of hydrocephalus following SAH. Cerebrospinal fluid (CSF) dynamics is of importance for the development of the hydrocephalus [6–8]. Hydrocephalus in patients with SAH may be induced mainly by obstructive mechanisms after the sudden increase in the arterial blood into the subarachnoid space lead to resist CSF outflow or block CSF circulation within the ventricular system in acute stage [9–11] and also may be caused mainly by absorptive problems attributable to impaired CSF absorption in chronic stage [12, 13]. Acute hydrocephalus is more frequent in patients with poor clinical grade and higher Fischer Scale scores [14, 15]. The insertion of external ventricular drainage (EVD) has been the standard care in the management of the acute post-SAH hydrocephalus, purposing primarily immediate improvement of the clinical condition, leading more suitable objection for curative surgical or endovascular intervention. Chronic hydrocephalus influences cognitive deficits and poorer neurological outcome and requires permanent CSF diversion [16, 17].

T. Murata, M.D. • T. Horiuchi, M.D. • K. Hongo, M.D. (✉)
Department of Neurosurgery, Shinshu University School of Medicine,
Matsumoto 390-8621, Japan
e-mail: khongo@shinshu-u.ac.jp

© Springer International Publishing AG 2017   191
A. Ammar (ed.), *Hydrocephalus*, DOI 10.1007/978-3-319-61304-8_14

Spontaneous or primary intracerebral hemorrhage (ICH) often penetrates the cerebral ventricular system and results in intraventricular hemorrhage (IVH) which is an independent predictor of poor functional outcome and higher mortality [18, 19]. Extension of hemorrhage into ventricles can impede normal CSF flow and make direct mass effects of ventricular blood, which results in acute obstructive hydrocephalus. The obstructive hydrocephalus caused by ventricular clots in the third and fourth ventricles develops in up to 50% of patients with IVH following ICH [20]. Acute obstructive hydrocephalus can be a life-threatening condition and lead to further neurological deterioration and poor outcome [18]. Insertion of EVD can be a lifesaving procedure, relief of acute obstruction, and subsequent herniation in ICH patients with IVH [21]. Recently, alternative treatment options include less-invasive endoscopic surgical evacuation of the hematoma and intraventricular thrombolytic therapy [22, 23]. On the other hand, chronic communicating hydro-cephalus can sometime develop with impairment of CSF absorption by IVH [24].

Here, we describe the management of hydrocephalus following subarachnoid hemorrhage and intracerebral hemorrhage with illustrative cases to attempt the CSF diversion procedure.

14.2   Illustrative Cases

Case 1 A 68-year-old man, with pacemaker placement and warfarinization for sick sinus syndrome, suffered sudden-onset severe headache followed by consciousness disturbance. A CT scan showed SAH with acute hydrocephalus and right acute sub-dural hematoma (ASDH) with midline shift (Fig. 14.1a, b). His Glasgow Coma Scale (GCS) score was 6 (E1V1M4), and anisoconia was found. Immediate three-dimensional CT angiography (3D-CTA) revealed no vascular anomaly associated with SAH and ASDH. He underwent small right frontal craniotomy for emergency removal of ASDH followed by insertion of EVD. Postoperatively, a CT scan showed resolution of hydrocephalus and midline shift (Fig. 14.1c), and his consciousness disturbance improved. Follow-up 3D-CTA demonstrated right posterior cerebral artery distal aneurysm (Fig. 14.2a). He underwent right occipital craniotomy and clipping of the aneurysm via the posterior interhemispheric approach. Postoperative CT scan and 3D-CTA disclosed complete clipping and no hydrocephalus (Fig. 14.2b, c). He was ambulant after rehabilitation.

Case 2 A 78-year-old healthy woman suddenly became unconscious (GCS score: 8). A CT scan revealed SAH with acute hydrocephalus (Fig. 14.3a, b), and 3D digi-tal subtraction angiography (DSA) showed left internal carotid artery bifurcation aneurysm (Fig. 14.3c). She underwent urgent placement of EVD and simultaneous surgery of clipping of the aneurysm. Her consciousness improved but deteriorated again 10 days after onset. A CT scan demonstrated the SAH well washed-out and relief of ventriculomegaly (Fig. 14.4a, b). A DSA disclosed complete clipping of the aneurysm and delayed cerebral vasospasm (Fig. 14.4c). Unfortunately, she expired from large cerebral infarctions due to severe vasospasm.

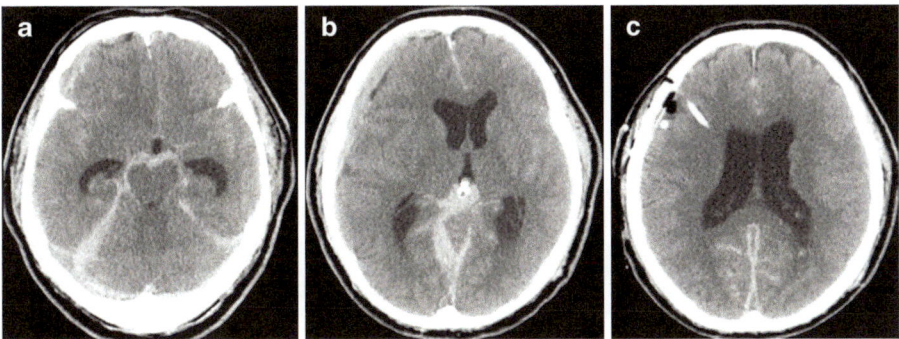

Fig. 14.1 (**a**, **b**) CT scans showing subarachnoid hemorrhage (SAH) with acute hydrocephalus and right acute subdural hematoma (ASDH) with midline shift. (**c**) Postoperative CT scan showing resolution of hydrocephalus and midline shift

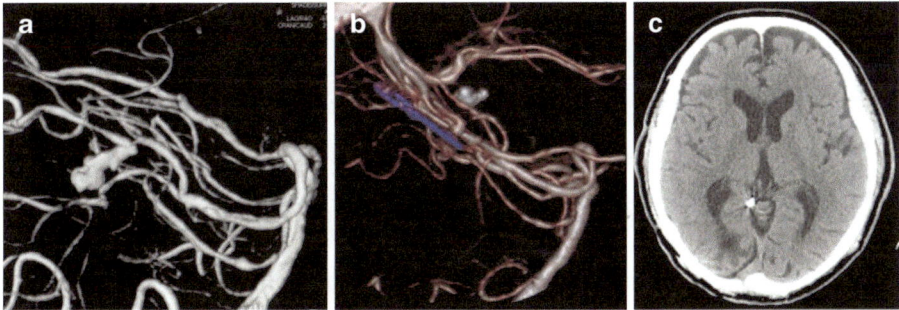

Fig. 14.2 (**a**, **b**) Follow-up and postoperative 3D-CTA showing right posterior cerebral artery distal aneurysm and complete clipping of it. (**c**) Follow-up CT scan showing no hydrocephalus

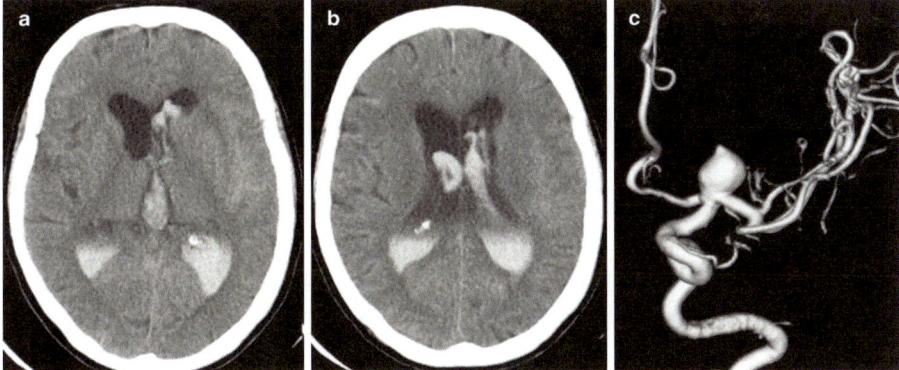

Fig. 14.3 (**a**, **b**) Initial CT scans revealing SAH in the left Sylvian fissure and much intraventricular hematoma with acute hydrocephalus. (**c**) Three-dimensional digital subtraction angiography (DSA) showing left internal carotid artery bifurcation aneurysm

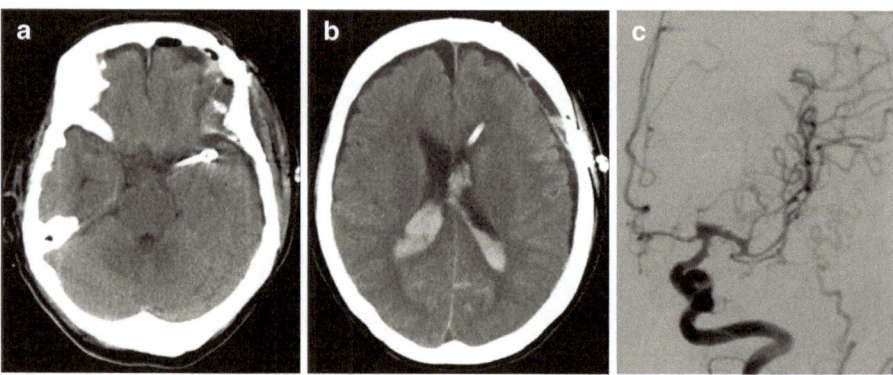

Fig. 14.4 (**a, b**) Postoperative CT scans demonstrating well washed-out SAH and relief of ventriculomegaly. (**c**) Anteroposterior view of left internal carotid artery injection (DSA) disclosing clipping of the aneurysm and delayed cerebral vasospasm

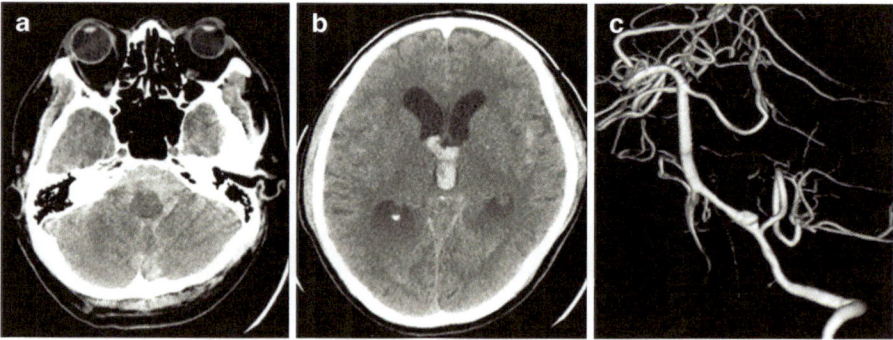

Fig. 14.5 (**a, b**) Initial CT scans revealing a lot of SAH in the posterior fossa and refluxed intraventricular hematoma with acute hydrocephalus. (**c**) Three-dimensional digital subtraction angiography (DSA) showing left vertebral artery dissecting aneurysm

Case 3 A 60-year-old healthy man suddenly presented with severe occipitalgia followed by consciousness disturbance with respiratory disorder. He was intubated and transferred to our hospital, and CT scans showed SAH with acute hydrocephalus (Fig. 14.5a, b). 3D-DSA demonstrated a left vertebral artery dissecting aneurysm (Fig. 14.5c). Urgent bilateral EVD was performed followed by endovascular coil embolization (Fig. 14.6a, b). A CT scan after 1 month from onset disclosed slight ventriculomegaly that remained (Fig. 14.6c), but he fully recovered from consciousness disturbance without any symptoms of chronic hydrocephalus.

Case 4 A 78-year-old healthy woman suffered severe headache and was transferred to our hospital. She presented somnolent which gradually worsened (GCS score: 12). A CT scan and 3D-CTA showed SAH suspecting rupture of an anterior communicat-

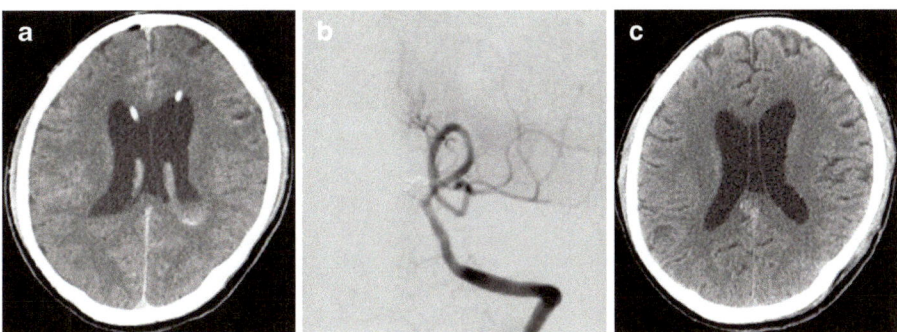

Fig. 14.6 (**a**) Immediate postoperative CT scans demonstrating relief of acute hydrocephalus. (**b**) Left oblique view of left vertebral artery injection (DSA) disclosing coil embolization of left vertebral artery. (**c**) Followed CT scan showing slight ventriculomegaly

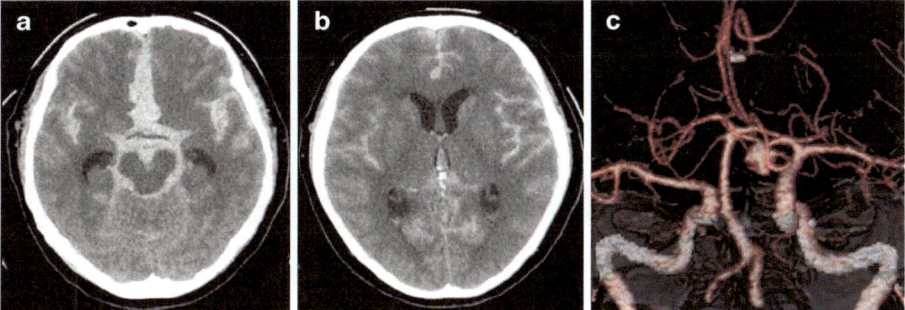

Fig. 14.7 (**a, b**) Initial CT scans revealing a lot of SAH in the basal cistern as well as anterior interhemispheric fissure and refluxed intraventricular hematoma with ventriculomegaly. (**c**) Initial 3D-CTA showing anterior communicating artery (ACoA) aneurysm

ing artery (ACoA) aneurysm (Fig. 14.7a–c). She underwent urgent clipping surgery, and postoperative 3D-CTA demonstrated complete clipping of the ACoA aneurysm (Fig. 14.8a). One month later, she presented with recent memory disturbance, urinary incontinence, and gait disturbance. Ventriculoperitoneal (VP) shunt was placed, and her symptoms disappeared. CT scans before and after VP shunt revealed the presence and absence of ventriculomegaly with periventricular lucency (Fig. 14.8b, c).

Case 5 A 43-year-old man, with untreated hypertension, presented with consciousness disturbance of sudden onset. He was deteriorated to GCS score 12 (E2V4M6) and showed moderate right hemiparesis. CT scan and MR imaging revealed IVH penetrated from hemorrhage of left caudate nucleus with acute obstructive hydrocephalus (Fig. 14.9a, b). MR angiography showed no abnormal vascular structure (Fig. 14.9c). He underwent endoscopic hematoma removal followed by

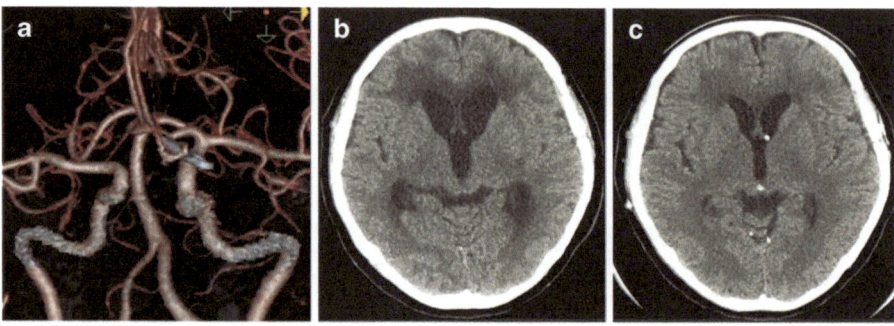

Fig. 14.8 (**a**) Postoperative 3D-CTA showing complete clipping of the ACoA aneurysm. (**b**) CT scan after 1 month from onset demonstrating chronic communicating hydrocephalus with periventricular hypodensity. (**c**) Postoperative CT scan showing resolution of hydrocephalus

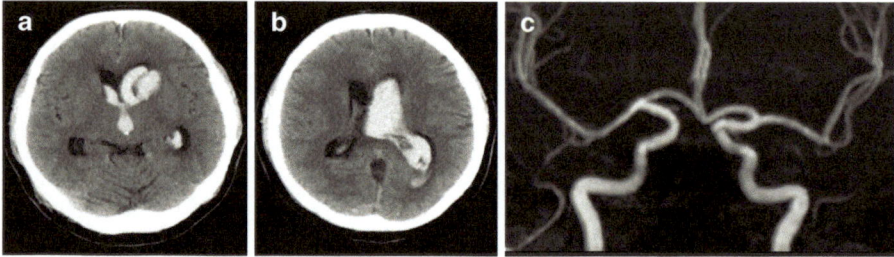

Fig. 14.9 (**a, b**) CT scans showing intraventricular hemorrhage (IVH) penetrated from hemorrhage of left caudate nucleus with acute obstructive hydrocephalus. (**c**) MR angiography showing no abnormal vascular structure

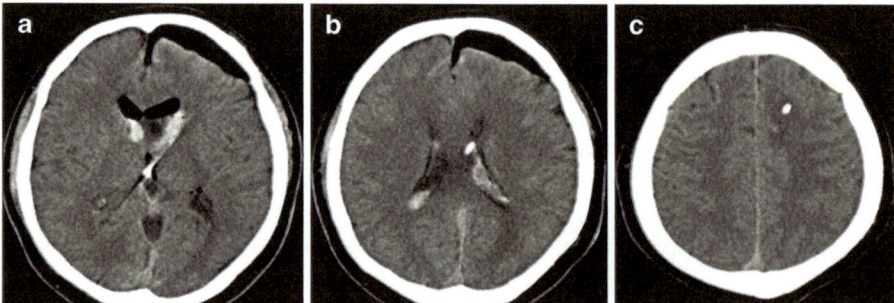

Fig. 14.10 (**a–c**) Postoperative CT scans demonstrating relief of hydrocephalus with partial removal of IVH, subtotal removal of intracerebral hemorrhage, and small trajectory along the external ventricular drainage

EVD via left frontal burr hole under the general anesthesia. Postoperative CT scan demonstrated relief of hydrocephalus with partial removal of IVH and subtotal removal of ICH (Fig. 14.10a, b) and small trajectory (Fig. 14.10c). He fully recovered and returned to the previous work 3 months later.

14.3   Discussion

14.3.1 Hydrocephalus Following SAH

An SAH is a neurological emergency with significant potential for long-term morbidity and mortality. Lives of patients with SAH are significantly affected by the severity of the initial hemorrhage and of the potential neurological complications such as rebleeding and hydrocephalus. After the patient is medically stabilized to prevent rebleeding and secondary brain injuries, evaluations of aneurysm and hydrocephalus are required. The decision of whether clipping or coiling for ruptured aneurysm has been made on the basis of angiography from various modalities and patient's general condition. The decision of whether or not to proceed with EVD also has been based on ventricular size from neuroimaging and patient's neurological condition. Acute hydrocephalus following SAH occurs in up to 20–30% of patients [14, 15]. The clinical significance of acute ventriculomegaly following SAH is uncertain because many patients did not deteriorate and present any symptoms but with ventriculomegaly [25]. However, in patients with a diminished level of consciousness, 40–80% had some degree of improvement after EVD [25–27]. In 1973, Kusske et al. reported one of the earliest placements of EVD in treatment of acute hydrocephalus following SAH significantly improved outcome [28]. They described the rationale of the placement of EVD included that patients with acute hydrocephalus are generally poor surgical candidates and the procedure could enhance their overall fitness to undergo surgery. Since then, the insertion of EVD has become the standard approach in management of acute hydrocephalus. In addition, EVD provided an accurate, reliable, and expeditious way to monitor and control for intracranial pressure (ICP). ICP management can prevent secondary cortical damage, improve neurological status, and facilitate dissecting maneuver during clipping surgery. In our Case 1, consciousness disturbance recovered after emergent removal of ASDH and placement of EVD, and curative clipping surgery was performed. However, the use of EVD has in itself been associated with increased risk of rebleeding due to abruptly decreased transmural aneurysmal pressure [5, 25], infection [29], and shunt dependency [30]. The most serious complication is acute rebleeding; on the basis of two small series, the placement of EVD may [31] or may not [32] be associated with rebleeding. More recent study of 546 prospectively patients with SAH showed no difference between the incidence of rebleeding with preoperative EVD placement and controls [33]. In our Cases 2 and 3, urgent EVD insertion followed by clipping or endovascular surgery was performed without rebleeding. The placement of EVD and simultaneous clipping surgery or urgent coil embolization are recommended. There are no guidelines for the placement and management of EVD including optimal insertion condition and timing, techniques to reduce the risk of rebleeding and infection, as well as optimal removal timing. Multi-institutional, prospective, large clinical trials are required for resolving these problems.

Chronic ventriculomegaly requiring permanent CSF diversion is reported at rate of 18–26% of survivors [15, 34, 35] and is predicted in SAH patients with elder age, early ventriculomegaly, intraventricular hemorrhage, poor clinical condition on onset, and female gender [2, 16, 36]. Incidence rates are not different in patients undergoing clipping or endovascular coiling treatment of their aneurysm [34, 35]. In our Case 4, she presented typical symptoms of normal pressure hydrocephalus in chronic stage following onset of SAH and fully recovered without any symptoms after VP shunt. Poorer neurological outcomes and cognitive deficits have been known as some of the adverse outcomes of hydrocephalus; therefore, CSF diversion such as ventriculoatrial, ventriculoperitoneal, and lumboperitoneal shunts can improve clinical status in symptomatic patients with chronic hydrocephalus after SAH.

14.3.2 Hydrocephalus Following ICH

A number of studies have demonstrated involvement of the third and fourth ventricles in patients with IVH following ICH to be predictive of adverse outcome and high mortality [18, 20]. Several possible mechanisms that explain the pathophysiology of this unfavorable event include acute obstructive hydrocephalus, the mass effect exerted by the clot, the toxicity of blood-resolving products to the periventricular parenchyma, and the development of chronic hydrocephalus. The purpose of treatment for acute obstructive hydrocephalus in ICH patients with IVH is to evacuate the intraventricular hematoma, thus relieving the obstruction to CSF flow, reversing ventricular dilatation, and restoring normal intracerebral pressure. EVD is recommended for the treatment of acute obstructive hydrocephalus in patient with consciousness disturbance secondary to its hydrocephalus; however, it is often occluded by clots, leading not to drain enough CSF. A systematic review from literatures showed EVD not only significantly reduced the mortality compared to conservative therapy but also did not improve the functional outcome [37]. EVD is a well ICP control but not well clot-clearance procedure which means it is not to resolve the mass effect and toxicity from erupted blood to the ventricles. Since then, for resolution of these problems, a number of clinical trials have been carried out for fast and effective clearance of blood clot from the ventricles. Recently, like our Case 5, the most effective and least invasive procedure seems to be endoscopic intraventricular hematoma evacuation followed by placement of EVD. Unfortunately, there is no evidence of efficacy in endoscopic procedure of large randomized trials. Chen et al. indicated randomized trial of IVH after ICH that compared 24 endoscopic IVH removals to 24 EVDs alone showed no differences between the two groups in terms of mortality or functional outcome [38]. The only benefit of endoscopic removal was a significant reduction of incidence of chronic communicating hydrocephalus. Furthermore, endoscopic hematoma evacuation is limited to use at a specialized center. On the other hand, intraventricular thrombolytic therapy can be instilled into the ventricles to maintain EVD functionality and promote fast clearance of the ventricles [23, 39]. The CLEAR IVH (Clot Lysis: Evaluating Accelerated Resolution of Intraventricular Hemorrhage) phase 2 trial indicated that low-dose recombinant tissue plasminogen activator (rtPA) can be safe and may increase the lysis rate [23], and CLEAR-3 randomized trial is ongoing. Chronic communicating hydrocephalus can develop in some of IVH patients with impairing CSF absorption and will require permanent CSF diversion. The less-invasive lumbar drainage might decrease in the need of permanent CSH shunting [24, 40], and randomized trial is ongoing.

Conclusions

Acute and chronic hydrocephalus is well-documented complication of aneurysmal SAH and spontaneous ICH. The placement of EVD in patients with acute hydrocephalus following SAH, particularly in patients of poor clinical grade, is recommended. This procedure can improve the clinical grade of these patients, make them better surgical candidates, and allow definite clipping surgery or endovascular intervention. However, there are no definitive practice guidelines for the insertion and management of EVD. Multi-institutional randomized controlled trial is needed, and evidence-based guidelines will help improvement of clinical outcome in patients with SAH. In case of acute hydrocephalus following ICH with IVH, EVD is recommended but has been proven often insufficient; the concomitant use of intraventricular antifibrinolytics such as rtPA or urokinase seems to be beneficial. Less-invasive endoscopic hematoma removal may be an effective option. In the management of chronic hydrocephalus following SAH and IVH, permanent CSF diversion such as ventriculoperitoneal or lumboperitoneal shunt is recommended.

References

1. Bagley C Jr. Blood in the cerebrospinal fluid. Resultant functional and organic alterations in the central nervous system. A. Experiment data. Arch Surg. 1928;17:18–38.
2. Vale FL, Bradley EL, Fisher WS. The relationship of subarachnoid hemorrhage and the need for postoperative shunting. J Neurosurg. 1997;86:462–6.
3. Black PM. Hydrocephalus and vasospasm after subarachnoid hemorrhage from ruptured intracranial aneurysms. Neurosurgery. 1986;18:12–6.
4. Spallone A, Gaglliardi FM. Hydrocephalus following aneurysmal subarachnoid hemorrhage. Zentralbl Neurochir. 1983;44:141–50.
5. Vassilouthis J, Richardson AE. Ventricular dilation and communicating hydrocephalus following spontaneous subarachnoid hemorrhage. J Neurosurg. 1979;51:341–35.
6. Auer LM, Mokry M. Disturbed cerebrospinal fluid circulation after subarachnoid hemorrhage and acute aneurysm surgery. Neurosurgery. 1990;26:804–9.
7. Doczi T, Nemessanyi Z, Szegvary Z, Huszka E. Disturbances of cerebrospinal fluid circulation during the acute stage of subarachnoid hemorrhage. Neurosurgery. 1983;12:435–8.
8. Yasargil MG, Yonekawa Y, Zumstein B, Stahl H. Hydrocephalus following spontaneous subarachnoid hemorrhage. J Neurosurg. 1973;39:474–9.
9. van Gijn J, Hijdra A, Wijdicks EF, Vermeulen M, van Crevel H. Acute hydrocephalus after aneurysmal subarachnoid hemorrhage. J Neurosurg. 1985;63:355–62.
10. Donauer E, Reif J, al-Khalaf B, Mengedoht EF, Faubert C. Intraventricular hemorrhage caused by aneurysms and angiomas. Acta Neurochir. 1993;122:23–31.
11. Hasan D, Herve L, Tanghe J. Distribution of cisternal blood in patients with acute hydrocephalus after subarachnoid hemorrhage. Ann Neurol. 1992;31:374–8.
12. Blasberg R, Johnson D, Fenstermacher J. Absorption resistance of cerebrospinal fluid after subarachnoid hemorrhage in the monkey: effects of heparin. Neurosurgery. 1981;9:686–91.
13. Brydon HL, Bayston R, Hayward R, Harkness W. The effect of protein and blood cells in the flow-pressure characteristics of shunts. Neurosurgery. 1996;38:498–505.
14. Suarez-Rivera O. Acute hydrocephalus after subarachnoid hemorrhage. Surg Neurol. 1998;49:563–5.
15. Sheehan JP, Polin RS, Sheehan JM, Baskaya MK, Kassell NF. Factors associated with hydrocephalus after aneurysmal subarachnoid hemorrhage. Neurosurgery. 1999;45:1120–7.
16. Steinke D, Weir B, Disney L. Hydrocephalus following aneurysmal subarachnoid hemorrhage. Neurol Res. 1987;9:3–9.
17. Yoshioka H, Inagawa T, Tokuda Y, Inokuchi F. Chronic hydrocephalus in elderly patients following subarachnoid hemorrhage. Surg Neurol. 2000;53:119–25.

18. Bhattathiri PS, Gregson B, Prasad KS, Mendelow AD, and the STICH Investigators. Intraventricular hemorrhage and hydrocephalus after spontaneous intracerebral hemorrhage: results from the STICH trial. Acta Neurochir Suppl (Wien). 2006;96:65–8.
19. Hanley DF. Intraventricular hemorrhage: severity factor and treatment target in spontaneous intracerebral hemorrhage. Stroke. 2009;40:1533–8.
20. Diringer MN, Edwards DF, Zazulia AR. Hydrocephalus: a previously unrecognized predictor of poor outcome from supratentorial intracerebral hemorrhage. Stroke. 1998;29:1352–7.
21. Sumer MM, Açikgöz B, Akpinar G. External ventricular drainage for acute obstructive hydrocephalus developing following spontaneous intracerebral haemorrhages. Neurol Sci. 2002;23:29–33.
22. Yadav YR, Mukerji G, Shenoy R, Basoor A, Jain G, Nelson A. Endoscopic management of hypertensive intraventricular haemorrhage with obstructive hydrocephalus. BMC Neurol. 2007;7:1.
23. Morgan T, Awad I, Keyl P, Lane K, Hanley D. Preliminary report of the clot lysis evaluating accelerated resolution of intraventricular hemorrhage (CLEAR-IVH) clinical trial. Acta Neurochir Suppl (Wien). 2008;105:217–20.
24. Huttner HB, Nagel S, Tognoni E, Köhrmann M, Jüttler E, Orakcioglu B, Schellinger PD, Schwab S, Bardutzky J. Intracerebral hemorrhage with severe ventricular involvement: lumbar drainage for communicating hydrocephalus. Stroke. 2007;38:183–7.
25. Hasan D, Vermeulen M, Wijdicks EF, Hijdra A, van Gijn J. Management problems in acute hydrocephalus after subarachnoid hemorrhage. Stroke. 1989;20:747–53.
26. Rajshekhar V, Harbaugh RE. Results of routine ventriculostomy with external ventricular drainage for acute hydrocephalus following subarachnoid haemorrhage. Acta Neurochir. 1992;115:8–14.
27. Milhorat TH. Acute hydrocephalus after aneurysmal subarachnoid hemorrhage. Neurosurgery. 1987;20:15–20.
28. Kusske JA, Turner PT, Ojemann GA, Harris AB. Ventriculostomy for the treatment of acute hydrocephalus following subarachnoid hemorrhage. J Neurosurg. 1973;38:591–5.
29. Lozier AP, Sciacca RR, Romagnoli MF, Connolly ES Jr. Ventriculostomy-related infections: a critical review of the literature. Neurosurgery. 2002;51:170–81.
30. Hirashima Y, Kurimoto M, Hayashi N, Umemura K, Hori E, Origasa H, Endo S. Duration of cerebrospinal fluid drainage in patients with aneurysmal subarachnoid hemorrhage for prevention of symptomatic vasospasm and late hydrocephalus. Neurol Med Chir (Tokyo). 2005;45:177–82.
31. Pare L, Delfino R, Leblanc R. The relationship of ventricular drainage to aneurysmal rebleeding. J Neurosurg. 1992;76:422–7.
32. McIver JI, Friedman JA, Wijdicks EF, Piepgras DG, Pichelmann MA, Toussaint LG III, McClelland RL, Nichols DA, Atkinson JL. Preoperative ventriculostomy and rebleeding after aneurysmal subarachnoid hemorrhage. J Neurosurg. 2002;97:1042–4.
33. Hellingman CA, van den Bergh WM, Beijer IS, van Dijk GW, Algra A, van Gijn J, Rinkel GJ. Risk of rebleeding after treatment of acute hydrocephalus in patients with aneurysmal subarachnoid hemorrhage. Stroke. 2007;38:96–9.
34. Gruber A, Reinprecht A, Bavinzski G, Czech T, Richling B. Chronic shunt-dependent hydrocephalus after early surgical and early endovascular treatment of ruptured intracranial aneurysms. Neurosurgery. 1999;44:503–9.
35. Sethi H, Moore A, Dervin J, Clifton A, MacSweeney JE. Hydrocephalus: comparison of clipping and embolization in aneurysm treatment. J Neurosurg. 2000;92:991–4.
36. Dorai Z, Hynan LS, Kopitnik TA, Samson D. Factors related to hydrocephalus after aneurysmal subarachnoid hemorrhage. Neurosurgery. 2003;52:763–9.
37. Nieuwkamp DJ, de Gans K, Rinkel GJ, Algra A. Treatment and outcome of severe intraventricular extension in patients with subarachnoid or intracerebral hemorrhage: a systematic review of the literature. J Neurol. 2000;247:117–21.
38. Chen CC, Liu CL, Tung YN, Lee HC, Chuang HC, Lin SZ, Cho DY. Endoscopic surgery for intraventricular hemorrhage (IVH) caused by thalamic hemorrhage: comparisons of endoscopic surgery and external ventricular drainage (EVD) surgery. World Neurosurg. 2011;75:264–8.
39. Naff NJ, Hanley DF, Keyl PM, Tuhrim S, Kraut M, Bederson J, Bullock R, Mayer SA, Schmutzhard E. Intraventricular thrombolysis speeds blood clot resolution: results of a pilot, prospective, randomized, double-blind, controlled trial. Neurosurgery. 2004;54:577–83.
40. Staykov D, Huttner HB, Struffert T, Ganslandt O, Doerfler A, Schwab S, Bardutzky J. Intraventricular fibrinolysis and lumbar drainage for ventricular hemorrhage. Stroke. 2009;40:3275–80.

Endoscopic Management of Hydrocephalus and Choroid Plexus Cauterization

15

Chima Oluigbo and Robert Keating

15.1 Introduction

The challenges of endoscopic hydrocephalus surgery include decreased visibility (often the result of minimal bleeding), a limited surgical corridor, difficult intraprocedural orientation, and the challenges of effecting action at the target site with current endoscopic instruments. Thus, a safe approach to endoscopic hydrocephalus surgery remains dependent upon high-quality imaging, an intimate knowledge of one's endoscopic equipment, and its limitations. Frameless stereotactic neuronavigation may offer additional anatomical guidance to intraventricular localization.

In the following paragraphs, we will define the spectrum of endoscopic hydrocephalus surgery at present, evidence for their efficacy, and possible future indications. These include:

- Endoscopic third ventriculostomy (ETV)
- Endoscopic third ventriculostomy and choroid plexus cauterization (ETV + CPC)
- Septum pellucidotomy and fenestration of intraventricular septations
- Endoscopic-assisted catheter placement

15.2 Endoscopic Third Ventriculostomy (ETV)

The use of endoscopic third ventriculostomy (ETV) in the management of hydrocephalus dates back to its use in 1923 by Mixter to treat a 9-year-old girl with hydrocephalus [2]. Since then, the advancements in optical technology and instrumentation design have helped to establish endoscopic third ventriculostomy as an effective mainstay treatment for children with obstructive hydrocephalus (Fig. 15.1).

C. Oluigbo, M.D. • R. Keating, M.D. (✉)
Department of Pediatric Neurosurgery, Children's National Medical Center, George
Washington University School of Medicine, Washington, DC, USA
e-mail: rkeating@childrensnational.org

© Springer International Publishing AG 2017
A. Ammar (ed.), *Hydrocephalus*, DOI 10.1007/978-3-319-61304-8_15

Fig. 15.1 Principle of endoscopic third ventriculostomy (ETV)

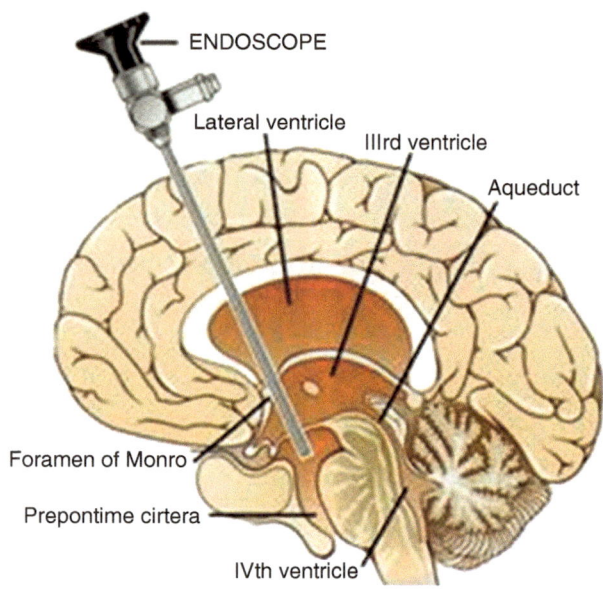

Table 15.1 Calculation of ETV success score (ETVSS) (*Reproduced from Kulkarni et al. 2010*)

Score	Age	Etiology	Prior shunt
0	<1 month	Postinfectious	Previous shunt
10	1 month to <6 months		No previous shunt
20		Myelomeningocele, IVH, nontectal brain tumor	
30	6 months to <1 year	Aqueductal stenosis, tectal tumor, others	
40	1 year to <10 years		
50	≥10 years		

At present, the principal indication for ETV is the treatment of patients with obstructive hydrocephalus due to different pathologies including congenital aqueductal stenosis, third ventricular obstruction from pineal region tumors, and tectal gliomas (may be done in conjunction with an endoscopic biopsy). Some practitioners have also applied ETV to the management of hydrocephalus associated with posterior fossa tumors [3]. Although ETV has been used for the treatment of communicating (nonobstructive) hydrocephalus, this is often associated with lower success rates.

The predictors of success following ETV have been studied extensively. Factors have been identified improving the chance of success and include obstructive etiology for hydrocephalus, older age at the time of the ETV, the presence of previous cerebrospinal fluid shunt, and preoperative assessments of prepontine cisternal scarring as well as thickness/bowing of the floor of the third ventricle. Kulkarni et al. developed an ETV success score (ETVSS), which predicts the success of ETV procedures when compared to a standard ventriculoperitoneal shunt. Relevant factors to this score include the etiology of hydrocephalus, age at the time of ETV surgery, and the previous presence of a CSF shunt (Table 15.1). An ETVSS of greater than 80 predicts the best chance of ETV success, while a score of less than 40 predicts a poor likelihood of ETV success [11]. Confirmation of the accuracy of the ETVSS has been independently confirmed by other investigators [6].

15.3   Operative Procedure

Access to the ventricular system is via a frontal precoronal (1 cm) burr hole 3 cm lateral to the midline. A 12-French cannula with a peel-away sheath is then introduced into the frontal horn of the respective lateral ventricle with a 0° endoscope being introduced from the respective lateral ventricle through the foramen of Monro to visualize the floor of the third ventricle (Fig. 15.2).

The landmarks identified on the floor of the third ventricle are the mammillary bodies (MB) and the tuber cinereum (TC) (Fig. 15.3). Perforation of the floor of the third ventricle is performed between the tuber cinereum and the mammillary bodies. In our practice, this perforation is usually made with a blunt instrument (e.g., biopsy probe, endoscope, Bugbee wire) and then expanded by inflation of a 3-French gauge Fogarty balloon. The use of ultrasonic aspirators, diathermy, and lasers for perforation of the floor of the third ventricle has been described by others. However, these may be associated with an increased risk of injury to surrounding critical structures as well as vascular injury to the basilar artery (and perforators) which may be associated with catastrophic bleeding and possibly mortality.

Fig. 15.2 The foramen of Monro as seen during the ETV procedure

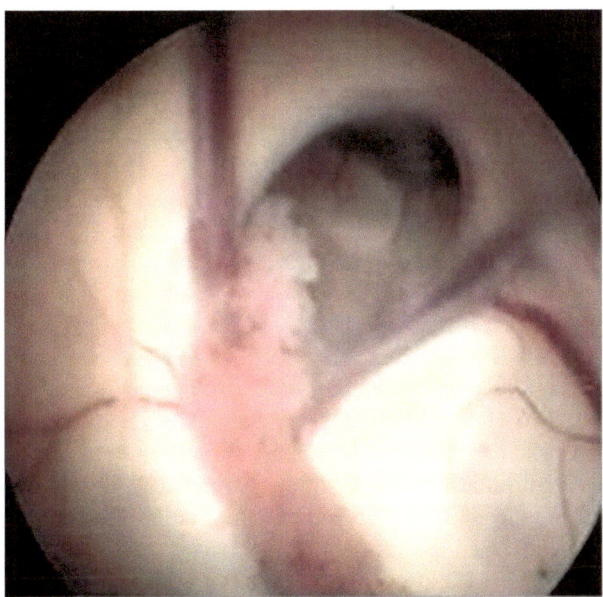

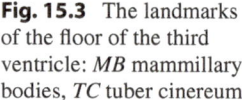

Fig. 15.3 The landmarks of the floor of the third ventricle: *MB* mammillary bodies, *TC* tuber cinereum

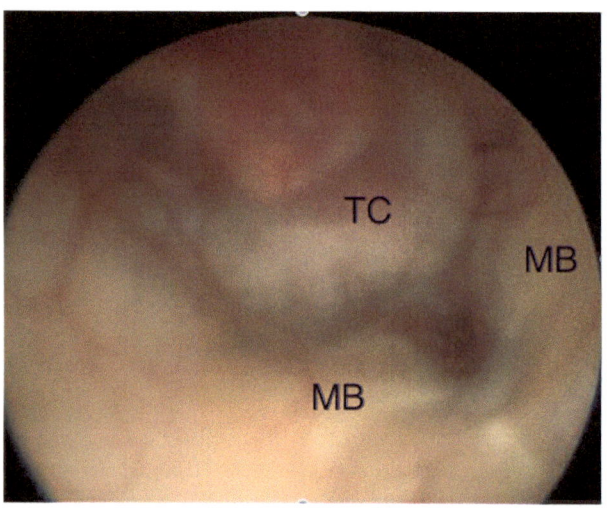

Other potential complications of ETV include CSF leak, wound infection, ventriculitis, intracranial hemorrhage, subdural hematoma, memory dysfunction, and hypothalamic/endocrine dysfunction including diabetes insipidus, menstrual irregularities, and hyperphagia. The most feared complication is catastrophic bleeding related to injury of the basilar artery with an incidence of 3.7% intraoperative hemorrhage and death 0.2% [1].

In the latest prospective multicenter results from the Hydrocephalus Clinical Research Network (HCRN) authored by Kulkarni et al., 336 eligible children who underwent initial ETV were assessed [13]. Of these, 141 patients (42%) had documented failure of the ETV requiring repeat hydrocephalus surgery during follow-up. ETV success rates at 30 days, 90 days, 6 months, 1 year, and 2 years were 73.7%, 66.7%, 64.8%, 61.7%, and 57.8%, respectively. Preoperative ETVSS was associated with ETV success ($p < 0.001$) as well as the intraoperative ability to visualize a "naked" basilar artery. It should be noted that this study was exclusively based on North American programs and demonstrated a relatively conservative approach to the use of ETV in these patients as 83% of the patients had ETV scores of 70 or greater.

In a systematic literature review and evidence-based guideline on cerebrospinal fluid shunt or ETV for the treatment of hydrocephalus in children, Limbrick et al. noted that in selected patients, CSF shunts and ETV demonstrated equivalent outcomes [14]. The greatest challenge to ETV success has been the patient under 1 year of age as well as patients with post-hemorrhagic hydrocephalus. A retrospective study by Jernigan et al. evaluated the success of an ETV in 5416 patients <1 year old at 41 institutions [9]. At 1 year, the success rate of the ETV was 36% vs. 60.4% for a ventriculoperitoneal shunt. Numerous other investigators have found similar results in the younger patients, particularly those <6 m of age [5, 8, 10, 15]. As a result, the addition of choroid plexus coagulation has been investigated to improve the likelihood of avoiding a shunt over the long term.

Assessment of the function of the ETV is evaluated clinically and by post-op imaging. While ventricular dilatation may persist over a significant period of time, other parameters can be helpful in determining the chance of success. Reduction of periventricular edema, expanded CSF cisterns/sulcal patterns, and reduction of third ventricular floor bowing are all consistent with a working ETV. CSF flow studies on MRI offer additional visualization and confirmation of a patent third ventriculostomy. However, flow across the floor of the third ventricle does not guarantee resolution of hydrocephalus, especially when there is a significant component of communicating hydrocephalus or prepontine cisternal scarring which effectively limits the egress of CSF from the third ventricle. Thus, the importance of visually ensuring that the ETV has perforated the membrane of Liliequist (allowing visualization of the basilar artery) and that there are no obvious limitations of CSF being able to reach the basal cisterns.

15.4   ETV and Choroid Plexus Cauterization

Open cauterization of the choroid plexus, which produces cerebrospinal fluid, was an early, understandably logical attempt at managing hydrocephalus. However, most of the early attempts were unsuccessful. Dandy performed an endoscopic cauterization of the choroid plexus in 1922, but this effort was also unsuccessful [2].

Subsequently, choroid plexus cauterization fell out of favor, and only sporadic reports in the literature by Scarff in 1952 and Pople in 1995 are noted [16, 18]. However, there has been a resurgence of interest in choroid plexus cauterization following the description of Warf et al. of his experience with the use of ETV and CPC in a large sub-Saharan African population where there were challenges with long-term follow-up after CSF shunting procedures [19]. Warf reported that the use of CPC substantially improved the chance of ETV success from 47 to 66%. For this procedure, he utilized a flexible endoscope and was able to achieve cauterization of at least 90% of the choroid plexus of the lateral and third ventricles as well as the temporal horn. He also used a septostomy for access to the contralateral ventricular choroid plexus. While it has been difficult for other centers to duplicate this degree of success, it is apparent that the degrees of choroid plexus cauterization, as well as the use of a flexible endoscope, are critical components of success in this endeavor.

The Hydrocephalus Clinical Research Network retrospectively reviewed the combination of an ETV and CPC at seven of their member centers and found that 50% of their patients failed at 1 month. This was worse than that seen for the cohort that had shunts placed [12]. They also found the success was significantly higher with greater degrees of choroid cauterization. Unfortunately, the use of a flexible scope and reaching back into the temporal/occipital horns require considerable experience and expertise. A recent meta-analysis of combined ETV and CPC in 2016 by Weil et al. demonstrated an efficacy of 63% with morbidity of 3.7% and 30-day mortality of 0.4% [20].

15.5 Septum Pellucidotomy and Fenestration of Intraventricular Septations

Intracranial endoscopy may also be used in the management of complex hydrocephalus with intraventricular septations as seen following ventriculitis (especially after gram-negative post-infection hydrocephalus) and intraventricular hemorrhage. This endoscopic application may obviate the need for multiple ventricular catheters to drain septated CSF collections in these challenging cases. However, in these cases, it should be noted that the use of stereotactic neuronavigation is strongly recommended because the septations often result in loss of normal intraventricular landmarks.

This may also be used in the setting of isolated lateral ventricular hydrocephalus following unilateral foramen of Monro obstruction or following deviation of the septum pellucidum following shunting of the contralateral ventricle. The results of this technique are mixed but in many situations may help to avoid the need for another shunt in the contralateral lateral ventricle.

15.6 Endoscopic-Assisted Catheter Placement

Endoscopy is routinely used in some centers as an adjunct in the placement of CSF shunts. The rationale for their use is that visual confirmation of good catheter placement may be associated with increased longevity of the shunt. A typical intra-catheter endoscope is the Neuropen endoscope which can be introduced through the lumen of the standard ventricular shunt catheter and extended through a cut made at the tip of the catheter to visualize structures beyond the tip of the shunt (Fig. 15.4). The endoscope is then withdrawn after visual confirmation of satisfactory placement of the ventricular catheter.

Although, in principle, the use of endoscopes for ventricular catheter placement seems rationale, evidence-based studies have not supported better outcomes with its use. A Class 1 randomized blinded study by Riva-Cambrin et al. in which 393 patients

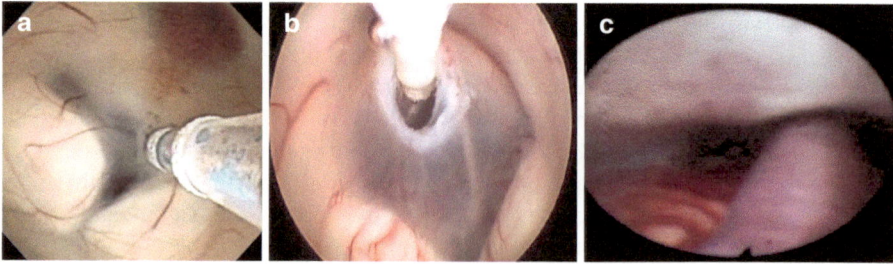

Fig. 15.4 Demonstrate operating views. (**a**) Insertion perforation of the floor of the third ventricle is usually made with a blunt instrument (e.g., biopsy probe, endoscope, Bugbee wire) and expanded by insertion of 3-French gauge Fogarty balloon. (**b**) Inflation of Fogarty Balloon. (**c**) The stoma as seen in the end of the procedure

requiring a shunt were randomized to either an endoscopy or standard insertion did not show any significant differences in the overall shunt survival rate [17].

A systematic literature review and evidence-based guideline by Flannery et al. reported that based on the present evidence, there's insufficient evidence to support the use of endoscopic guidance for routine ventricular catheter placement [4].

Conclusion

The use of the endoscope has transformed the treatment of hydrocephalus over the past 75 years and is now considered an indispensable part of the neurosurgeon's armamentarium for this often challenging condition. Better optics, lighting, instrumentation, and stereotaxic localization have all contributed to the successful avoidance of shunts in certain hydrocephalus cohorts. In the future, continued efforts will undoubtedly improve our approach and outcomes to the patients afflicted with hydrocephalus.

References

1. Bouras T, Sgouros S. Complications of endoscopic third ventriculostomy. J Neurosurg Pediatr. 2011;7:643–9.
2. Choudhri O, Feroze AH, Nathan J, Cheshier S, Guzman R. Ventricular endoscopy in the pediatric population: review of indications. Childs Nerv Syst. 2014;30:1625–43.
3. Di Rocco F, Juca CE, Zerah M, Sainte-Rose C. Endoscopic third ventriculostomy and posterior fossa tumors. World Neurosurg. 2013;79:S18.e15–9.
4. Flannery AM, Duhaime AC, Tamber MS, Kemp J, Pediatric Hydrocephalus Systematic Review and Evidence-Based Guidelines Task Force. Pediatric hydrocephalus: systematic literature review and evidence-based guidelines. Part 3: Endoscopic computer-assisted electromagnetic navigation and ultrasonography as technical adjuvants for shunt placement. J Neurosurg Pediatr. 2014;14(Suppl 1):24–9.
5. Fritsch MJ, Kienke S, Ankermann T, Padoin M, Mehdorn HM. Endoscopic third ventriculostomy in infants. J Neurosurg. 2005;103:50–3.
6. Garcia LG, Lopez BR, Botella GI, Paez MD, da Rosa SP, Rius F, et al. Endoscopic third ventriculostomy success score (ETVSS) predicting success in a series of 50 pediatric patients. Are the outcomes of our patients predictable? Childs Nerv Syst. 2012;28:1157–62.
7. Ishii M, Gallia GL. Application of technology for minimally invasive neurosurgery. Neurosurg Clin N Am. 2010;21:585–94.
8. Javadpour M, Mallucci C, Brodbelt A, Golash A, May P. The impact of endoscopic third ventriculostomy on the management of newly diagnosed hydrocephalus in infants. Pediatr Neurosurg. 2001;35:131–5.
9. Jernigan SC, Berry JG, Graham DA, Goumnerova L. The comparative effectiveness of ventricular shunt placement versus endoscopic third ventriculostomy for initial treatment of hydrocephalus in infants. J Neurosurg Pediatr. 2014;13:295–300.
10. Kadrian D, van Gelder J, Florida D, Jones R, Vonau M, Teo C, et al. Long-term reliability of endoscopic third ventriculostomy. Neurosurgery. 2008;62(Suppl 2):614–21.
11. Kulkarni AV, Drake JM, Kestle JR, Mallucci CL, Sgouros S, Constantini S, et al. Predicting who will benefit from endoscopic third ventriculostomy compared with shunt insertion in childhood hydrocephalus using the ETV success score. J Neurosurg Pediatr. 2010;6:310–5.
12. Kulkarni AV, Riva-Cambrin J, Browd SR, Drake JM, Holubkov R, Kestle JR, et al. Endoscopic third ventriculostomy and choroid plexus cauterization in infants with hydrocephalus: a retrospective hydrocephalus clinical research network study. J Neurosurg Pediatr. 2014;14:224–9.

13. Kulkarni AV, Riva-Cambrin J, Holubkov R, Browd SR, Cochrane DD, Drake JM, et al. Endoscopic third ventriculostomy in children: prospective, multicenter results from the hydrocephalus clinical research network. J Neurosurg Pediatr. 2016;18:423–9.
14. Limbrick DD Jr, Baird LC, Klimo P Jr, Riva-Cambrin J, Flannery AM, Pediatric Hydrocephalus Systematic Review and Evidence-Based Guidelines Task Force. Pediatric hydrocephalus: systematic literature review and evidence-based guidelines. Part 4: Cerebrospinal fluid shunt or endoscopic third ventriculostomy for the treatment of hydrocephalus in children. J Neurosurg Pediatr. 2014;14(Suppl 1):30–4.
15. Ogiwara H, Dipatri AJ Jr, Alden TD, Bowman RM, Tomita T. Endoscopic third ventriculostomy for obstructive hydrocephalus in children younger than 6 months of age. Childs Nerv Syst. 2010;26:343–7.
16. Pople IK, Ettles D. The role of endoscopic choroid plexus coagulation in the management of hydrocephalus. Neurosurgery. 1995;36:698–701. discussion 701-2
17. Riva-Cambrin J, Kestle JR, Holubkov R, Butler J, Kulkarni AV, Drake J, et al. Risk factors for shunt malfunction in pediatric hydrocephalus: a multicenter prospective cohort study. J Neurosurg Pediatr. 2016;17:382–90.
18. Scarff JE. Nonobstructive hydrocephalus: treatment by endoscopic cauterization of the choroid plexus long term results. J Neurosurg. 1952;9:164–76.
19. Warf BC. Comparison of endoscopic third ventriculostomy alone and combined with choroid plexus cauterization in infants younger than 1 year of age: a prospective study in 550 african children. J Neurosurg. 2005;103:475–81.
20. Weil AG, Westwick H, Wang S, Alotaibi NM, Elkaim L, Ibrahim GM, et al. Efficacy and safety of endoscopic third ventriculostomy and choroid plexus cauterization for infantile hydrocephalus: a systematic review and meta-analysis. Childs Nerv Syst. 2016;32:2119–31.
21. Zada G, Liu C, Apuzzo ML. "Through the looking glass": optical physics, issues, and the evolution of neuroendoscopy. World Neurosurg. 2013;79:S3–13.

Shunts and Shunt Complications

16

Yazid Maghrabi and Saleh Baeesa

16.1 Introduction

Hydrocephalus is a common entity in neurosurgical practice affecting both adult and pediatric populations [1, 2]. In the USA, it accounts for around 69,000 discharges/year [3]. Since their introduction in the 1950s, ventricular shunting especially ventriculoperitoneal (VP) shunts have been the standard of care for patients suffering from hydrocephalus, even with the presence of new therapeutic advances such as endoscopic ventriculostomy [1]. Ventricular shunting is a common procedure in the field of neurosurgery [1, 3]. It was estimated that in the USA, 36,000 shunt insertion procedures are done every year with 14,000 of shunts requiring revisions annually [3]. Such procedure costs around billions of dollars yearly, which make hydrocephalus such a big financial burden on health systems [4]. However, such burden is due to high rate of complications and revision surgery rather than technical issues of shunt placement procedures [5].

In this chapter, we intend to discuss shunts including shunt systems and hardware, mentioning different kinds of shunt types and their use. However, our main focus is to offer an updated view of shunt complications that can be faced by practicing neurosurgeons and offer an update from the literature on the diagnosis and management of such entities.

16.2 Shunt Hardware and Valve Systems

Shunt systems are composed of a proximal ventricular catheter that is located in the ventricular system of the brain, or lumber subdural space, a one-way valve system, and a distal catheter located in different body cavities, commonly the peritoneum or

Y. Maghrabi, M.B.B.S. • S. Baeesa, M.B.Ch.B., F.R.C.S.C. (✉)
Division of Neurosurgery, Department of Surgery, Faculty of Medicine, King Abdulaziz University, P.O. Box 80215, Jeddah 21589, Saudi Arabia
e-mail: salehbaeesa@gmail.com

© Springer International Publishing AG 2017
A. Ammar (ed.), *Hydrocephalus*, DOI 10.1007/978-3-319-61304-8_16

right atrium [6]. The valve system is considered to be a major part of the shunt system since it provides a narrow gate for cerebrospinal fluid (CSF) to be drained out of the enlarged ventricular system. Many valve subtypes are available in the market, but the most commonly used ones are either differential or programmable pressure valves, or flow-regulating valves [6, 7].

16.2.1 Differential Pressure Valves

Such valve system includes four subtypes, namely, slit, miter, diaphragm, and ball-in-cone. The main goal of such valve system is keeping the intraventricular pressure in a constant state that it does not become high or low. The mechanism by which such valves work is determined by their opening and closing pressure. Moreover, when the intraventricular pressure exceeds the opening pressure of the valve, this will lead to its opening, which will allow CSF to flow at a rate determined by other factors. Then, when intraventricular pressure decreases to a point below the valve's closing pressure, the valve will close stopping the flow of CSF [6].

For static nonprogrammable valves, many prototypes exist as regards opening pressures: very low (<1 cmH$_2$O), low (1–4 cmH$_2$O), medium (4–8 cmH2O), and high (>8 cmH$_2$O). However, we have to put in mind that no standard classification exists that governs the actual pressure number designation and it is manufacturer dependent [6, 8, 9].

16.2.1.1 Slit Valves

Single-slit valves are considered to be the oldest and simplest type of valves, where the slit resides in the wall of the valve, opening up for CSF to flow when the pressure exceeds the opening pressure of the valve [6, 7, 9, 10]. Opening pressure is exclusively set by the thickness of the tube itself [7, 9, 10].

Multiple modified prototypes are available, with different characteristics such as Codman Holter valve and Phoenix Holter-Hausner valve. Codman Holter valve is composed of two slits made of silicon: those slits attached to springs preventing the collapse of the tube. Those slits are housed within a silicon-made chamber. Moreover, Phoenix Holter-Hausner valve differs from Codman Holter valve that it has two silicon slits in cruciform shape [7, 9, 10]. The main advantages of these types are easy placement and accessibility, and it does not have to be connected directly to the ventricular catheter [9].

16.2.1.2 Miter Valves

These valve types are composed of two silicon-made leaflets, each one having a free end and an end connected to the wall of the valve: this will ensure the flow of CSF in a unidirectional way [9, 10]. A commercially available prototype of this type of valves is Heyer-Schulte in-line valve.

16.2.1.3 Diaphragm Valves

These valves are characterized by the presence of silicon diaphragm that resides on the base. With pressure differential, the diaphragm moves in a flexible manner for

CSF to flow in one direction [9, 10]. Unlike slit valves, diaphragm valves need to be connected directly to a ventricular catheter [9].

16.2.1.4 Ball-in-Cone Valves

These are types of valves in which spring is connected with a ruby ball. This ball closes the orifice in which the CSF flows when the pressure differential is low. Depending on many factors such as CSF flow and viscosity, the ball moves in both directions. Many prototypes exist such as Cordis-Hakim valve and Codman Medos Hakim valve [9, 10].

Cordis-Hakim valve is composed of two valves: one is located where the CSF enters coming from the ventricular system with ruby ball housed in stainless steel spring, and the other is located toward the outlet. Codman Medos Hakim, on the other hand, composed of a chamber that will lead to a ruby ball residing in a valve seat. Moreover, a reservoir dividing the system into two valves, proximal and distal [10].

16.2.2 Programmable Valves

Adjustable, programmable valves have similar design and follow similar mechanisms of static (nonprogrammable) pressure differential valves. The only difference is that the neurosurgeon can adjust the opening pressure without the need of revision surgery. However, due to their high cost, most neurosurgeons reserve the use of such valves to cases difficult to manage such as slit ventricles and subdural collection [6]. Multiple programmable valves are available in the market such as adjustable Codman Hakim valve, which works by ball-in-cone mechanism, Sophy programmable pressure valve by Sophysa, Strata by Medtronic, and Certas plus by Codman [6, 11].

16.2.3 Flow-Regulating Valves

Other types of valves, which work with a mechanism different than pressure differential, are flow-regulating valves. These valves are designed to keep a constant flow of CSF regardless of pressure differential. Flow-regulating valves are composed of a solid cylinder with a ring connected to pressure-sensitive membrane. Moreover, a gap is present between the ring and the cylinder, and such gap changes in size in response to high or low CSF pressure [6, 10].

16.2.4 Additional Features of Valve Systems

- *On-off device* In some cases, neurosurgeons are faced with cases in which drainage needs to be stopped temporarily due to overdrainage or for revision of shunt system. It is designed as a radiopaque titanium plug that can be pressed by slight pressure on the skin. Moreover, due to its radiopacity, the position of the device can be confirmed by X-ray [6, 9, 10].

- *Antisiphon device (ASD)* The main function of such device is to guard against over drainage in response to a difference between atmospheric and intraventricular pressures. This device is usually positioned under the scalp, and it has a small diaphragm that regulates the flow of CSF [6, 9, 10].
- *Shunt filter* is a filter that can be added to shunt system, made of cellulose with silicon mesh, which is housed within a silicon chamber. This filter is used in case of hydrocephalus where the obstruction resulted from malignancy. It prevents malignant cells from spreading throughout shunt system to the rest of the body. The main disadvantage is that the filter can be blocked by blood or tissue debris [9, 10].

16.3 Ventricular Catheter (Proximal Tubing)

These catheters are within direct access to the ventricular system. It is usually made of synthetic silicone (Silastic). For visualization under X-ray, such catheters are impregnated in barium or contain an intermittent radiopaque markings [9, 10]. Silicon or syntactic silicon is usually the most preferred material to be used because of its biocompatibility, flexibility, elasticity, and high tensile strength [10]. Multiple subtypes are present, straight catheters that have a length ranging from 15 to 23 cm. Moreover, pre-shaped catheters where the curvature of the catheter is pre-shaped for better fitness on burr hole, ranging in length from 5 to 13 cm [9, 10]. Furthermore, versatile designs of catheter tips are also present such as smooth-tipped catheter with multiple perforations and grooved design with multiple perforations [9, 10].

16.4 Peritoneal Catheter (Distal Tubing)

These catheters are of various lengths. They might be open or closed-ended. They are similar to ventricular catheters, made of silicon synthetic, and are either barium-impregnated or contain radiopaque marking for visualization under X-ray [9, 10].

16.5 Shunt Types

16.5.1 Ventriculoperitoneal Shunt (VP Shunt)

It is the most commonly used shunt type in modern neurosurgery for the treatment of hydrocephalus [6, 11, 12]. The presence of ventriculitis, intraventricular hemorrhage, and peritoneal infection are contraindications for VP shunting. In this type of shunt, the proximal catheter is positioned in the lateral ventricles and the distal catheter in the peritoneal cavity [12].

16.5.2 Ventriculoatrial Shunt (VA Shunt)

This type is not usually used as the first choice because it may be associated with the development of cor pulmonale and shunt nephritis [6, 13]. It is used when there are absolute contraindications to VP shunting in the presence of peritoneal pathology such as peritonitis [6, 11, 12]. It is utilized for the treatment of all types of hydrocephalus, pseudotumor cerebri, and subdural hygromas [12].

CSF infection is an absolute contraindication of VA shunting. Moreover, other diseases of cardiovascular origin such as congestive heart failure and pulmonary hypertension pose a contraindication for placement of such shunt type since it increases the fluid load and may be fatal to the patient [12]. The proximal catheter in this type is positioned in the lateral ventricles, similar to VP [9, 12]. The distal catheter is positioned in the superior vena cava (SVC), just above the tricuspid valve and through the jugular vein [6, 12].

16.5.3 Ventriculopleural Shunt (VPL Shunt)

It is considered as an option for the treatment of hydrocephalus when VP and VA shunts are contraindicated [11, 14]. It has been recommended that such type of shunt not be used in children less than 7 years of age to prevent the development of hydrothorax [11]. Patients with the history of chest surgery and adhesions in the chest cavity and patients with active chest disease are contraindicated to undergo VPL shunt placement [6]. The proximal catheter in this type is positioned in the lateral ventricles similar to VP and VA [6, 9, 12]. The distal catheter is positioned in the pleural space [6].

16.5.4 Less Common Types of Shunts

16.5.4.1   Lumboperitoneal Shunt (LP Shunt)
This type usually is reserved for use in communicating hydrocephalus, mostly in pseudotumor cerebri and CSF fistula [11]. It has been found in SINPHONI-2 study, an open-label randomized trial, that lumboperitoneal shunts can benefit patients suffering from idiopathic normal pressure hydrocephalus. [15].

16.5.4.2   Extremely Rare Shunt Types
There have been some reports of insertion of the distal catheter in multiple visceral sites such as the gallbladder, uterus, and bladder. Moreover, such types are rare due to the association of high complications' rate [11].

16.6   Complications and Incidence

Many authors proposed a wide variety of classifications of shunt complications, but yet a standard classification is lacking [16]. In the following section, we intended to classify complications into two categories: early complications occurring within the first year after insertion of shunt system and late complications that are occurring after 1 year of shunt system insertion (Table 16.1).

Shunts are the cornerstone of hydrocephalus treatment. However, complications are inevitable during patients' lifetime [16]. In a large population study conducted in California where 14,455 patients were included, it was found that the cumulative complication rate at 5 years was 35% [16, 17]. An observational study in pediatric patients done by Khan et al. found that the median time from shunt placement to shunt failure is approximately 68 days. Moreover, shunt blockage constituted 15.1%, while shunt infection was found in 8% of cases [18]. Concerning the adult population, Korinek et al. have found that the most common complication of shunts was shunt obstruction comprising of 15.9% of cases, followed by catheter malposition in 7.1% [19]. This shows that shunt complications differ between pediatric and adult populations, and this can be attributed to the process of growth in children [20, 21].

16.6.1 Early Complications (Up to 1 Year)

16.6.1.1 Shunt Infection: "A Neurosurgeon's Worst Nightmare"

Studies in the last 10 years have reported that the rate of shunt infection ranges from 8.5 to 15% [16–18, 20–22]. It is also the second to the third most common cause of shunt failure in children [18, 19]. In a large multicenter prospective cohort study, it has been found that age less than 6 months is considered as a risk factor for developing shunt infection. Moreover, intraventricular hemorrhage of prematurity is also a risk factor. Other risk factors present, which are also considered as the causes of hydrocephalus such as tumors, myelomeningocele, post-infectious meningitis, are predictors of the possibility of development of shunt infection. Gender, race, and post-conceptional age can be considered risk factors [23]. It is important to note that the most important risk factor of them all is the length of the procedure since most of the shunt infections occur in the perioperative period [11, 16, 22].

Shunt infections usually occur in the early perioperative period due to intraoperative contamination of shunt system due to breach in sterility, poor surgical technique, imperfect surgical materials, or long operations [16, 24, 25]. Moreover, other rare sources of contamination the might lead to shunt infection are meningitis that can contaminate the proximal catheter and peritonitis that can contaminate the distal

Table 16.1 Classification of complications based on shunt insertion period

Early complications (*within 1 year after shunt insertion*)	Late complications (*1 year after shunt insertion*)
Shunt infection	Shunt fracture
Shunt obstruction	Secondary craniosynostosis
Shunt disconnection	Meningeal fibrosis
Shunt malposition and migration	Acquired cranioencephalic disproportion
Slit ventricle syndrome (SVS)	Entrapment of the fourth ventricle (EFV)
	Pneumocephalus

catheter [16, 24]. Late shunt infection is rare and usually results from contamination of the distal tubing by colonic flora [16, 25].

Contamination of shunt system is usually by skin flora such as *S. aureus* and coagulase-negative staphylococci that include *S. epidermidis* [16, 22, 25]. Gram-negative bacteria have been isolated in from infected VP shunts in other studies [16, 22, 24]. A cohort of 224 pediatric patients found that most common organisms isolated from infected shunts were negative staphylococci (45.7%), followed by *S. aureus* (22.9%); other organisms were identified but not with a great frequency such as *K. pneumoniae* and *E. coli* [22]. Shunt infection presents clinically with multiple signs and symptoms such as local inflammatory reaction around the shunt system manifested with classic features of inflammation: swelling, hotness, redness, and loss of shunt function. Patients can also present with lethargy, irritability, nausea, and vomiting. Moreover, seizures can occur as sequelae of the infectious process [16, 24]. Some authors argue that fever can be present or absent. However, Lee and his colleagues found that fever was the most common presentation of patients with shunt infection [16, 22, 24].

Laboratory testing can be utilized as the initial step for diagnosing shunt infection. This includes complete blood count (CBC) to check for white blood cells (WBC), which are increased in most cases of shunt infection. Moreover, acute phase reactants such as C-reactive protein (CRP) and erythrocyte sedimentation rate (ESR) can aid in the diagnosis of infection [11, 22]. The most important diagnostic test is CSF analysis and culture, a shunt tap from valve reservoir allowing isolation of causative factors, which is more sensitive and specific than the lumbar and ventricular since they might yield sterile cultures and might, by itself, cause other iatrogenic complications [11, 16, 24].

Imaging studies such as computer tomography (CT) and magnetic resonance imaging (MRI) are not usually utilized for the diagnosis of shunt infection [11]. Meningitis can be seen in CT and MRI as irregular leptomeningeal enhancement. Moreover, ventriculitis is demonstrated in CT and MRI as irregular ependymal enhancement [16, 26]. It should be noted that shunt infection can happen concurrently with peritonitis in VP shunts, or be coupled with endocarditis or septic emboli in VA shunts, and empyema in VPL shunts [11, 16, 26].

Most authors recommend the removal of shunt system in the context of suspected shunt infection with the insertion of external ventricular drain (EVD) with simultaneous administration of antibiotics [11, 22, 27]. EVDs have many advantages that they enable surgeons to monitor CSF and take CSF sample to see response to treatment [11]. In some cases, the proximal catheter might be left in place with externalization of the distal catheter [11, 22, 27].

As initial antibiotic therapy, vancomycin usually is used either alone or with another antimicrobial agent. Third-generation cephalosporin and meropenem can be used, and then the therapy can be adjusted according to culture and sensitivity [11, 22, 27]. Nevertheless, antibiotic therapy alone without shunt removal has the low rate of success and might carry the high risk of secondary infections; thus, it is not usually recommended [11, 22]. Usually antibiotics are used for approximately 10 days after three sterile CSF samples, 1 day apart. This marks also the point where EVD has to be removed and new shunt system to be inserted [11, 27].

Shunt infection causes high morbidity and mortality. Thus, prevention of such complication will prevent subsequent catastrophic events [28]. Since the main mode of infection transmission is intraoperative by surgeon's gloves, decreasing the frequency of manual contact between the surgeon's hands will also decrease shunt infection rates [22, 28, 29]. Decreasing the duration of the procedure and double gloving would also contribute to decrease the infection rates [22, 29]. A retrospective cohort study by Rehman et al. have found that changing the gloves before handling the shunt hardware reduces the infection rate dramatically [28].

A randomized controlled trial found that injecting vancomycin and gentamicin around VP shunt causes a significant decrease in shunt infection rate [30]. Furthermore, meta-analysis of observational studies has found that antibiotic-impregnated catheters (AIC) are effective in reducing the rate of shunt infections [31].

16.6.1.2   Shunt Obstruction (Malfunction)

Shunt obstruction is one of the most common complications of shunt systems, and it has been extensively reported in the literature [18, 19, 32]. It has been estimated that half of the patients will suffer from symptoms related to shunt obstruction at least once in 12 years following shunt insertion [32, 33]. Moreover, the rate of shunt obstruction has been estimated to be around 0.5% and 5% annually [16, 24, 32, 34].

Obstruction can occur in three sites: (1) the tip of the proximal catheter due to ependymal cellular reaction, (2) valve with blood or tissue debris, or (3) distal tubing, which rarely occurs [24]. Regardless of the site, shunt obstruction presents clinically with signs and symptoms suggestive of increased intracranial pressure (ICP) such as a headache, vomiting, and drowsiness [16, 25, 32]. Presentation of shunt obstruction in infants includes nausea, vomiting, and irritability [16, 24]. There are also reports stating that shunt obstruction can be presented with atypical presentations such as cranial nerve palsies, ataxia, and seizure [32, 35].

Laboratory tests have a limited role in diagnosing shunt obstruction. However, clinical presentation with imaging studies such as CT is the gold standard for diagnosing such entity [16, 24, 32]. Shunt obstruction can be diagnosed with CT or conventional radiography by comparison of direct postoperative images with images obtained after the appearance of symptoms [16, 24]. Several authors proposed many treatment options such as tube removal, but this option is sometimes not feasible as it might cause profuse bleeding. Others suggested doing endoscopic third ventriculostomy, nevertheless the newer trend is shunt freeing through endoscopy [36].

16.6.1.3   Shunt Disconnection

Several authors consider shunt disconnection as the second most common cause of shunt malfunction after shunt obstruction [16, 26]. The rate of shunt disconnection in the literature has been estimated to range from 4 to 15%, however, not all shunt disconnection results to shunt failure [37]. Literature reports that disconnection usually occurs with multiple pieces' shunt systems and the usage of one-piece system can reduce the rate of such complication [37–39]. Other contributing factors to shunt disconnection are the growth of the patient and extreme neck movement [37, 38].

The most problematic sites where disconnection can occur are in the neck and abdomen [38]. Shunt disconnection is manifested with symptoms of increased ICP, as in shunt obstruction, but some patients can be asymptomatic and the diagnosis is made incidentally [16, 37]. The gold standard to diagnose such phenomenon is by obtaining "shunt X-ray series" where gaps between shunt tube segments are seen and occasionally areas of classifications are observed [16, 26]. Laparoscopy is considered as a gold standard in the retrieval of a disconnected distal catheter and insertion of new one to ensure proper functioning of shunt system [40]. As regards proximal tubing and shunt valve, revision surgery is indicated [11].

16.6.1.4 Shunt Malposition and Migration

This is considered as a rare complication of shunt insertion and usually occurs with proximal or distal catheters [16, 24]. There have been many reports on the migration of distal tubing into many sites such as the scrotum, stomach, chest wall, bladder, and colon [41–46]. Such complication necessitates emergency revision surgery [47].

16.6.1.5 Slit Ventricular Syndrome (SVS)

SVS is a group of symptoms that are suggestive of raised ICP that are present intermittently coupled with the slit-like appearance of ventricles in imaging studies [16, 25]. Two-thirds of shunted patients present with the radiographic finding of slit-like ventricles. However, only 11% of these patients become symptomatic [48, 49]. The most accepted theory of the pathogenesis of such syndrome is rapid overdrainage of CSF that results in the obstruction of the proximal catheter's tip leading to intermittent shunt malfunction [16, 25]. Diagnosis of such syndrome is mostly clinical and imaging studies can reveal small ventricles [16, 26].

Many treatment options have been advocated by multiple authors such as cortisone therapy, adjustment of programmable valves, cranioplasty, and even removal of shunt system [50]. A cohort follow-up study over 25 years has found that implanting antisiphon device (ASD) as a prophylactic measure would reduce the incidence of SVS [50].

16.6.2 Late Complications (More Than 1 Year)

16.6.2.1 Shunt Fracture

It is considered as a late complication since it is usually seen in older children and adolescent age group. It results from repeated mechanical stress, calcification, and growth of the patient [16, 33]. Most common sites for catheter breakage are in the neck and upper chest, and fractures manifested clinically by signs and symptoms of raised ICP coupled with pain, swelling, or erythema surrounding shunt tract. The diagnosis is made by plain radiography, which has high sensitivity in detecting discontinuation in shunt tract [16, 51]. The treatment, in this case, is similar to treatment options for shunt disconnection that has been mentioned in the previous section, in which revision surgery is indicated [11, 40].

16.6.2.2 Secondary Craniosynostosis

In the case of chronic overdrainage, an osteoclastic/osteoblastic activity starts to arise, leading to the formation of new bony tissue and subsequent fusion of skull sutures [16, 52]. The incident of such complication ranges from 10 to 15% [16, 52]. This entity is diagnosed best by imaging studies such as CT; 3D constructed images are widely used nowadays for such cases. Sometimes parenchymal brain changes are needed to be checked; this is usually done with MRI imaging [16]. Such complication is treated surgically by craniotomy and cranioplasty [16].

16.6.2.3 Meningeal Fibrosis

A rare complication is also resulting from chronic overdrainage leading to collagen and granulation tissue deposition and subsequently fibrosis, which appears as hyperintense areas in contrast MRI and hyperdense areas in contrast CT [16]. This entity is treated conservatively [53].

16.6.2.4 Acquired Cranioencephalic Disproportion

This complication is diagnosed in cases where brain volume is growing and cannot be accommodated by the skull. Such disease is not well studied, and the incident of such entity is not well known. Patients suffering from such complication experience severe headaches, irritability, ataxia, dizziness, and lethargy [54]. In skull X-ray, the skull appears thickened and might be associated with craniosynostosis [54]. Moreover, CT might show "crowded posterior fossa" that would show more tonsil herniations (acquired Chiari type 1 malformation) [16, 54].

Strategies to treat such complications are either to drain the CSF or to decrease ICP by either shunt revision surgery, endoscopic third ventriculostomy, or externalization of the shunt [54–56]. Other surgical techniques have been proposed to expand the size of the cranial vault such as subtemporal craniotomy [54, 57].

16.6.2.5 Entrapment of the Fourth Ventricle (EFV)

Trapping of the fourth ventricle happens when the aqueduct of Sylvius, foramina of Luschka and Magendie are occluded leaving the fourth ventricle isolated from the rest of ventricular system [58]. The occlusion results from posthemorrhagic or postmeningitic states. But in the case of shunts, it can be functional occlusion due to chronic over drainage [58–60]. Incident of EFV ranges from 2 to 3%, and it has been estimated that time interval from shunt insertion to EFV was 3.1 years [58, 61]. EFV presents clinically with signs and symptoms that overlap with other shunt complications such as shunt obstruction and shunt disconnection. Therefore, clinical features have to be correlated with imaging studies like CT and MRI [58]. MRI is superior to CT in a way that it delineates in great details the occlusion of the aqueduct [16, 58].

Placement of separate shunts for the fourth ventricle has been the cornerstone of treatment for several years, but it has been reported to be associated with many complications [58]. Other options have been developed and proved to be safe and efficient such as endoscopic aqueductoplasty, aqueductal stent placement, and fenestration of EFV via craniotomy.

16.6.2.6 Pneumocephalus

Pneumocephalus is an extremely rare complication of shunt system and can only occur in the presence of bony defects in the skull base and with a negative pressure of shunt system, air might enter into the cranial vault. Imaging studies are the gold standard for diagnosis [16].

Conclusion

Shunt systems are widely used in the treatment of hydrocephalus. However, they are associated with high level of complications, which patients will face at least once in their lifetime. Complications usually occur at the early postoperative period, while the rare late complications occur more than 1 year after shunt insertion, which can be a direct result of shunt overdrainage. Neurosurgeons should have a high degree of intuition in order to identify and treat shunt complications the earliest to decrease morbidity and mortality.

References

1. Stone JJ, Walker CT, Jacobson M, Phillips V, Silberstein HJ. Revision rate of pediatric ventriculoperitoneal shunts after 15 years. J Neurosurg Pediatr. 2013;11(1):15–9.
2. Laurence KM, Coates S. The natural history of hydrocephalus. Detailed analysis of 182 unoperated cases. Arch Dis Child. 1962;37:345–62.
3. Massimi L, Paternoster G, Fasano T, Di Rocco C. On the changing epidemiology of hydrocephalus. Childs Nerv Syst. 2009;25(7):795–800.
4. Patwardhan RV, Nanda A. Implanted ventricular shunts in the United States: the billion-dollar-a-year cost of hydrocephalus treatment. Neurosurgery. 2005;56:139–45.
5. Gottfried ON, Binning MJ, Sherr G, Couldwell WT. Distal ventriculoperitoneal shunt failure secondary to Clostridium Difficile colitis. Acta Neurochir. 2005;147:335–8.
6. Cinalli G, Maixner WJ, Sainte-Rose C. Pediatric Hydrocephalus. Milano: Springer; 2005.
7. Drake JM, Sainte-Rose C. The shunt book. New York: Blackwell Scientific; 1995.
8. Albright AL, Pollack IF, Adelson P. Principles and practice of pediatric neurosurgery. 3rd ed. New York: Thieme; 2014.
9. Post EM. Currently available shunt systems: a review. Neurosurgery. 1985;16(2):257–60.
10. Ramamurthy B, Sridhar K, Vasudeva MC. Textbook of operative neurosurgery. New Delhi: BI Publications Pvt Ltd; 2005.
11. Greenberg MS. Handbook of neurosurgery. 8th ed. New York: Thieme; 2016.
12. Rengachary SS, Wilkins RH. Neurosurgical operative atlas. Chicago: AANS Publications Committee; 1993.
13. Lundar T, Langmoen IA, Hovind KH. Fatal cardiopulmonary complications in children treated with ventriculoatrial shunts. Childs Nerv Syst. 1991;7:215–7.
14. Jones RFC, Currie BG, Kwok BCT. Ventriculopleural shunts for hydrocephalus: a useful alternative. Neurosurgery. 1988;23:753–5.
15. Kazui H, Miyajima M, Mori E, Ishikawa M, SINPHONI-2 Investigators. Lumboperitoneal shunt surgery for idiopathic normal pressure hydrocephalus (SINPHONI-2): an open-label randomised trial. Lancet Neurol. 2015;14(6):585–94.
16. Di Rocco C, Turgut M, Jallo G, Martínez-Lage JF. Complications of CSF shunting in hydrocephalus. Cham: Springer; 2015.
17. Wu Y, Green NL, Wrensch MR. Ventriculoperitoneal shunt complications in California: 1990 to 2000. Neurosurgery. 2007;61:557–62.

18. Khan F, Shamim MS, Rehman A, Bari ME. Analysis of factors affecting ventriculoperitoneal shunt survival in pediatric patients. Childs Nerv Syst. 2013;29(5):791–802.
19. Korinek AM, Fulla-Oller L, Boch AL, Golmard JL, Hadiji B, Puybasset L. Morbidity of ventricular cerebrospinal fluid shunt surgery in adults: an 8-year study. Neurosurgery. 2011;68(4):985–94.
20. Park MK, Kim M, Park KS, Park SH, Hwang JH, Hwang SK. A retrospective analysis of ventriculoperitoneal shunt revision cases of a single institute. J Korean Neurosurg Soc. 2015;57(5):359–63.
21. Lee L, Low S, Low D, Ng LP, Nolan C, Seow WT. Late pediatric ventriculoperitoneal shunt failures: a Singapore tertiary institution's experience. Neurosurg Focus. 2016;41(5):E7.
22. Lee JK, Seok JY, Lee JH, Choi EH, Phi JH, Kim SK, et al. Incidence and risk factors of ventriculoperitoneal shunt infections in children: a study of 333 consecutive shunts in 6 years. J Korean Med Sci. 2012;27(12):1563–8.
23. Simon TD, Butler J, Whitlock KB, Browd SR, Holubkov R, Kestle JR, et al. Risk factors for first cerebrospinal fluid shunt infection: findings from a multi-center prospective cohort study. J Pediatr. 2014;164(6):1462–8. e2
24. Sivaganesan A, Krishnamurthy R, Sahni D, Viswanathan C. Neuroimaging of ventriculoperitoneal shunt complications in children. Pediatr Radiol. 2012;42:1029–46.
25. Di Rocco C, Massimi L, Tamburrini G. Shunts vs. endoscopic third ventriculostomy in infants: are there different types and/or rates of complications? A review. Childs Nerv Syst. 2006;22:1573–89.
26. Wallace AN, McConathy J, Menias CO, Bhalla S, Wippold FJ II. Imaging evaluation of CSF shunts. Am J Roentgenol. 2014;202:38–53.
27. Kestle JR, Garton HJ, Whitehead WE, Drake JM, Kulkarni AV, Cochrane DD. Management of shunt infections: a multicenter pilot study. J Neurosurg. 2006;105(3 Suppl):177–81.
28. Rehman AU, Rehman TU, Bashir HH, Gupta V. A simple method to reduce infection of ventriculoperitoneal shunts. J Neurosurg Pediatr. 2010;5(6):569–72.
29. Kulkarni AV, Drake JM, Lamberti-Pasculli M. Cerebrospinal fluid shunt infection: a prospective study of risk factors. J Neurosurg. 2001;94:195–201.
30. Moussa WM, Mohamed MA. Efficacy of postoperative antibiotic injection in and around ventriculoperitoneal shunt in reduction of shunt infection: a randomized controlled trial. Clin Neurol Neurosurg. 2016;143:144–9.
31. Thomas R, Lee S, Patole S, Rao S. Antibiotic-impregnated catheters for the prevention of CSF shunt infections: a systematic review and meta-analysis. Br J Neurosurg. 2012;26(2):175–84.
32. Barnes NP, Jones SJ, Hayward RD, Harkness WJ, Thompson D. Ventriculoperitoneal shunt block: what are the best predictive clinical indicators? Arch Dis Child. 2002;87(3):198–201.
33. Sainte-Rose C, Piatt JH, Renier D, et al. Mechanical complications in shunts. Pediatr Neurosurg. 1991;17:2–9.
34. Rekate HL. Shunt revision: complications and their prevention. Pediatr Neurosurg. 1991;17:155–62.
35. Lee TT, Uribe J, Ragheb J, et al. Unique clinical presentation of pediatric shunt malfunction. Pediatr Neurosurg. 1999;30:122–6.
36. Singh D, Saxena A, Jagetia A, Singh H, Tandon MS, Ganjoo P. Endoscopic observations of blocked ventriculoperitoneal (VP) shunt: a step toward better understanding of shunt obstruction and its removal. Br J Neurosurg. 2012;26(5):747–53.
37. Lee YH, Park EK, Kim DS, Choi JU, Shim KW. What should we do with a discontinued shunt? Childs Nerv Syst. 2010;26(6):791–6.
38. Aldrich EF, Harmann P. Disconnection as a cause of ventriculoperitoneal shunt malfunction in multicomponent shunt systems. Pediatr Neurosurg. 1990;16:309–11.
39. Epstein F. How to keep shunts functioning, or "the impossible dream". Clin Neurosurg. 1985;32:608–31.
40. Yu S, Bensard DD, Partrick DA, Petty JK, Karrer FM, Hendrickson RJ. Laparoscopic guidance or revision of ventriculoperitoneal shunts in children. JSLS. 2006;10(1):122–5.

41. Albala DM, Danaher JW, Huntsman WT. Ventriculoperitoneal shunt migration into the scrotum. Am J Surg. 1989;55(11):685–8.
42. Alonso-Vanegas M, Alvarez JL, Delgado L, Mendizabal R, Jimenez JL, Sanchez-Cabrera JM. Gastric perforation due to ventriculo-peritoneal shunt. Pediatr Neurosurg. 1994;21:192–4.
43. Borkar SA, Satyarthee GD, Khan RN, Sharma BS, Mahapatra AK. Spontaneous extrusion of migrated venticuloperitoneal shunt through chest wall: a case report. Turk Neurosurg. 2008;18(1):95–8.
44. Burnette DJ. Bladder perforation and urethral catheter extrusion: an unusual complication of cerebrospinal fluid-peritoneal shunting. J Urol. 1982;127:543–4.
45. Fischer G, Goebel H, Latta E. Penetration of the colon by a ventriculo-peritoneal drain resulting in an intra-cerebral abscess. Zentralbl Neurochir. 1983;44:155–60.
46. Glatstein MM, Roth J, Scolnik D, Haham A, Rimon A, Koren L. Late presentation of massive pleural effusion from intrathoracic migration of a ventriculoperitoneal shunt catheter: case report and review of the literature. Pediatr Emerg Care. 2012;28(2):180–2.
47. Ammar A, Nasser M. Intraventricular migration of VP shunt. Neurosurg Rev. 1995;18(4):293–5.
48. Baskin JJ, Manwaring KH, Rekate HL. Ventricular shunt removal: the ultimate treatment of the slit ventricle syndrome. J Neurosurg. 1998;88(3):478–84.
49. Walker ML, Fried A, Petronio J. Diagnosis and treatment of the slit ventricle syndrome. Neurosurg Clin N Am. 1993;4:707–14.
50. Gruber RW, Roehrig B. Prevention of ventricular catheter obstruction and slit ventricle syndrome by the prophylactic use of the Integra antisiphon device in shunt therapy for pediatric hypertensive hydrocephalus: a 25-year follow-up study. J Neurosurg Pediatr. 2010;5(1):4–16.
51. Browd S, Ragel B, Gottfried O. Failure of cerebrospinal fluid shunts. Part I: obstruction and mechanical failure. Pediatr Neurol. 2006;34:83–92.
52. Albright AL, Tyler-Kabara E. Slit-ventricle syndrome secondary to shunt-induced suture ossification. Neurosurgery. 2001;48:764–9.
53. Bhatia V, Panda P, Sharma S, Sood RG. Post shunt meningeal fibrosis: role of contrast enhanced MRI in differentiation from chronic subdural hematoma. IJNS. 2012;1(2):158–60.
54. Sandler AL, Goodrich JT, Daniels LB III, Biswas A, Abbott R. Craniocerebral disproportion: a topical review and proposal toward a new definition, diagnosis, and treatment protocol. Childs Nerv Syst. 2013;29(11):1997–2010.
55. Cinalli G, Salazar C, Mallucci C, Yada JZ, Zerah M, Sainte-Rose C. The role of endoscopic third ventriculostomy in the management of shunt malfunction. Neurosurgery. 1998;43:1323–7.
56. Gil Z, Siomin V, Beni-Adani L, Sira L, Constantini S. Ventricular catheter placement in children with hydrocephalus and small ventricles: the use of a frameless neuronavigation system. Childs Nerv Syst. 2002;18:26–9.
57. Epstein FJ, Fleischer AS, Hochwald GM, Ransohoff J. Subtemporal craniectomy for recurrent shunt obstruction secondary to small ventricles. J Neurosurg. 1974;41:29–31.
58. Udayakumaran S, Biyani N, Rosenbaum DP, Ben-Sira L, Constantini S, Beni-Adani L. Posterior fossa craniotomy for trapped fourth ventricle in shunt-treated hydrocephalic children: long-term outcome. J Neurosurg Pediatr. 2011;7(1):52–63.
59. Oi S, Matsumoto S. Isolated fourth ventricle. J Pediatr Neurosci. 1986;2:282–6.
60. Oi S, Matsumoto S. Pathophysiology of aqueductal obstruction in isolated IV ventricle after shunting. Childs Nerv Syst. 1986;2:282–6.
61. Eder HG, Leber KA, Gruber W. Complications after shunting isolated IV ventricles. Childs Nerv Syst. 1997;13:13–6.

Strategies to Minimize Shunt Complications and Optimize Long-Term Outcomes

17

Dominic Venne

17.1 Introduction and General Concepts

It is a fact that high percentage of patients with hydrocephalus will require a CSF shunt. These shunts, mainly ventriculoperitoneal (VP shunt) and ventriculoatrial (VA shunt), have saved millions of life since their introduction in 1952 [1]. Improvements in shunt technology and surgical techniques have decreased the morbidity and mortality of these procedures. Unfortunately, CSF shunts are still associated with high complication rates, mainly, when compared with similar procedures where implants are used (cardiac pacemaker, DBS, intrathecal pumps). In fact, CSF shunts are associated with a 2-year failure rate of greater than 40% and more worrisome, an infection rate of 3–15% [2, 3]. Moreover, CSF shunt infections are associated with a long-term risk of mortality greater than 30%, which is nearly twice the rate for children without infection [4].

Among the multiple risk factors associated with CSF shunt infection, the cause of the hydrocephalus [5], the young age of the patient [6], the presence of a previous shunt system [7], the duration of the shunt operation [8, 9], and the presence of a postoperative CSF leak [10, 11] have been identified as critical ones.

Neurosurgeons are well aware of these complications and always seek for ways to improve their surgical outcomes. Most recommendations, like the ones listed below, are based on historical reports, retrospective studies, personal observations, and very few randomized studies. This chapter aims at raising the awareness and the understanding of simple details that might have significant positive effects on the outcome of shunted patients.

D. Venne, M.D., M.Sc., F.R.C.S.C.
Department of Neurosurgery, Cleveland Clinic Abu Dhabi, Abu Dhabi, United Arab Emirates
e-mail: dominic_venne@yahoo.com

© Springer International Publishing AG 2017
A. Ammar (ed.), *Hydrocephalus*, DOI 10.1007/978-3-319-61304-8_17

17.1.1 Patient Selection

This concept applies to any kind of surgery. When selected properly, patients always have better outcomes. Concerning hydrocephalus patients, as mentioned by H. L. Rekate, [12] not all patients with enlarged ventricles need treatment. The old saying mentioning, "when in doubt, shunt," has probably resulted in the overusage of shunts. Several children with "enlarged ventricles" have been shunted although their intracranial "homeostasis" was probably normal. In the absence of clear signs or symptoms of raised intracranial pressure, the final decision for CSF shunt insertion will depend on clinical factors such the rapid enlargement of the head circumference, radiological evidence of ventricular enlargement, ancillary tests such as transcranial Doppler including the measurement of pulsatility index and resistance index [13, 14], and obviously the physician's clinical experience.

17.1.2 Team Selection

Shunt placements have been considered for many years as an excellent surgical procedure for young residents to learn and develop independently their surgical skills. Unfortunately, this concept is fundamentally wrong since CSF shunt surgeries can be associated with severe complications that can lead to grim outcomes. Although technically simple, the clinical outcome depends greatly on the experience of the surgeons and the attention to fine details. Table 17.1 lists only few of the numerous complications related with shunt surgery.

17.2 Reducing Infectious Complications

CSF shunt infection is a disastrous complication that is associated with poor functional outcomes and even significant mortality. The vast majority of infections appear to be acquired at the moment of surgery (intraoperative contamination) or in the immediate postoperative period (usually from CSF leakage or suboptimal wound closure). Most of the infections will occur within 8 weeks and 90% within the first 6 months after surgery [15]. The most frequent pathogens found in CSF shunt infection are coagulase-negative staphylococcus, followed by *Staphylococcus aureus*. Reducing the rate of infection to zero percent should be a constant quest. Achievable or not, the rate of CSF shunt infections can certainly be decreased through a series of small measures that contribute all together to the success of this procedure.

17.2.1 Preoperative Measures

17.2.1.1 Preoperative Skin Preparation

Several studies have looked at the infection source. While infection can originate from the surgical team members [16], there is clear evidence that a large percentage of infected CSF shunts originate from the patient's own skin flora [17]. In fact, the high

Table 17.1 Lists of the complications related with shunt surgery

Complications	Clinical effects
Proximal catheter misplacement (extraventricular/intraparenchymal insertion)	• Unresolved and sustained intracranial hypertension • Lobar, basal ganglia, brainstem hemorrhage leading to permanent neurologic deficits
Inverted valve direction	• Unresolved and sustained intracranial hypertension
Distal catheter misplacement	• Unresolved and sustained intracranial hypertension • Bowel perforation, peritonitis
Hemorrhage (extra-axial, intra-axial, and intraventricular)	Mass effect causing raised intracranial pressure, neurologic deficits, seizures, shunt obstruction, death
Postoperative CSF leak	Increased risk of shunt infection, meningitis, seizures, decreased intellectual performance, death
Proximal shunt infection	Meningitis, empyema, brain abscess leading to epilepsy, decreased mental and cognitive performance, septicemia, death
Distal shunt infection	Intra-abdominal pseudocysts, peritonitis, septicemia
Improper valve or pressure setting selection	Unresolved and sustained intracranial hypertension, overdrainage
Improper catheter length	Early malfunction due to proximal migration of distal catheters
Suboptimal suturing and fixation	Catheter disconnection from the valve, connectors
Suboptimal skin closure	CSF leak, hardware exposure leading to shunt infection. Poor cosmetic results leading to emotional/psychological distress

prevalence of gastrointestinal bacteria (vancomycin-resistant enterococci) and methicillin- or vancomycin-resistant *Staphylococcus aureus* (MRSA and VRSA) found in CSF shunt infections in long-stay ICU patients highlights the importance of patient skin colonization. In a similar manner, neonates born with spinal dysraphism (open myelomeningocele) have a higher incidence of gastrointestinal tract bacteria [18]. Teenagers have also a higher rate of *Propionibacterium acnes* found in infected VP shunts. Therefore, reducing the number of bacteria at the surgical site prior the procedure itself would seem to be a logical practice that could decrease the infection rate. As a matter of fact, the Centers for Disease Control and Prevention (CDC) has recommended this practice to prevent surgical site infections (SSI) [19, 20]. In terms of agent to be used, chlorhexidine has been shown to reduce the number of organisms at the incision site more efficiently than povidone-iodine or soap and water [21]. Therefore, our practice has been to prescribe a hair shampoo and body wash or bathing with chlorhexidine soap in the hours before surgery in order to decrease the bacterial count and skin crusts and even remove dust and debris as often seen in road traffic accident patients. For ICU patients where chlorhexidine bathing is not feasible, chlorhexidine-impregnated cloth is an excellent alternative to reduce SSI as shown in the orthopedic literature [22–24].

17.2.1.2 Preoperative Nursing Care

Nursing care is of paramount importance in the prevention of SSI. Many patients have highly colonized wounds (gastrostomy, ileostomy, tracheostomy sites) that should be kept cleaned and isolated from the planned surgical site.

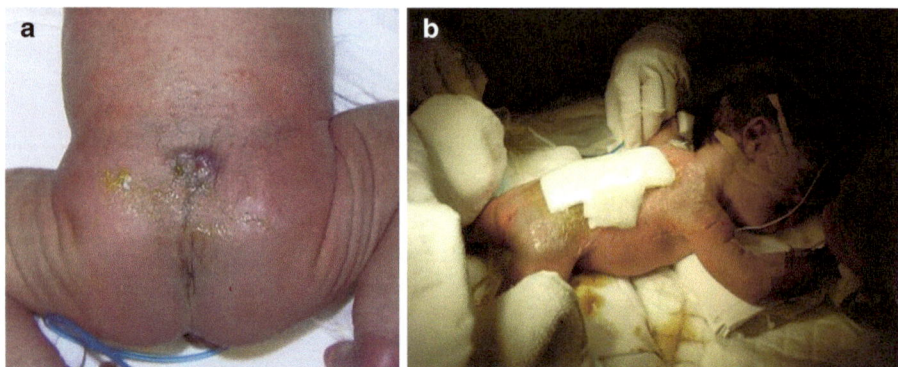

Fig. 17.1 (**a**) Stool contamination of an open neural tube defect (**b**) Isolation technique

In neonates born with open neural tube defects (myelomeningocele), stools should be contained to the perianal area and often cleaned (Fig. 17.1a). While the patient is kept in prone or lateral position, the open neural tube defect (NTD) should be kept warm and covered with a nonadhesive sterile wet dressing [25]. These wet gauzes can then be covered with an adhesive membrane to keep the spinal dysraphism physically isolated from the adjacent perianal area (Fig. 17.1b). Obviously, open NTD should be repaired as early as possible, and CSF shunt insertion should be delayed as long as the child can tolerate the condition [18].

17.2.1.3 Intravenous Preoperative Antibiotic Prophylaxis
The current recommendations for clean elective procedures are to administer intravenous antibiotics one hour prior the incision and to repeat the dose intraoperatively for long procedures. Although not proven statistically superior, it is a common practice to extend the coverage for 24 hours after surgery [26, 27].

17.2.1.4 Hair Clipping and Shaving
Cochrane Database Reviews have shown no differences in the SSI for non-neurosurgical procedures when hair was removed prior to surgery compared to those who have not [28]. In fact, there is no report of increased infection for procedure done without hair clipping or shaving as long as the scalp and hair are cleaned and prepped properly with antiseptic soap. Retrospective studies have shown a higher rate of infection for patients who were shaved several hours prior to surgery compared to those whose hair were not removed [29]. If it is necessary to remove hair, clipping might result in fewer SSI compared to shaving with a razor. However, in our experience, shaving has not lead to increased infection rate when done immediately preoperatively using a gentle technique in order to avoid skin abrasions.

17.2.2 Intraoperative Measures

17.2.2.1   Skin Preparation

Skin preparation solution is an important factor in the prevention of surgical site infections. Although some studies have possibly shown povidone-iodine-based compounds to be superior to chlorhexidine for general surgery procedures, there is no clear evidence to confirm this finding in neurosurgery [30]. Other studies have shown chlorhexidine to be superior where long action is required [31]. A three-step prep technique including 3 minutes chlorhexidine scrub followed by alcohol-based cleaning and, finally, with a 10% povidone-iodine paint allowed to dry before the application of sterile drapes is a very efficient way for skin preparation.

17.2.2.2   Draping Techniques

Drapes should be fixed and not allowed to move intraoperatively to unsterile areas mainly near tracheostomy and gastrostomy sites (Fig. 17.2).

17.2.2.3   Iodophor-Impregnated Adhesive Drapes

Iodophor-impregnated adhesive drapes are commonly used in neurosurgical procedures. They offer the advantage of constant iodine release, which theoretically decreases the bacterial count and proliferation throughout the surgical procedure. Studies have shown reduced bacterial sampling count of the wound at the end of the procedure from 15 to 1.6% with the use of such adhesive drapes [32].

17.2.2.4   Antibiotic Irrigating Solution

Disturbing reports have shown positive cultures for coagulase-negative *Staphylococcus* in the physical operating room environment such as drapes, instrument tables, and patient skin [17]. The highest concentration being near the surgical

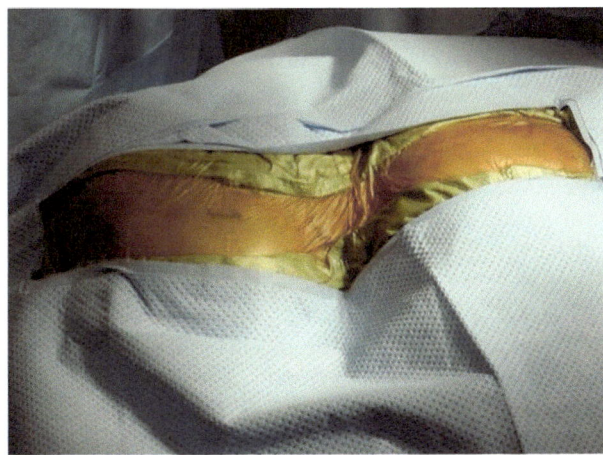

Fig. 17.2 Draping technique with the usage of iodophor-impregnated adhesive drapes

team suggests that bacteria associated with shunt infections might be airborne [33]. In the same study, positive environmental cultures were not correlated with the surgeon, length of the procedure, and time of day, but interestingly were more likely to occur in a room other than the designated neurosurgical operating room. There was also a correlation between the occurrence of positive environmental cultures and positive cerebrospinal fluid cultures. Taken into consideration these findings, the intraoperative usage of antibiotic irrigating solution (such vancomycin and gentamycin) seems to be a safe practice to keep valves, connectors, and non-impregnated antibiotic catheters from potential contamination during a CSF shunt procedure. However this practice is not recommended for antibiotic-impregnated catheters since it can accelerate the release of antibiotics from the catheter tubing.

17.2.2.5   Intrathecal Antibiotics
The usage of a single dose of intraventricular antibiotic (vancomycin and gentamicin) at the end of the procedure has been described as a highly efficient measure to reduce CSF shunt infection. The combination of intraventricular gentamicin and vancomycin with systemic antibiotic therapy had the greatest effect in decreasing the incidence of perioperative shunt infection [34]. Several protocols are available in the literature with different antibiotic dosage.

17.2.2.6   Antibiotic-Impregnated Catheters
Review of the literature demonstrates that antibiotic-impregnated shunt (AIS) catheters are associated with significant reduction incidence of CSF shunt infection compared to non-AIS catheters. Moreover these AIS catheters do not appear to be associated with an increased incidence of antibiotic-resistant microorganisms [35, 36]. By reducing the incidence of infection, the utilization of AIS catheters has been associated also with significant hospital cost savings [37].

17.2.2.7   Double Gloving Technique and Non-touch Technique
Kulkarni and Drake [38] have estimated the overall incidence of at least one hole in a surgical glove used by a member of the operating team to be at 33.4% during CSF shunt procedures. This percentage was similar to other studies of orthopedic, gynecological, and general surgeries. This extremely high rate of glove perforation highlights the importance of double gloving. In addition since the majority of glove holes are not recognized by surgeons, it is recommended to minimize manual handling of valves and catheters during CSF shunt surgeries [39]. Non-touch technique implies also to avoid the shunt hardware touching the patient' skin.

17.2.2.8   Skin Handling
Meticulous care should be given to skin handling with avoidance to crush tissue with forceps or to use monopolar cautery at the surface of the skin (Fig. 17.3).

17.2.2.9   Skin Closure
Skin closure is extremely important in CSF shunt surgery. In fact, one of the most important factors identified in shunt infection is the presence of postoperative CSF leakage [38]. Therefore, whenever possible, a multilayer skin closure is

Fig. 17.3 Demonstrate depicting skin handling where forceps are used to retract, avoiding tissue crushing

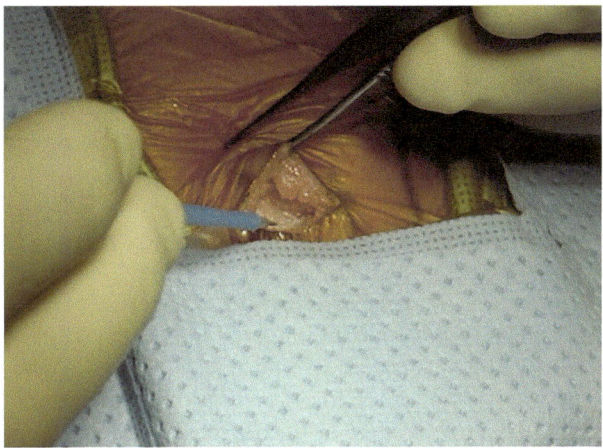

recommended with watertight suturing of the cranial incision(s) that can be achieved with an intradermal subcuticular suture. Also incisions should be planned in a way that most of the hardware will not be placed directly under the incisions. Absorbable monofilament suturing material is favored since it is associated with lower infection rate [40].

17.2.3 Postoperative Measures

17.2.3.1   Postoperative Wound Care and Dressing
Dressings should be changed at any moments if they are soaked with stools or any other contaminated fluids.

17.2.3.2   Postoperative CSF Fistula
Postoperative CSF leakage is associated with a high risk of shunt infection [38]. As soon as recognized, the wound should be inspected and additional sutures should be added if a leakage site is identified. Although not proven scientifically, the addition of intravenous antibiotics for few days should be considered in order to cover this period of potential shunt contamination.

17.3   Surgical Strategies to Optimize Long-Term Outcome

Ventricular catheter placement is associated with a high rate of malpositioning due to several factors. First of all, it is a blind technique that relies on poorly defined external landmarks often covers or not visible during surgery. Secondly, the ventricular system is a complex tridimensional structure, occasionally asymmetrical, that can be distorted due to acute or subacute events. Therefore, minute changes in the location of the burr hole or the trajectory of the catheter can result easily in an off-target catheter. In order to minimize such unfortunate outcomes, neurosurgeons can either choose "easier" targets or take advantage of existing technology such as stereotactic navigation.

17.3.1 Option Sites for Ventricular Placement

Ventricular catheters can be inserted in the frontal, parietal or occipital regions. Several studies have shown that with non-stereotactic techniques (freehand technique), the rate of malpositioned catheters is significantly higher in the occipital region. In fact, ventricular catheters are correctly placed in 85% of parietal and 64% of frontal shunts (this difference is not statistically significant) and only in 42% for occipital shunts ($p < 0.01$) [41]. The difficulty with occipital catheter placement seems related with the location of the burr hole. Therefore, parietal and frontal catheters are more likely to be placed successfully in the targeted ventricle. As expected, the revision rate for occipital catheters is also much higher than for frontal placed catheters [42]. However, shunt survival appears to be related to the final destination of the catheter tip in relation to the choroid plexus [43]. Also attention should be made to place the valve subcutaneously, away from the skin incision to avoid wound breakage and possible hardware exposure (Fig. 17.4a, b).

17.3.2 Techniques for Accurate Ventricular Catheter Placement

17.3.2.1 Intraoperative Image-Guided Shunt Placement
Traditional anatomical landmarks are generic guides that are often misleading since they do not take into consideration the specific anatomy of patients. The utilization of image-guided shunt placement (navigational guidance) has the advantage of using specific preoperative patient's data. This technology provides high accuracy and allows first-pass ventricular catheterization even in slit ventricles [44]. Moreover, since image guidance shunt placement reduces poor shunt placement, it decreases significantly shunt revision rate [45].

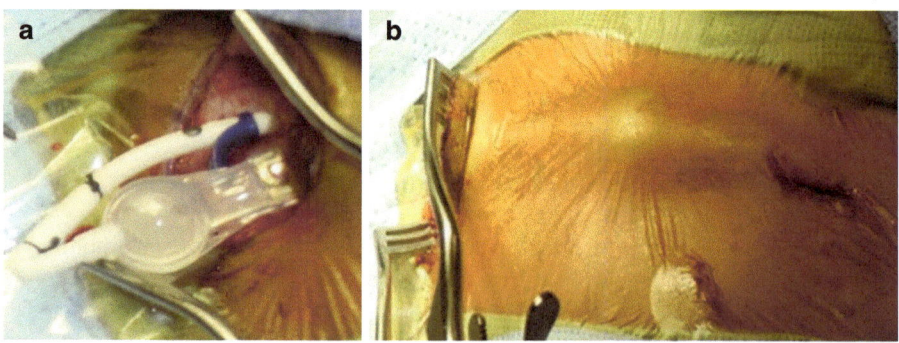

Fig. 17.4 (**a, b**) Subcutaneous placement of the valve

17.3.2.2   Endoscopic-Assisted Shunt Placement

Endoscopic-assisted shunt placement provides the operator the ability to position the catheter in a precise ventricular area that might be less prone for obstruction (for instance, choroid plexus). Compared to conventional techniques, it provides increased positioning accuracy [46, 47].

17.4   Techniques for Optimal Placement of Distal Catheter

17.4.1   Open Technique

The open technique, commonly used during CSF shunt insertion, is a simple approach in children or slim patients but can sometimes be more challenging in obese or patients who underwent previously several abdominal surgeries. Special attention should be paid in identifying clearly the pre-peritoneal space and the peritoneal cavity. In cases where there is doubt, it is always a wise decision to ask a colleague general surgeon to help at this stage of surgery. Once inserted, the catheter should be fixed to the abdominal wall usually with a purse-string suture using the posterior sheath of the rectus abdominis muscle/transverse abdominis muscle. The purse-string suture should be tight enough to prevent movement of the catheter but not to strangulate and reduce the lumen of the distal catheter.

17.4.2   Laparoscopic Insertion

Laparoscopic-assisted CSF shunt placement is associated with a low incidence of distal catheter malfunction. In fact, laparoscopic placement of the distal shunt into the peritoneal cavity allows the surgeon to cut adhesions when necessary, to choose the optimal location of the catheter, and, finally, to demonstrate the patency of the shunt by direct visualization of CSF flowing spontaneously at the distal tip. Moreover, it reduces the risk of visceral complications, CSF pseudocysts, or extra-peritoneal placement of the catheter [48, 49].

17.5   Minimizing the Rate of Hardware Malfunction and Revision

17.5.1   Valve Choice and Placement

So far, there have been no prospective randomized studies that were able to demonstrate a clear advantage for any specific shunt component, mechanism, or valve design over another. Programmable or non-programmable and flow- or pressure-regulated valves have the same survival, malfunction, and infection rates [50].

17.5.2 Avoiding Disconnection and Hardware Breakage

Disconnection and breakage of catheters in the vicinity of valve or connectors is a common finding among adolescent patients who were shunted in early age. This complication results most probably from catheter tension during the physical development in adolescence and also from repetitive neck movements. Therefore, during the primary insertion of a CSF shunt, the operator should secure tightly all connectors with nonabsorbable sutures but also avoid tension between the anchoring points that are usually located at the level of the abdominal wall and the retro-mastoid region where the sternocleidomastoid, trapezius, and splenius capitis muscles have their attachments.

17.5.3 Avoiding Proximal and Distal Migration (Fig. 17.5a, b)

Distal catheter movement or migration can be described as proximal retraction of the distal catheter outside of the peritoneal cavity. This finding is usually the result of physical growth during adolescence in a patient where a suboptimal catheter length was inserted in the peritoneal cavity. When inserting a VP shunt in a neonate, the operator should anticipate the patient physical development and insert sufficient catheter length to compensate with the abdominal and torso development. Usually the insertion of 90 cm of distal catheter within the peritoneal cavity is sufficient and well tolerated even in neonates.

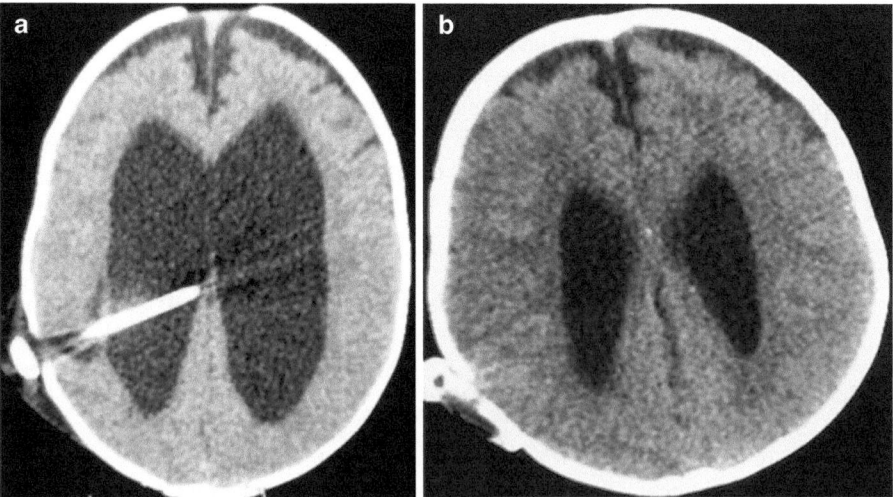

Fig. 17.5 (**a**) Showing immediate postoperative placement of a right-sided parietal catheter and catheter. (**b**) Showing movement of the catheter after several months

Proximal catheter movement happens less frequently and is more likely to occur in young patients (<6 months) and with a thin cortical mantle [51]. Although an anterior shunt entry (frontal point) is associated with a higher accuracy rate placement, it is unfortunately associated a higher incidence of shortening most probably due to greater displacement of the burr hole adjacent to the coronal suture. Therefore, in the neonatal population, a parieto-occipital route might be considered in order to provide a longer shunt function [52, 53].

Conclusion
Several important measures should be taken in order to avoid shunt complications. These measures should start with proper patient selection (correcting the cause of hydrocephalus rather than inserting a shunt whenever feasible), peri-operative skin care (reducing the bacterial count), minimizing contact with the CSF shunt hardware (avoiding shunt contamination), and the respect of basic surgical principles (anatomical reconstruction and skin closure concepts).

References

1. Nulsen FE, Spitz EB. Treatment of hydrocephalus by direct shunt from ventricle to jugular vein. Surg Forum. 1952;2:399–403.
2. Drake JM, Kestle JRW, Milner R. Randomized trial of cerebrospinal fluid shunt valve design in pediatric hydrocephalus. Neurosurgery. 1998;43:294–305.
3. Drake JM, Kulkarni AV. Cerebrospinal fluid shunt infections. Neurosurg Q. 1993;3:283–94.
4. Walters BC, Hoffman HJ, Hendrick EB, et al. Cerebrospinal fluid shunt infection. Influences on initial management and subsequent outcome. J Neurosurg. 1984;60:1014–21.
5. Ammirati M, Raimondi AJ. Cerebrospinal fluid shunt infections in children. A study on the relationship between the etiology of the hydrocephalus, age at the time of shunt placement, and infection rate. Childs Nerv Syst. 1987;3:106–9.
6. Serlo W, Fernell E, Heikkinen E, et al. Functions and complications of shunts in different etiologies of childhood hydrocephalus. Childs Nerv Syst. 1990;6:92–4.
7. Renier D, Lacombe J, Pierre-Kahn A, et al. Factors causing acute shunt infection. Computer analysis of 1174 operations. J Neurosurg. 1984;61:1072–8.
8. Kestle JRW, Hoffman HJ, Soloniuk D, et al. A concerted effort to prevent shunt infection. Childs Nerv Syst. 1993;9:163–5.
9. Kontny U, Hofling B, Gutjahr P, et al. CSF shunt infections in children. Infection. 1993;21:89–92.
10. Welch K. Residual shunt infection in a program aimed at its prevention. Z Kinderchir. 1979;28:374–7.
11. Abhaya AV, Drake JM, et al. Cerebrospinal fluid shunt infection: a prospective study of risk factors. J Neurosurg. 2001;94:195–201.
12. Rekate HL. Chapter 7: Treatment of hydrocephalus. In: Albright AL, Pollack IF, Adelson PD, editors. Principles and practice of pediatric neurosurgery. New York: Thieme. p. 94–130.
13. Jindal A, Mahapatra AK. Correlation of ventricular size and transcranial Doppler findings before and after ventricular peritoneal shunt in patients with hydrocephalus: prospective study of 35 patients. J Neurol Neurosurg Psychiatry. 1998;65:269–71.
14. Rainov NG, Weise JB, Burkert W. Transcranial Doppler sonography in adult hydrocephalic patients. Neurosurg Rev. 2000;23:34–8.

15. McGirt MJ, Leveque JC, Wellons JC III, et al. Cerebrospinal fluid shunt survival and etiologies of failures: a seven-year institutional experience. Pediatr Neurosurg. 2002;36(5):248–55.
16. Shapiro S, Boaz J, et al. Origin of organisms infecting ventricular shunts. Neurosurgery. 1988;22:868–75.
17. Duhaime AC, Bonner K, McGowan KL, et al. Distribution of bacteria in the operating room environment and its relation to ventricular shunt infections: a prospective study. Childs Nerv Syst. 1991;7:211–4.
18. Caldarelli M, Di Rocco C, La Marca F. Shunt complications in the first postoperative year in children with meningomyelocele. Childs Nerv Syst. 1996;12(12):748–54.
19. Centers for Disease Control and Prevention. Guideline for Prevention of Surgical Site Infection. 1999. http://www.cdc.gov/ncidod/dhqp/pdf/guidelines/SSI.pdf.
20. Climo MW, Sepkowitz KA, Zuccotti G, et al. The effectiveness of daily bathing with chlorhexidine on the acquisition of methicillin-resistant Staphylococcus aureus, vancomycin-resistant Enterococcus, and healthcare-associated bloodstream infections: results of a quasi-experimental multicenter trial. Crit Care Med. 2009;37:1858–65.
21. Darouiche RO, Wall MJ, Itani KMF, et al. Chlorhexidine–alcohol versus povidone–iodine for surgical-site antisepsis. N Engl J Med. 2010;362:18–26.
22. Vernon MO, Hayden MK, et al. Chlorhexidine gluconate to cleanse patients in a medical intensive care unit: the effectiveness of source control to reduce the bioburden of vancomycin-resistant enterococci. Arch Intern Med. 2006;166(3):306–12.
23. Zywiel MG, Daley JA, Delanois RE, et al. Advance pre-operative chlorhexidine reduces the incidence of surgical site infections in knee arthroplasty. Int Orthop. 2011;35(7):1001–6.
24. Eiselt D. Presurgical skin preparation with a novel 2% chlorhexidine gluconate cloth reduces rates of surgical site infection in orthopaedic surgical patients. Orthop Nurs. 2009;28(3):141–5.
25. McLone DG. Care of the neonate with a myelomeningocele. Neurosurg Clin N Am. 1998;9(1):111–20.
26. Ratilal B, Costa J, Sampaio C. Antibiotic prophylaxis for surgical introduction of intracranial ventricular shunts. Cochrane Database Syst Rev. 2006;(3):CD005365.
27. Ratilal B, Costa J, Sampaio C. Antibiotic prophylaxis for surgical introduction of intracranial ventricular shunts: a systematic review. J Neurosurg Pediatr. 2008;1(1):48–56.
28. Tanner J, Woodings D, Moncaster K. Preoperative hair removal to reduce surgical site infection. Cochrane Database Syst Rev. 2006;(2):CD004122.
29. Horgan MA, Piatt JH Jr. Shaving of the scalp may increase the rate of infection in CSF shunt surgery. Pediatr Neurosurg. 1997;26:180–4.
30. Swenson BR, Hedrick TL, Metzger R, et al. Effects of preoperative skin preparation on post-operative wound infection rates: a prospective study of 3 skin preparation protocols. Infect Control Hosp Epidemiol. 2009;30(10):964–71.
31. Macias JH, Arreguin V, Munoz JM, et al. Chlorhexidine is a better antiseptic than povidone iodine and sodium hypochlorite because of its substantive effect. Am J Infect Control. 2013;41(7):634–7.
32. Fairclough JA, Johnson D, Mackie I. The prevention of wound contamination by skin organisms by the pre-operative application of an iodophor impregnated plastic adhesive drape. J Int Med Res. 1986;14(2):105–9.
33. Edminston CE Jr, Seabrook GR, Cambria RA, et al. Molecular epidemiology of microbial contamination in the operating room environment: is there a risk for infection? Surgery. 2005;138(4):573–9. discussion 579-82
34. Ragel BT, Browd SR, Schmidt RH. Surgical shunt infection: significant reduction when using intraventricular and systemic antibiotic agents. J Neurosurg. 2006;105(2):242–7.
35. Parker SL, Anderson WN, Lilienfield S, et al. Cerebrospinal shunt infection in patients receiving antibiotic-impregnated versus standard shunts. J Neurosurg Pediatr. 2011;8(3):259–65.
36. Klimo P Jr, Thompson CJ, Baird LC, et al. Pediatric hydrocephalus: systematic literature review and evidence-based guidelines. Part 7: antibiotic-impregnated shunt systems versus conventional shunts in children: a systematic review and meta-analysis. J Neurosurg Pediatr. 2014;14(Suppl 1):53–9.

37. Attenello FJ, Garces_Ambrossi GL, Zaidi HA, et al. Hospital costs associated with shunt infections in patients receiving antibiotic-impregnated shunt catheters versus standard shunt catheters. Neurosurgery. 2010;66(2):284–9. discussion 289
38. Kulkarni AV, Drake JM, Lamberti-Pasculli M. Cerebrospinal fluid shunt infection: a prospective study of risk factors. J Neurosurg. 2001;94:195–201.
39. Faillace WJ. A no-touch technique protocol to diminish cerebrospinal fluid shunt infection. Surg Neurol. 1995;43:344–50.
40. Alexander JW, Solomkin JS, Edwards MJ. Updated recommendations for control of surgical site infections. Ann Surg. 2011;253(6):1082–93.
41. Lind CR, Tsai AM, Law AJ, et al. Ventricular catheter placement accuracy in non-stereotactic shunt surgery for hydrocephalus. J Clin Neurosci. 2009;16(7):918–20.
42. Price S, Santarius T, Richards H, et al The accuracy of ventricular catheter placement: does it influence shunt revision rates? Annual Meeting of the Society for Research into Hydrocephalus and Spina Bifida Cambridge, UK. 30 August–2 September 2006.
43. Dickerman RD, McConarthy WJ, Morgan J, et al. Failure rate of frontal versus parietal approaches for proximal catheter placement in ventriculoperitoneal shunts: revisited. J Clin Neurosci. 2005;12(7):781–3.
44. Lind CR, Tsai AM, Law AJ, et al. Ventricular catheter trajectories from traditional shunt approaches: a morphometric study in adults with hydrocephalus. J Neurosurg. 2008;108(5):930–3.
45. Hayhurst C, Beems T, Jenkinson MD, et al. Effect of electromagnetic-navigated shunt placement on failure rates: a prospective multicenter study. J Neurosurg. 2010;113(6):1273–8.
46. Theodosopoulos PV, Abosch A, McDermott MW. Intraoperative fiber-optic endoscopy for ventricular catheter insertion. Can J Neurol Sci. 2001;28(1):56–60.
47. Wilson TJ, Stetler WR Jr, Al-Holou WN, Sullivan SE. Comparison of the accuracy of ventricular catheter placement using freehand placement, ultrasonic guidance, and stereotactic neuronavigation. J Neurosurg. 2013;119:66–70.
48. Sekula RF Jr, Marchan EM, Oh MY, et al. Laparoscopically assisted peritoneal shunt insertion for hydrocephalus. Br J Neurosurg. 2009;23(4):439–42.
49. Sosin M, Sofat S, Felbaum DR, et al. Laparoscopic-assisted peritoneal shunt insertion for ventriculoperitoneal and lumboperitoneal shunt placement: an institutional experience of 53 consecutive cases. Surg Laparosc Endosc Percutan Tech. 2015;25(3):235–7.
50. Baird LC, Mazzola CA, Auguste KI, et al. Pediatric hydrocephalus: systematic literature review and evidence-based guidelines. Part 5: effect of valve type on cerebrospinal fluid shunt efficacy. J Neurosurg Pediatr. 2014;14(Suppl 1):35–43.
51. Whitehead WE, Riva-Cambrin J, Wellons JC, et al. Factors associated with ventricular catheter movement and inaccurate catheter location: post hoc analysis of the hydrocephalus clinical research network ultrasound-guided shunt placement study. J Neurosurg Pediatr. 2014;14(2):173–8.
52. Kemp J, Flannery AM, Tamber MS, Duhaime AC. Pediatric hydrocephalus: systematic literature review and evidence-based guidelines. Part 9: effect of ventricular catheter entry point and position. J Neurosurg Pediatr. 2014;14(Suppl 1):72–6.
53. Nakahara K, Shimizu S, Utsuki S, et al. Shortening of ventricular shunt catheter associated with cranial growth: effect of the frontal and parieto-occipital access route on long-term shunt patency. Childs Nerv Syst. 2009;25(1):91–4.

Psychological Management of Hydrocephalic Patients and Its Parents

18

Lujain A.S. Ammar and Ahmed Ammar

18.1 Prenatal Diagnosis and Parental Responses

Baile et al. in 2000 composed SPIKES, a six step protocol for empathically and effectively delivering an undesirable diagnosis [1]. Previous chapters in this book have explained the likely outcomes of congenital hydrocephalus and the likelihood of survival and continual deficits over the lifetime of the patient. A study conducted by Garne and colleagues in 2010, explored the prevalence and outcomes of the condition in four European regions. Between 1996 and 2003, 87 cases were identified. 42 of the identified cases (48%) resulted in termination of pregnancy for fetal abnormality. Of the 41 live births, 14 infants (34%) died within the first year of life, most of them within the first week [4]. A study by Chaplin et al. in 2005 on the experiences of parents who received a prenatal diagnosis of spina bifida or hydrocephalus and decided to proceed with the pregnancy sheds light onto the particular psychological stresses that parents undergo when they receive such a diagnosis [2]. All participants in the study reported feelings of disbelief, distress, or sadness upon hearing the diagnosis. One participant described it as the "worst feeling in the world." The primary and immediate thoughts of the parents were concern for their child's future and dismay of having a "sick" child [3]. However, practitioners must be mindful of the fact that not all people will react in the same manner. When actually receiving the diagnosis, the reactions of parents may vary. Some are likely to reject the diagnosis in order to maintain hope and stave off the fear of the unknown. After receiving the diagnosis, people will cope in different ways and experience a

L.A.S. Ammar (✉)
Middlesex University, The Burroughs, London NW4 4BT, UK
e-mail: lujiiammar@gmail.com

A. Ammar, M.D., M.B.Ch.B., D.M.Sc.
Department of Neurosurgery, King Fahd University Hospital, Imam Abdulrahman Bin Faisal University, Al Khobar, Saudi Arabia
e-mail: ahmed@ahmedamma.com

© Springer International Publishing AG 2017
A. Ammar (ed.), *Hydrocephalus*, DOI 10.1007/978-3-319-61304-8_18

Table 18.1 Outline of what parents stated as their more positive or negative experiences of receiving the diagnosis as mentioned in Chaplin et al. [3]

Positive experiences	Negative experiences
Parents reported that it was particularly helpful when a clinician delivered the diagnosis "gently and nicely"	Some professionals were reported to express negative attitudes toward the fetus's condition and recommended termination
Practitioners who delivered the diagnosis calmly, clearly, and honestly, giving all the facts and naming the condition	Practitioners who refused to give diagnosis information, despite being perceived as having the expertise to do so
Calling attention to the possible positive outcomes of the condition and treatment possibilities	Parents reported professionals withholding information and not providing a clear picture of the condition
Referring parents onto specialist centers or clinicians	Not following up on referrals to other professionals or centers
	The use of negative and insensitive language

variety of emotions at different stages. Some parents reported that the most trying period of the pregnancy was immediately after receiving the diagnosis (Table 18.1). The rest of the pregnancy was characterized by feelings of concern, but also feelings of hope and strength. Other parents reported that the latter half of their pregnancy was the most stressful, using terms such as "traumatizing" and "debilitating." They reported feelings of isolation. Others still expressed their emotional states as fluctuating between confidence and distress. The most common emotion exhibited across the parents was acute shock. Stages of the pregnancy that caused the most distress and anxiety were the period of delay between being tested and receiving the diagnosis, when waiting to be referred to a specialist, and when trying to gain clear balanced information [3, 4].

According to Chaplin et al. [3], typically a parent is inspired to collect as much information as possible, in order to prepare for any foreseeable outcome. For these particular parents, understanding the "worst-case scenario" was perceived as more comforting than the uncertainty they would be facing otherwise [3]. Obviously, medical practitioners provide a valuable source of support and information at this stage. However, parents will also often turn to other sources. Some contacted local hydrocephalus organizations, counselors, social workers, and the media. The Internet is widely accessed as it provides a wealth of information and allows parents to guide their own search for information. For specific information, parents reported accessing literature written by specialist personnel.

18.2 Breaking the News

Providing parents with a full, accurate view of their fetus's diagnosis is vital in supporting their general psychological well-being and global functioning. As shown above, the manner in which a healthcare professional greatly impacts the way in which the parent perceives the fetus' condition, and their decision on whether to continue with the pregnancy or terminate. Table 18.2 lists practices and techniques

Table 18.2 details the possible good and bad practices professionals may perform while delivering a diagnosis (*Reproduced from Zheng G. 2011*)

Good practice	Bad practice
Take the parents into a private room	Delivering a diagnosis in a ward or crowded area is a violation of the patient's privacy
Before presenting the diagnosis, a practitioner should ask some open-ended questions, to gain an understanding of the parents' views their fetus' medical situation	Being blunt, and "straight to the point," is potentially extremely distressing to those receiving the diagnosis
Using comparisons or metaphors may assist in explaining the condition	Using medical jargon or terms the patient is unlikely to understand

that may assist in delivering bad news to patients, and some that are likely to cause more distress [5].

People tend to fixate on the diagnosis as soon as it is delivered, and so are unlikely to absorb any other information if the diagnosis is delivered bluntly, and at the very beginning of the conversation [5]. Because of this, and the general differences in understanding and medical knowledge in people, healthcare professionals should ensure that they check the mother's comprehension of the diagnosis and that she has absorbed enough information. As mentioned above, making a comparison or providing a metaphor may assist comprehension. Mothers should also be given the freedom to ask any questions they may have, and to discuss their concerns. If they become too emotional to continue with the conversation, offering them empathy and understanding is crucial. Once they sufficiently recover and ask for more information, then the practitioner may continue delivering the diagnosis [5].

A six-step protocol in delivering "bad news" to patients was formed by Dr. Baile and colleagues in 2000. This is commonly referred to as the SPIKES protocol, which incorporates a step-by-step guide to delivering a difficult diagnosis, as well as the management of the patient's stress.

18.2.1 S: Setting Up the Interview

1. Setting up a quiet and private space to talk to the mother is very important. Sometimes, it is not possible but steps should be taken to make sure the area is as private as possible.
2. Involving the mother's significant others while delivering the diagnosis can help them feel more supported. This could be the fetus's other parent or anyone else the mother would like.
3. Making a connection with the mother will help facilitate communication.
4. Before you begin, make sure the others are aware of any time constraints you may have, and if you may have to leave. If possible, ensure you won't be interrupted for the duration of the interview.

18.2.2 P: Assessing the Patient's Perception

Asking a parent why they think you conducted the tests you did, and what they think is going on, is a good way to get a clear idea of what the patient understands. That way you can tailor the explanation to match their level of understanding. It can also give you an idea if the patient is utilizing maladaptive processes such as denial and ignoring important facts or has unreasonable expectations of the outcome.

18.2.3 I: Obtaining the Patient's Invitation

Most parents will want full disclosure on the fetus's health and diagnosis. However, this cannot be assumed, and they should be asked if they want the results and how much they want to know. Rejecting negative information is a common, natural psychological defense mechanism, and as such, parents may not be mentally prepared to hear devastating news. If they do not want the information, offer to answer their questions in the future or whether you can ask if you can talk to a friend or family member of theirs, so they can get the information in the future.

18.2.4 K: Giving Knowledge and Information to Patients

1. Warning the parents that bad news is coming may help them process the coming information. Phrases like "Unfortunately …" or "I am sorry to tell you …" may serve this function.
2. Avoid using medical jargon, abbreviations, or terms the parents are unlikely to understand. You may use metaphors and comparisons to assist. A possible metaphor for hydrocephalus could be: "Imagine there is water in a bag. This bag has a limited amount of space, but it does have an opening to drain the water. Now imagine that opening got blocked, the water would fill the bag and would cause a lot of pressure, and compress anything else in the bag." Possible treatment methods such as shunts could be explained using the same imagery: "To decrease the pressure in the bag, we can create another opening for the water to drain out of."

18.2.5 E: Addressing Emotions and Empathic Responses

1. People all react to bad news in different ways. Some will react with tears, silence, anger, disbelief, or hysteria. Knowing how to accommodate all of these reactions is challenging.
2. If you cannot discern the emotion the patient is feeling, possibly because of a black expression on their face, asking open-ended questions about how they are

feeling or what they are thinking may help. Allow the parent to have time to express their feelings. Allow them to know you understand that the emotional reaction is a response to the diagnosis perhaps by saying something like "I know this isn't what you wanted to hear …."
3. Using body language to communicate your empathy is very important. Leaning forward in your chair, moving the chair closer to the mother, perhaps light physical contact such as touching the arm of the person you are speaking to, and maintaining eye contact all communicate empathy and sincerity.

18.2.6 S: Strategy and Summary

1. Having a clear plan of action will help regulate a parent's anxiety levels.
2. Again, it is important to gain consent before moving on to drafting a treatment plan.
3. Giving treatment options is not only a legal and ethical mandate, but will also make the parent feel as if you value their opinion in the matter. Summarizing all the key aspects of the conversation is vital, as you can once again assess the other party's understanding of what is happening.

18.3 Making the Decision on Termination of Pregnancy for Fetal Abnormality

Deciding whether or not to terminate a pregnancy due to fetal abnormality is an incredibly difficult and personal decision for parents to come to. Healthcare practitioners play a huge role in coming to the decision and should be mindful of this fact. Lyus et al. in 2014 explained that the care of patients carrying fetus' with abnormalities is particularly demanding due to the fact that clinicians and midwives need to be heedful of a range of psychological, emotional, and clinical issues the patient may be experiencing. The patients must receive nondirective and nonjudgmental support in order to make their decision [6].

Below is the outline on the influencing factors that a parent must take into account when making the decision on whether to continue with or terminate a pregnancy for fetal abnormality [3]:

A parent will consider:

- The value they place on human life
- The extent and severity of the disability or condition of the fetus, most specifically:
  - The potential quality of life of their child
  - Whether the infant will be able to interact with his/her environment
  - Whether there is any chance at all of survival
  - And any physical or intellectual deficits the child may have
- The opinions of the medical professionals the parents have interacted with

- The parents' beliefs in their own ability to cope. This encompasses:
  - The confidence they gained from their own skills, relationships, and resources
  - Their feelings of self-worth and self-confidence
  - Their religious beliefs and influence

The age and ethnicity of the parents, as well as the potential effects of the quality of the child's life may have on their existing family life, are also considerations that affect the decision [6].

18.4   Child Outcomes

If the parents decide to proceed with the pregnancy and deliver, the main concern at the time is the treatment of the infant. However, even if treatment is successful and the infant survives into childhood, other psychological concerns may arise. As explained in previous chapters, one third of infants with hydrocephalus will have neurological deficits and potentially disabilities. Lavigne and Faier-Routman in 1993 performed a meta-analysis to examine correlates of pediatric physical disability and adjustment in general life. They concluded that children with sensory or neurological disorders tend to have decreased self-esteem and have more trouble adjusting [10]. Children with hydrocephalus, particularly, seem to have an identifiable pattern of personality development. Reportedly, children with hydrocephalus are at increased risk for various problems relating to psychological adjustment, while their parents are at increased risk for psychological distress [4]. Having a chronic disability can have an impact of various facets of everyday life. There are the demands of the illness itself, such as self-care and hospitalizations, as well as the impact it has on academic performance, employment, family, and social life [7].

There is also the matter of facial abnormalities. If the child has a permanent facial abnormality as a result of the hydrocephalus, this can lead to more psychological adjustment issues. Facial appearance is a critical aspect of social interaction. Therefore, young people with facial abnormalities tend to have more adjustment problems [8]. Patients with congenital facial anomalies tend to present with chronic depression and social withdrawal, and anxiety. The study conducted by Lim et al. in 2010, however, showed no significant differences between the control group and those with congenital facial anomalies in respect to anxiety, depression, and quality of life. Nonetheless, they did find that a patient with congenital facial anomalies had fewer narcissistic tendencies and was more avoidant than the control group [8].

Furthermore, the social and economic status of the family of the patient, along with the healthcare available for them, has an impact of the child's outcome of hydrocephalus. Children with hydrocephalus living in countries without universal healthcare presented with worse child health outcomes, greater unmet healthcare needs, and poorer cognitive and social outcomes. They also found that even in countries with a universal healthcare system, such as Canada, families with the top 20% of income have significant advantages in terms of healthcare. Families that function poorly tend to have worse child outcomes than other families, and parents with higher education predict better child outcomes, both in terms of health and cognition [9].

18.5 Advice for Parents

Receiving a diagnosis that your child may have deficits may grow up with difficulties or worse is such a traumatizing and incredibly stressful event. Whatever decision you make regarding your family's future is yours alone, and you have the right to expect nonjudgmental treatment from your healthcare practitioners. If you would like to direct your own search for information regarding hydrocephalus, and its outcomes and treatments, we recommend the following:

- Stay away from sites such as Web MD, etc. Sites such as these will give you the bare minimum of information.
- If you want specific information look at published articles and books. These tend to be filled with facts and evidence.
- Look for charities and organizations that work with hydrocephalus. They may be able to provide information and support for you.
- Online forums and support groups for parents of children with hydrocephalus can be a great way to meet and swap stories with parents in the same situation as you.
- Seek support. It can feel overwhelming and very isolating to be in the position you are in. Seeking support from your friends, family, and professionals such as therapists can help you share your burden and not feel so alone.
- Don't be afraid to ask questions and to expect answers to your questions.
- Most importantly, remember that nothing you could have done could have prevented this.

References

1. Baile WF, Buckman R, Lenzi R, Glober G, Beale EA, Kudelka AP. SPIKES—a six-step protocol for delivering bad news: application to the patient with cancer. Oncologist. 2000;5:302–11.
2. Garne E, Loane M, Addor M, Boyd PA, Barisic I, Dolk H. Congenital hydrocephalus- prevalence, prenatal diagnosis & outcome of pregnancy in 4 european regions. Eur J Paediatr Neurol. 2010;14:150–5.
3. Chaplin J, Schweitzer R, Perkoulidis S. Experiences of prenatal diagnosis of spina bifida or hydrocephalus in parents who decide to continue with their pregnancy. J Genet Couns. 2005;14:151–62.
4. Donders J, Rourke BP, Canady AI. Emotional adjustment of children with hydrocephalus and their parents. J Child Neurol. 1992;7:365–80.
5. Zheng G. Delivering bad news to patients—the necessary evil. J Med Coll PLA. 2011;26:103–8.
6. Lyus R, Creed K, Fisher J, McKeon L. Termination of pregnancy for fetal abnormality. BJM. 2014;22:332–7.

7. Pit-ten Cate IM, Kennedy C, Stevenson J. Disability and quality of life in spinabifida and hydrocephalus. Dev Med Child Neurol. 2002;44:317–22.
8. Lim SY, et al. Concealment, depression and poor quality of life in patients with congenital facial anomalies. J Plast Reconstr Aesthet Surg. 2010;63:1982–9.
9. Kulkarni AV, Cochrane DD, McNeely D, Shams L. Medical, social and economic factors associated with health-related quality of life in canadian children with hydrocephalus. J Pediatr. 2008;153:689–95.
10. Lavigne JV, Faier-Routman J. Correlates of psychological adjustment to pediatric physical disorders: a meta analytic review and comparison with existing models. J Dev Behav Pediatr. 1993;14:117–23.

Follow-Up Strategies for Hydrocephalus Patients

19

Ta-Chih Tan and Ahmed Ammar

19.1 Introduction

The traditional classification of hydrocephalus (HC) as communicating (a-resorptive) and noncommunicating (obstructed) hydrocephalus does not stand anymore and has been strongly challenged. There are cases in which the cause of hydrocephalus is due to noncommunicating causes and a-resorptive causes as well, which is known as a complex hydrocephalus.

The causes of infantile communicating (a-resorptive) hydrocephalus (HC) are posthemorrhagic, post-infectious, and other causes. In noncommunicating (obstructed) hydrocephalus, in children age group, the causes are tumor or congenital aqueduct stenosis, Dandy-Walker syndrome (DWS), Chiari malformation, and some other congenital anomalies. However, categorizing the long-term challenges is facilitated by dividing the hydrocephalus into infant, childhood, adolescent, and adult hydrocephalus, as well as into different methods of management such as (a) shunts, ventriculoperitoneal shunt (VP shunt) and ventriculoatrial shunt (VA shunt), and (b) endoscopic third ventriculostomy (ETV) with or without Choroid Plexus Cauterization (CPC).

T.-C. Tan (✉)
Department of Pediatrics, Paediatric Neurosurgery, Helios-Kliniken Wiesbaden, Ludwig-Erhardstr 100, Wiesbaden 65199, Germany
e-mail: icttan@yahoo.co.uk

A. Ammar, M.D., M.B.Ch.B., D.M.Sc.
Department of Neurosurgery, King Fahd University Hospital, Imam Abdulrahman Bin Faisal University, Al Khobar, Saudi Arabia
e-mail: ahmed@ahmedammar.com

© Springer International Publishing AG 2017
A. Ammar (ed.), *Hydrocephalus*, DOI 10.1007/978-3-319-61304-8_19

In order to design an efficient follow-up plan during this age, certain facts have to be taken into consideration:

- During this period the growth of the head is the fastest in human's life, especially in preterm infants.
- The fontanelle is still present, providing the opportunity for ultrasound examination of the ventricles, avoiding unnecessary radiations.
- As long as the fontanelle is open and the sutures of the skull unclosed, raised intracranial pressure will be accommodated to certain degree and counterbalanced by protrusion of the fontanelle as well as the rapid progressive growth of the head circumference.
- The skin of the infant is still very fragile, thin, and vulnerable. The brain tissues are still very soft. Therefore, to avoid bedsores and pressure sores over shunt's valves and materials extra care should be taken considering the positioning of the infant, especially of the ears during OR. It is also recommendable to provide a fixation at the intracranial entry point of the intracranial catheter. Due to the softness of the brain and the relative rigidity of the catheter, the risk of dislocation is higher.
- Preterm infants present themselves more often with an underdeveloped omentum. CSF ascites may result as CSF accumulates in the small peritoneal cavity which occurs and, without being absorbed, will result in shunt malfunction. In such cases, VP shunt should not be used.
- Infants with posthemorrhagic and post-infectious hydrocephalus are at serious risk to develop loculated/multicystic hydrocephalus, which can complicate the therapy considerably.
- Preterm infants with a bodyweight below 1000 grams are more at risk for NEC when treated with a VP shunt. Therefore, it is preferable to use a CSF reservoir for reducing the intracranial pressure by tapping the CSF through punction. After reaching bodyweight of 1000 grams, when a the hydrocpehalus is not ameliorating, a permanent shunt can be considered.

19.2   Infant and Children Follow-Up Plans and Strategies

The neurosurgical team should be clear with what they should follow up and examine in every visit. They should know what to expect, what to maintain, and what to avoid. It is advisable that such plan should be made before discharging the patients and discussed with patient's parents. The plan should be designed to each individual case taking in consideration these conditions:

1. Patient's general condition such as age, full term, premature, associated anomalies, chromosomal abnormalities, and neurological condition
2. Causes and type of hydrocephalus
3. Method of treatment and type of shunt
4. The available and rapid access to the nearby neurosurgical unit. Legibility for treatment and follow-up and medical insurance
5. Parents and family education and social status

19.2.1 Neonate and Children Follow-Up Plan and Strategy

19.2.1.1 Infant/Children's Parents' Education

Parents, close relatives, and the patients, if possible, should be informed and taught how to observe and to detect in an early stage any unwanted signs and symptoms:

1. The parents should be informed if the cause of hydrocephalus is genetic. They should know about the sequelae and consequences of such finding and they should be examined genetically as well.
2. Shunt malfunction. They should be trained to test the valve and should be clearly instructed to avoid frequent and excessive pumping of the valve.
3. Signs of shunt infection or malfunction.
4. Avoid leaving the head in one position all the time to prevent posterior plagio-cephaly and secondary craniosynostosis. The latter can also be induced by chronic overdrainage. Avoid any pressure sores along the shunt system.
5. Observe any CSF collection along the shunt tube.
6. Observe possible abdominal distention.
7. Awareness about possible neurological and psychological disorders that may appear such as epilepsy or behavioral disorders.
8. They should be aware about the possible delays in some of milestones.
9. They should know the type and pressure of the valve.
10. They should be aware and agree about the follow-up plans and schedules.
11. They should have round-the-clock access to the neurosurgical team.
12. The parents and close relatives may be advised to go for counseling and psychological support.

19.2.1.2 General, Neurological, and Cognitive Function Development

Follow-up examination plan should be planned according to the order and frequency of each visit:

1. In the first visit postoperative, particular attention should be given to:
   (a) Wound healing. Signs of fresh wound gaping over the valve should be detected and dealt with immediately.
   (b) Subdural hygroma may be caused by overdrainage.
   (c) Headache, epilepsy, or any signs of neurological deficit.
   (d) Early signs of infections.
   (e) Early signs of shunt malfunction.
   (f) The response of the ventricles to the performed procedure (ventricle's volume)
2. In follow-up visits:
   The treating team should carefully carry on complete neurological examination and psychological considerations, and determine the pattern of the development of the child, which include:
   (a) Growth rate, especially of the head cinrcumference, and neurological abnormalities

(b) The milestones and cognitive functions and social, cognitive, and psychological development
3. The shunt:
    The inserted shunts should be examined in every visit to ensure:
    (a) Shunt's patency
    (b) Valve's function
    (c) No signs of skin inflammation along the tube or CSF collection around the tube and no shunt pressure sores over the valve
4. ETV, with or without CPC:
    Special attention should be given in every visit that:
    (a) Third ventricle fenestration (the stoma) is opened, and the CSF flow through the stoma is adequate [4].
    (b) Ventricular size and volumes [3, 4, 5].
    (c) No signs of intraventricular bleeding or ventriculitis.
    (d) The ICP is reduced; in some cases ICP may increase directly after ETV, due to pending adjustment but should be reduced not long after that. A closed external ventricular catheter can be left in place for emergency.
5. Radiological examinations:
    The treating team should use the available facilities in the hospital to secure proper follow-up and good patient care. The plan should be made to determine:
    (a) Frequency of ultrasound with additional RI measurement (Resistive Index)/CT scan/MRI/shunt survey
    (b) Indications for urgent CT scan and shunt survey
    (c) CT perfusion
    (d) MRI/DTI
    (e) Transcranial Doppler (TCD) to follow IP changes
6. ICP:
    ICP follow-up after insertion of the shunt or ETV is crucial. However, it is not easy. Attempts should be made to estimate the ICP in each visit and draw the pattern of ICP progress [6, 7]. TCD and fundus examination may be helpful.
7. Psychological and social services:
    The child and parents should be provided with social and psychological support and aid, which is a vital pillar in the development of the child [8–12].
8. Rehabilitation and physiotherapy:
    These services should be given as early as possible. Parents should learn how to perform certain exercises at home to provide regular therapy for their child to ensure optimal stimulation of the sensory, cognitive, and motor functions.
9. Plans for follow-up in different medical services:
    The attending neurosurgical team should be careful not to develop a narrowed view of the patient, discarding all other possible medical disciplines except concerning hydrocephalus and shunt. Vaccinations and regular pediatric follow-ups are necessary as well as any other necessary department, for example, urology, cardiology, and orthopedics.
10. Plans for the next OPD visits:
    It is strongly recommended that at the end of each outpatient clinical visit, the family and patient leave the clinic with a clear plan and schedule regarding their next visit.

19.2.2 Adolescence and Adult Follow-Up Plans

In general, the follow-up strategy should involve the family and close relatives of such patients. They should feel that they are part of the treating team.

1. Patient's education
   The patient should be educated, trained, and be aware of the following:
   (a) Early signs of shunt infection or malfunction.
   (b) The ETV procedure and possible complications and sequels of events.
   (c) Signs and symptoms of increased ICP.
   (d) The type and location of the shunt and the pressure of the valve.
   (e) What is hydrocephalus? What are the associated anomalies? What are the expected long-term outcomes and possible complications?
2. Motivation, encouragement, and inspiration
   The patient should be aware of the fact that a neither shunt nor ETV will hinder the chances to have a normal life. They should be motivated to be socially active and to reach for all possible personal goals like a good education, jobs, and family life and have an optimistic view for the future.
   Annual or periodic meeting with all hydrocephalus and myelomeningocele patients and their family members is a very valuable form of social and psychological support. The therapeutic team should inquire about the progress of adolescence in the educational establishments or the adult's job satisfaction and performance and family life. Support should be given by all means. Our experience with such meeting is very positive.
3. Follow-up plans
   The treating team should perform neurological examinations, psychological tests, and radiological examination to detect any alteration of the neurological or psychological status and ensure the patency of the shunt or third ventricular fenestration.
4. ICP and adjusting the valve pressure
   In cases of programmable valves and cases of NPH, checking the ICP and adjusting the valve's pressure are mandatory. Beware, however, of adjusting too often and/or too soon, especially after implantation of a new valve with a new resistance. This will reduce the adaptive ability of the brain and will enhance the risk of creating a stiffness of the brain with chronic therapy-resistant cephalgia as a result.

19.3 The Recommended Tools for Follow-Up Children and Adults

1. Head circumference. It gives the physician an inclination of the function of the shunt. One should be aware though that a "shunted" head tends to grow along a minor percentile after shunting. Remaining at the same percentile could even indicate a malfunction.
2. Observation of consistent development of the infant: arrested development, regression, and small "insignificant" signs like temporary strabismus can be signs of malfunction.

3. Ultrasound of the head in young children, when still possible, abdomen, and, if needed, sutures. Especially in young children, a regular and thorough follow-up can be provided, avoiding the elaborate preparations including sedation for an MRI or the radiation of a CT scan. The ventricular width and volume can be measured, as well as the amount of present intraperitoneal free fluid and the resistive index through TCD as an extra parameter for assessing raised intracranial pressure. The current data shows the usefulness to use it to detect any changes of ICP.
4. The ophthalmologic follow-up, to detect fundus changes and papilledema.
5. MRI, especially for infants with multicystic hydrocephalus, to provide a detailed follow-up, considering that the drainage of one cyst could lead to the "growth" of another cyst by reducing the local pressure.
6. CT scan is a useful examination method. However frequent and not clinically indicated CT scan should be avoided in general and especially in infants and children because of the risk of secondary malignancy.
7. Psychological and behavior examination.

The long-term management of hydrocephalus does not end with shunt insertion or ETV performance. In fact the surgical procedure, either insertion of a shunt or ETV, is the start of a long plan of management, which will continue as long as the patient lives. The follow-up of such patients is the cornerstone of proper and efficient management! It plays a very important role in shaping the whole life and future of such infants. A successful follow-up program definitely will help them to have an acceptable or even high quality of life. Therefore, the follow-up should expand ways beyond testing the shunt or the patency of the third ventricle fenestration. Strategy for the follow-up of each patient should be designed very individually according to the age, cause of hydrocephalus, methods of treatment, associated CNS anomalies such as spina bifida, associated other anomalies, complications, and predicting consequences of mental, cognitive function, and psychological development.

19.4   Frequently Asked Questions by the Parents

The treating team should be prepared to answer the questions of the relatives in a simple and comprehensible way, informing them with facts without scaring them or giving them false hopes.

The most frequently asked questions are:

1. Will my child be a healthy child?
2. Will he/she grow up normally?
3. Will he/she go to a regular school?
4. What are the complications of the hydrocephalus?
5. What are the complications of surgery?

6. What can we do to provide him/her with the best care?
7. What are the chances of getting hydrocephalic fetus on next pregnancy? Is there any way to prevent it?

19.5 Follow-Up Strategy

The aims of the follow-up:

1. To ensure that the signs and symptoms of hydrocephalus are treated
2. To provide the best possible services and facilities to allow the child to have the best possible quality of life
3. To detect any new problems which may appear as complications of surgery or consequences of the hydrocephalus

19.6 Common Complications of Hydrocephalus

The complications of hydrocephalus could be divided into:

1. *Complications related to the nature progressive of the disease*:
   (a) Delayed milestones.
   (b) Cerebral ischemia due to enlarged ventricles and paraventricular edema or due to the occlusion of the posterior cerebral artery due to tentorial herniation as consequences of uncontrolled increased transtentorial herniation.
   (c) Visual changes or blindness due to papilledema and optic nerve and chiasm damage.
   (d) The progressive of dilatation of third ventricle is very serious as it causes an alteration in level of consciousness, impairment of midbrain functions, compression of optic chiasm, and hormonal changes.
   (e) Empty sella syndrome as has been reported in cases of long-standing hydrocephalus [13].
   (f) Impairment or defect of some of the cognitive functions as memory.
   (g) Incontinence and gait disturbances, due to frontal lobe compression.
   (h) Behavioral disorders and alteration of cognitive functions.
   (i) Epilepsy.
2. *Complications related to genetic abnormalities and associated anomalies* such as spina bifida, DWS, and cerebral and corpus coliseum anomalies [14].
3. *Complications related the medical treatment* such as dehydration, allergies, renal failure, electrolyte imbalance, acidosis, abnormal liver functions, and others.
4. *Complications related to the surgical procedures and the inserted devices* such as infection or skin laceration and necrotized wounds along the shunt valve and tubes. Migration of the distal or proximal catheters are reported complications [13, 15–19].
5. *Complications related to ETV*.

19.7   Notes for Shunt Follow-Up

There is a general agreement that the ideal shunt does not exist yet. It is believed that once a shunt is inserted, this patient will be under your care for the rest of his life. VP shunt has the advantage of the possibility of deposing more length in the intraperitoneal space to compensate for the first years of growth. Often, enough for the first 6–7 years. Also, free fluid in the intraperitoneal space, displayed through ultrasound, can be easily used as a diagnostic sign of a well-functioning shunt [1, 2, 3, 18, 20]. VA shunt is preferred to be the first line of treatment in some centers. It does work properly and avoid the malresorptive problems in the peritoneal cavity. It may have low risk of thrombosis or emboli. Additionally, the distal part of the shunt will have to be revised more often than a distal peritoneal catheter.

19.8   Notes for ETV Follow-Up

Many neurosurgeons gain increasing confidence in using ETV as the first line of treatment for hydrocephalus. ETV was found to be effective and useful method of treatment in children, adolescences, and adults when the cause is an obstructive and noncommunicating hydrocephalus. There is the impression that the success rate of ETV is less optimal considering the high rate of reclosing of the stoma in the first year of an infant. However, there has been no evidence so far that shunts have an advantage over ETV. ETV can be the only method to deal with cases of a loculated/multicystic hydrocephalus by reducing the required proximal catheters through shunting of the individual cysts.

19.9   Hydrocephalus-Spina Bifida Multidisciplinary Pediatric Outpatient Clinics

Due to the diversity of the medical, neurological, and surgical problems, we strongly recommend to establish special clinic, which is called "Hydrocephalus-Spina Bifida Multidisciplinary Pediatric Outpatient Clinic."

19.9.1 Mission of the Clinic

To provide a comprehensive and efficient all-inclusive medical service in one visit

19.9.2 Vision

To draw short-term and long-term plans for the management of each case

To help patients, parents, and relatives to have convenient, professional, and productive visits as all the needed services from different disciplines will be available in just one visit

To give a chance to the different consultant to have time for professional and scientific discussion for each case and make the proper decisions without any delay

19.9.3 The Structure of the Clinics

The clinic is consistent of neurosurgeon/pediatric neurosurgeons, pediatric neurologist, genetic specialist/consultant, pediatrician, pediatric orthopedic, pediatric urologist (they are required in cases of spina bifida), psychologist, physiotherapist, trained nurses, social worker, and secretary. Such multidisciplinary clinics proved to be very efficient and will answer to most of the patient's needs. The patients and families truly appreciate that service. Patient satisfaction index of these clinics is remarkably high.

19.10 Prediction of the Outcome

The prediction of the outcome of the hydrocephalus patient in general is based on several factors, such as:

1. Chromosomal abnormalities and genetic profile of the patient
2. CNS associated anomalies and general associated anomalies
3. Efficacy of management and follow-ups
4. Family care

A considerable number of these patient's life and futures were seriously affected by the complications or errors of surgery or medical treatment.

In general, different surgical procedures produce no substantial differences in the outcome on the long haul as long as the treatment is successful. The point of entry of the ventricular catheter, either frontal or occipital, does not seem to affect the function of the shunt [21]. However, in premature and neonates, we prefer the frontal insertion of the shunt to avoid pressure sores. The types of valves in several studies had the same rate of success or failures [22]. ETV after the first year of life may have more or less equivalent results to shunt insertion [23].

19.11 Valve Type Identification

From time to time, a new patient appears in the OPD without a complete record on the detailed history of surgery and medical care. It is important to determine the proper type of the shunt, which can be obtained by a simple skull and valve X-ray.

Conclusion

Insertion of shunts or a successful ETV is not the final treatment for hydrocephalus. These procedures are only the initial treatment of lifelong management of hydrocephalus. These patients need comprehensive management for their neurological, surgical, and psychological condition. The aim is, and should always be, to provide all the facilities and expertise to allow these patients to have the highest quality of life possible. It is possible for some of these children to have a healthy development and a good quality of life with the proper care.

The parents can easily do the measurements of the head circumference at home, once every 2 weeks in the beginning up to only once every month when the condition is stable. Parents can also check the patency of the shunt tube.

The frequency of the ultrasound can vary between twice a month and once every 2 months, depending on the stability of the situation.

After the closing of the fontanelle, an ophthalmological and MRI examination can be carried out once or twice a year. In case of multiple cysts, we recommend a more frequent schedule, until the stability or regression of the cysts is established and confirmed.

It should be taken into consideration that shunt infection could be asymptomatic. The clinical presentations of a shunt infection in neonates include poor feeding; poor sucking; crying and irritability; vomiting; fever, either low-grade or high-grade fever; drowsiness; somnolence; and lethargy. Anterior fontanelle may be tense and bulging. The symptoms of older children, adolescents, and adults include blurring vision, headache, nausea and vomiting, and irritability. Distension and painful abdomen may occur as well. Also the development may slow down or even reverse and the achievements may deteriorate.

Giving the utmost care of such children is mandatory.

References

1. Colak A, Albright AL, Pollack IF. Follow-up of children with shunted hydrocephalus. Pediatr Neurosurg. 1997;27(4):208–10.
2. Lumenta CB, Skotarczak U. Long-term follow-up in 233 patients with congenital hydrocephalus. Childs Nerv Syst. 1995;11(3):173–5.
3. Sufianov AA, Sufianova GZ, Iakimov IA. Endoscopic third ventriculostomy in patients younger than 2 years: outcome analysis of 41 hydrocephalus cases. J Neurosurg Pediatr. 2010;5(4):392–401.
4. Lucic MA, Koprivsek K, Kozic D, Spero M, Spirovski M, Lucic S. Dynamic magnetic resonance imaging of endoscopic third ventriculostomy patency with differently acquired fast imaging with steady-state precession sequences. Bosn J Basic Med Sci. 2014;14(3):165–70.
5. St George E, Natarajan K, Sgouros S. Changes in ventricular volume in hydrocephalic children following successful endoscopic third ventriculostomy. Childs Nerv Syst. 2004;20(11–12):834–8.
6. Cinalli G, Spennato P, Ruggiero C, Aliberti F, Zerah M, Trischitta V, Cianciulli E, Maggi G. Intracranial pressure monitoring and lumbar puncture after endoscopic third ventriculostomy in children. Neurosurgery. 2006;58(1):126–36.

7. Czosnyka M, Pickard JD. Monitoring and interpretation of intracranial pressure. J Neurol Neurosurg Psychiatry. 2004;75(6):813–21.
8. Hommet C, Billard C, Gillet P, Barthez MA, Lourmiere JM, Santini JJ, de Toffol B, Corcia P, Autret A. Neuropsychologic and adaptive functioning in adolescents and young adults shunted for congenital hydrocephalus. J Child Neurol. 1999;14(3):144–50.
9. Iddon JL, Morgan DJ, Ahmed R, Loveday C, Sahakian BJ, Pickard JD. Memory and learning in young adults with hydrocephalus and spina bifida: specific cognitive profiles. Eur J Pediatr Surg. 2003;13(Suppl 1):S32–5.
10. Lacy M, Pyykkonen BA, Hunter SJ, Do T, Oliveira M, Austria E, Mottlow D, Larson E, Frim D. Intellectual functioning in children with early shunted posthemorrhagic hydrocephalus. Pediatr Neurosurg. 2008;44(5):376–81.
11. Larysz P, Larysz D, Mandera M. Radiological findings in relation to the neurodevelopmental outcome in hydrocephalic children treated with shunt insertion or endoscopic third ventriculostomy. Childs Nerv Syst. 2014;30(1):99–104.
12. Pelegrín Valero C, Tirapu Ustarroz J, Landa González N. Neuropsychological deficits in hydrocephalus associated spina bifida. Rev Neurol. 2001;32(5):489–97.
13. Ammar A, Sultan A, Mulhim F, Yousef A. Empty Sella syndrome does it exist in children? J Neurosurg. 1999;91:960–3.
14. Anderson EM. Cognitive deficits in children with spina bifida and hydrocephalus: a review of the literature. Br J Educ Psychol. 1973;43(3):257–68.
15. Ammar A, Ibrahim AWI, Nasser M, Rashid M. CSF hydrocele unusual complication of V-P shunt. Neurosurg Rev. 1991;14:141–3.
16. Ammar A, Nasser M. Longterm complications of buried valves. Neurosurg Rev. 1995;18:65–7.
17. Ammar A, Nasser M. Intraventricular migration of VP shunts. Neurosurg Rev. 1995;18:293–5.
18. Ammar A. Nasser M. Anaizi A. Farag M. Management of VP shunt complications. Proceeding of the 13th World Congress of Neurosurgical Surgeons, F6 19R038.507-516, 2005 June 19–24, Marrakesh, Morocco.
19. Hadzikarik N, Nasser M, Mashani A, Ammar A. CSF hydrothorax- VP shunt complication without displacement of a peritoneal catheter. Childs Nerv Syst. 2002;8:179–18.
20. Iglesias S, Ros B, Martín Á, Carrasco A, Segura M, Delgado A, Rius F, Arráez MÁ. Surgical outcome of the shunt: 15-year experience in a single institution. Childs Nerv Syst. 2016;32(12):2377–85.
21. Kemp J, Flannery AM, Tamber MS, Duhaime AC, Pediatric Hydrocephalus Systematic Review and Evidence-Based Guidelines Task Force. Pediatric hydrocephalus: systematic literature review and evidence-based guidelines. Part 9: effect of ventricular catheter entry point and position. J Neurosurg Pediatr. 2014;14(Suppl 1):72–6.
22. Baird LC, Mazzola CA, Auguste KI, Klimo P Jr, Flannery AM, Pediatric Hydrocephalus Systematic Review and Evidence-Based Guidelines Task Force. Pediatric hydrocephalus: systematic literature review and evidence-based guidelines. Part 5: effect of valve type on cerebrospinal fluid shunt efficacy. J Neurosurg Pediatr. 2014;14(Suppl 1):35–43.
23. Limbrick DD Jr, Baird LC, Klimo P Jr, Riva-Cambrin J, Flannery AM, Pediatric Hydrocephalus Systematic Review and Evidence-Based Guidelines Task Force. Pediatric hydrocephalus: systematic literature review and evidence-based guidelines. Part 4: cerebrospinal fluid shunt or endoscopic third ventriculostomy for the treatment of hydrocephalus in children. J Neurosurg Pediatr. 2014;14(Suppl 1):30–4.

Complex Hydrocephalus: Management by "Smart Shunt"

20

Nobuhito Morota

20.1 Introduction

Treatment of hydrocephalus is not necessarily straightforward. Despite the fact that surgical procedure for hydrocephalus is regarded as one of basic procedures in neurosurgery, some hydrocephalus stays intractable against surgical management. Multiloculated hydrocephalus, repeated shunt infection, slit ventricle syndromes, and some hydrocephalus with complicated anatomy due to associated brain anomalies can be tough to cure. Frequency of such complex hydrocephalus is not clear but estimated about 10–20% based on the recent report [1–3].

> "Complexö hydrocephalus implies those hydrocephalus which require multiple surgery for the treatment of hydrocephalus. However, its exact definition of what is complex is not clear enough. Several factors can be listed for "complexity".

1. Anatomical factor (e.g., multiloculated hydrocephalus, isolated fourth ventricle)
2. Pathological factor (e.g., post hemorrhagic hydrocephalus, post infectious hydrocephalus)
3. Mechanical factor (e.g., hydrocephalus associated with other intracranial lesions, normal pressure hydrocephalus)
4. Human factor (e.g., inexperienced surgeon, poor surgical tactics, insufficient protection against infection)

Electronic Supplementary Material The online version of this chapter (doi:10.1007/978-3-319-61304-8_20) contains supplementary material, which is available to authorized users.

N. Morota
Division of Neurosurgery, Tokyo Metropolitan Children's Medical Center,
2-8-29 Musashi-dai, Fuchu, Tokyo 183-8561, Japan
e-mail: nobu.m01@gmail.com

© Springer International Publishing AG 2017
A. Ammar (ed.), *Hydrocephalus*, DOI 10.1007/978-3-319-61304-8_20

More than one factor can lurk in a single case of complex hydrocephalus. Posthemorrhagic or infectious hydrocephalus often develops multiloculated hydrocephalus, which can lead to complex, tough-for-treatment hydrocephalus. Gastroenteritis that associates with a very low birth weight baby makes treatment of posthemorrhagic hydrocephalus extremely difficult since ventriculoperitoneal (VP) shunt is precluded from the surgical choice.

Complex hydrocephalus often requires multiple surgeries. However, hydrocephalus which needs multiple surgeries does not necessarily means they are a complex one. Surgical technique, choice of surgical procedures, shunt infection rate, and associated general condition of the patient, all influence the total number of surgery. It should be reminded that not a small number of complex hydrocephalus could be a result of failed or not properly managed hydrocephalus. Simple hydrocephalus does have potential to turn into complex one if anything unfavorable develops or being handled poorly in the clinical course. On the other hand, potentially complex hydrocephalus can be controlled by a single procedure if treated appropriately.

In this chapter, the author presents personal experience of treating complex hydrocephalus.

The author has witnessed so many cases of poorly handled simple hydrocephalus which turned to complicated one before referral. Figure 20.1 demonstrates one of such cases with three shunts left but any of them not functioning properly.

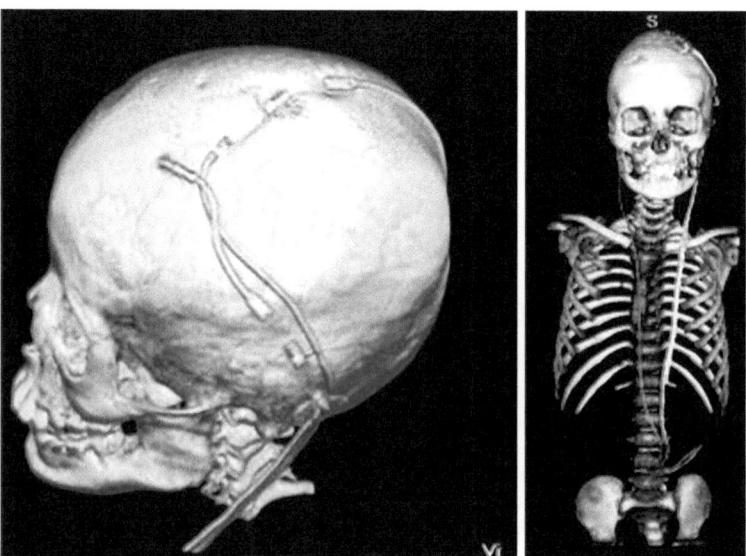

Fig. 20.1 Example of poorly managed hydrocephalus. The child received three shunts but none of them functioning properly

The author defines "complex hydrocephalus" as those which or which may require multiple surgeries for the treatment of hydrocephalus when treated traditional patterns. Basic concept of handling hydrocephalus in the author's hands is "the simpler, the better." Whatever the anatomy or background pathogenesis is complex, the simple surgical procedure often brings better surgical outcome.

20.2 Subjects

The author treated 434 patients with variety of hydrocephalus between April 2002 and March 2015. Age of the patients at the first treatment ranged from 1 day to 26 years old, but more than half of them (236 patients (53%)) were aged less than a year old. The total number of surgical procedures applied for the treatment was counted 1007.

There were 360 children who were newly diagnosed and treated from the beginning. One hundred eighty children had a single surgery, 26 had more than 5 surgeries, and five had more than 10 surgeries. Endoscopic surgery tended to control hydrocephalus after a single procedure. On the other hand, there were 74 children who were referred to my institute after several failed treatment of hydrocephalus at other hospitals. Despite possibly complicated hydrocephalus, 37 children received only one surgery, most of it was a single shunt revision through an elaborated surgical approach. Nine had more than five surgeries, and only one had more than ten.

Illustrative cases were from the abovementioned patients.

20.3 Classification and Treatment Policy for Complex Hydrocephalus

Classification of complex hydrocephalus is not straightforward. Several factors interfere or synergy to form complex hydrocephalus.

The author classified complex hydrocephalus simply as primary and secondary one. The primary complex hydrocephalus means complexity is inherent in hydrocephalus due to one or several factors. Original complexity can be enhanced further by human factors. The secondary complex hydrocephalus means hydrocephalus is originally simple one but turned complex during the clinical course following poor management or complications due to the human factor.

In the following paragraph, the author describes several selected cases of complex hydrocephalus and how to recover or escape from chain of complications. What is important is to sense and predict the possible complexity of hydrocephalus prior to surgery and to avoid complications. Surgical procedures should be aimed as simple as possible. "The simpler, the better" is the key concept for the treatment of hydrocephalus. When shunting the CSF, make the shunt system as simple as possible. The author prefers to call such shunt system as "smart shunt." Smart shunt can be achieved with the assistance of neuroendoscopic procedures.

20.4   Illustrative Cases

The author demonstrates representative cases of complex hydrocephalus and how they were treated. The goal of surgery is, after all, the control of hydrocephalus. It is also important to stabilize hydrocephalus with fewer postoperative complications, including shunt failure. In that purpose, simple surgical solution seems better than complex procedures.

20.4.1 Subdural Hematoma Following a VP Shunt

Case 1
A 1-year-old girl, hydrocephalus associated with myelomeningocele. The girl underwent a VP shunt after repair of myelomeningocele (Fig. 20.2a). A CT scan taken 3 months later revealed bilateral subdural fluid collection (hematoma) (Fig. 20.2b). Conservative observation showed increase of hematoma before her referral.

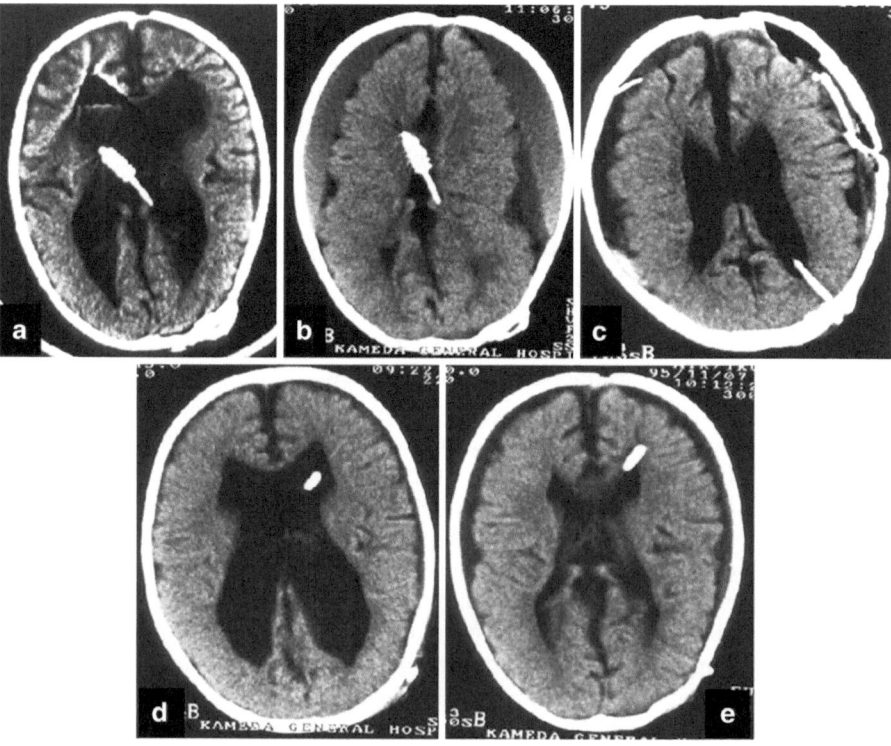

Fig. 20.2 Subdural hematoma following a VP shunt. The VP shunt peritoneal catheter was ligated and a bilateral SP shunt was placed. After reduction of subdural space, a new VP shunt with a programmable valve was installed

Classification: Secondary complex hydrocephalus.
Treatment:

1. Bilateral subduro-peritoneal (SP) shunt using a three-way connector was installed. The VP shunt peritoneal catheter was closed by ligation (Fig. 20.2c).
2. After confirming no subdural hematoma and the ventricle enlarged in size (Fig 20.2d), a new VP shunt was installed. A programmable valve was used for shunt revision through the anterior approach. The peritoneal catheter of the SP shunt was alternatively used for the new VP shunt. The remaining SP and old VP shunt systems were removed.

Outcome: The patient has remained stable with no shunt revision since then.

Case 2
A 4-month-old girl with congenital hydrocephalus. The baby had intrauterine diagnosis of congenital hydrocephalus and had a VP shunt at other hospital. After the shunt, the cerebral cortex sank by subdural fluid collection (hematoma). Her parents were told that the condition was unable to treat. They came to me for seeking if there is any possible treatment left for her.
Classification: Secondary complex hydrocephalus.
Treatment:

1. A burr hole was opened on the subdural hematoma.
2. Physiological saline was infused from the shunt valve so that the collapsed brain mantle can expand again.
3. The subdural hematoma was drained from the burr hole as the brain re-expanded (Fig. 20.3b).
4. The shunt valve was changed to a new pressure programmable one, which was set at 15 cm H2O. A postoperative CT taken next day showed dramatic solution for the post-shunted subdural fluid collection (Fig. 20.3c).

Outcome: The patient remained stable though severe developmental delay was present.

20.4.2 Multiloculated Hydrocephalus

Case 3
A 6-month-old girl with hydrocephalus due to brain abscess and ventriculitis. The baby had been treated for MRSA meningitis. She developed a seizure and was referred after a CT which showed multiple MRSA abscesses and ventriculitis. Loculated and isolated bilateral ventricular compartments were also presented (Fig. 20.4a).
Classification: Primary complex hydrocephalus.

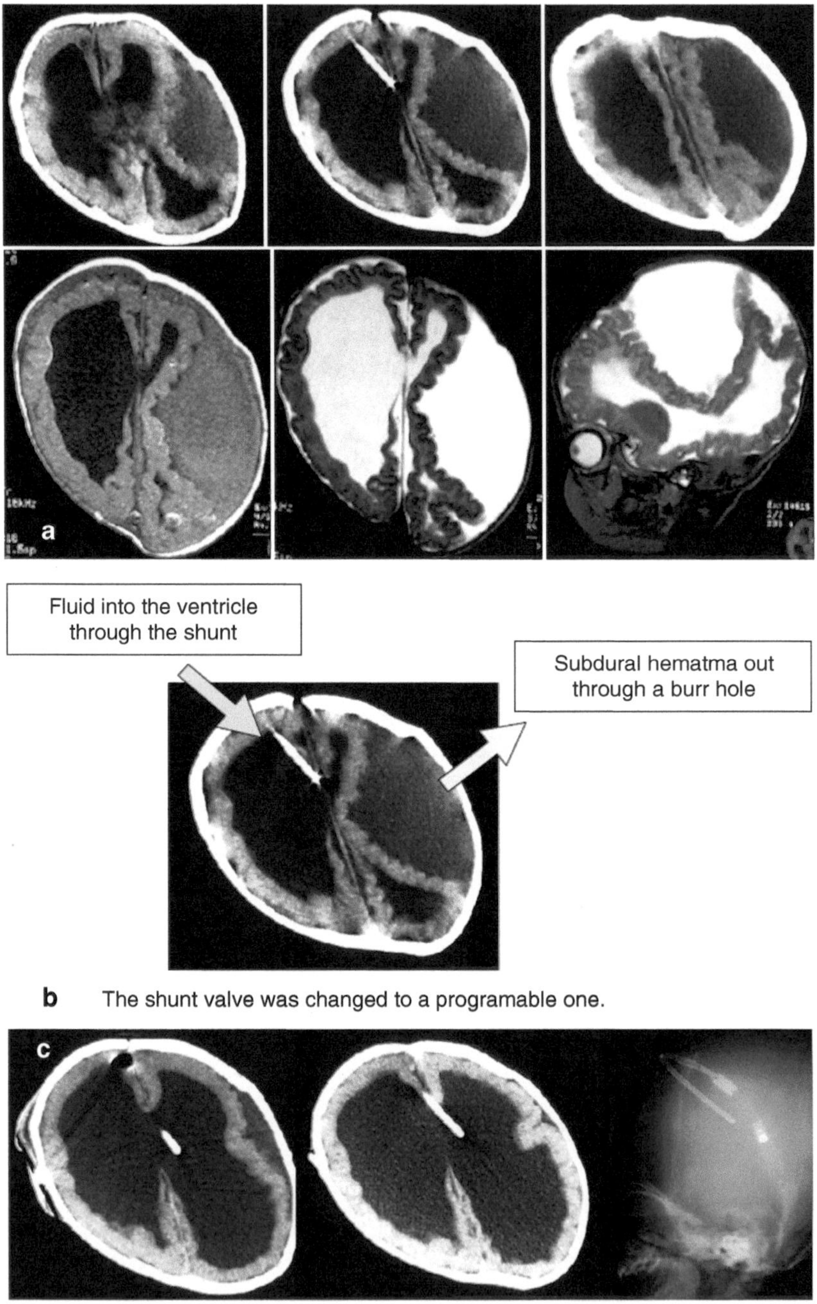

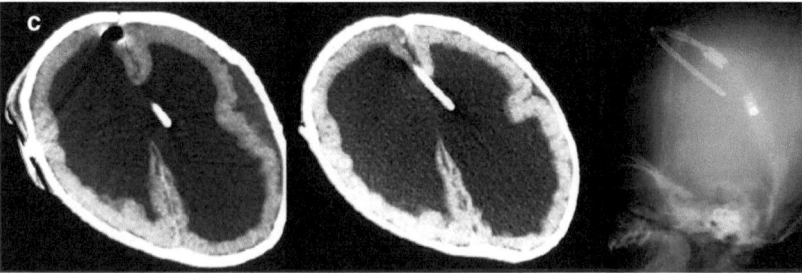

Fluid into the ventricle
through the shunt

Subdural hematma out
through a burr hole

b The shunt valve was changed to a programable one.

Fig. 20.3 (**a**) Subdural hematoma following a VP shunt. A thin cortex sunk after a VP shunt. A burr hole was opened on the subdural hematoma, and saline was infused through the shunt valve. (**b**) The cortex re-expanded and the subdural hematoma was drained. (**c**) The shunt valve was changed to a new one with programmable function that was set at high level to prevent re-accumulation of the subdural hematoma

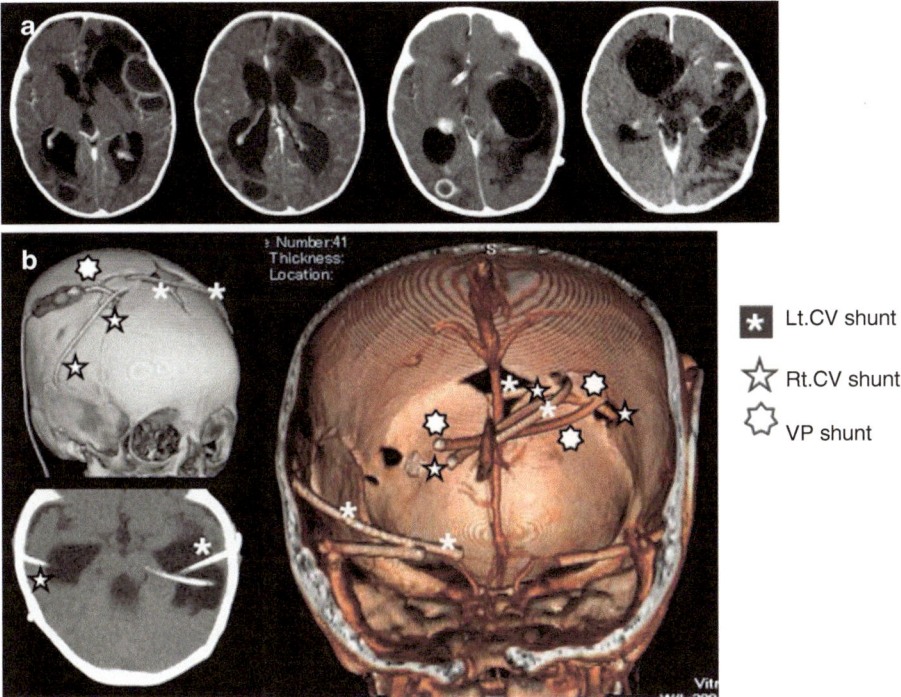

Lt.CV shunt

Rt.CV shunt

VP shunt

Fig. 20.4 Multiple abscesses and wide spread ventriculitis (**a**) were treated radically by surgery and direct injection of VCM into the ventricle. Multiloculated ventricles were partially connected by endoscopic fenestration. Finally, a single VP shunt system was installed with the use of two internal shunts (**b**) Removal and aspiration

Treatment:

1. Removal and aspiration of lt.frontotemporal abscesses. Endoscopic fenestration of the rt.anterior horn to the ventricle and external ventricular drainage (EVD).
2. Endoscopic fenestration of the trapped rt.temporal horn. Placement of a CSF reservoir to the trapped lt.temporal horn.
3. Removal of the occipital abscess. A cyst-ventricular internal shunt was placed between the lt.temporal horn and the ventricle body. EVD was changed to a new one.
4. Another cyst-ventricular internal shunt between the rt.temporal horn and the ventricle body.
5. VP shunt. Figure 20.4b shows the final condition after the shunt. Bilateral internal shunts between the temporal horn and ventricle body, and a single shunt system were installed. Total duration of treatment was three and a half months.

Outcome: The shunt remained stable without infection or malfunction for more than 10 years. The patient was severely delayed in neurodevelopment but happily with the family.

Case 4

A 2-year-old girl with posthemorrhagic hydrocephalus. The girl was born with very low birth weight and suffered intraventricular hemorrhage soon after the birth. She was treated at other hospital where she was born. The treatment was complicated by shunt infection and she received more than 20 surgeries before referral. When she was transferred, she had two shunts and two CSF reservoirs in each compartmented ventricle. A CT ventriculogram suggested that there were three compartments, one ventricle and two isolated cysts (Fig.20.5 upper).

Classification: Secondary complex hydrocephalus.
Treatment (Video 1):

1. The patient was placed in supine lateral position with the head rotated to the rt.side and was fixed in a Sugita head frame (Fig. 20.5 middle).
2. CT navigation was registered. A rigid endoscope was synchronized for navigation as a guiding instrument.

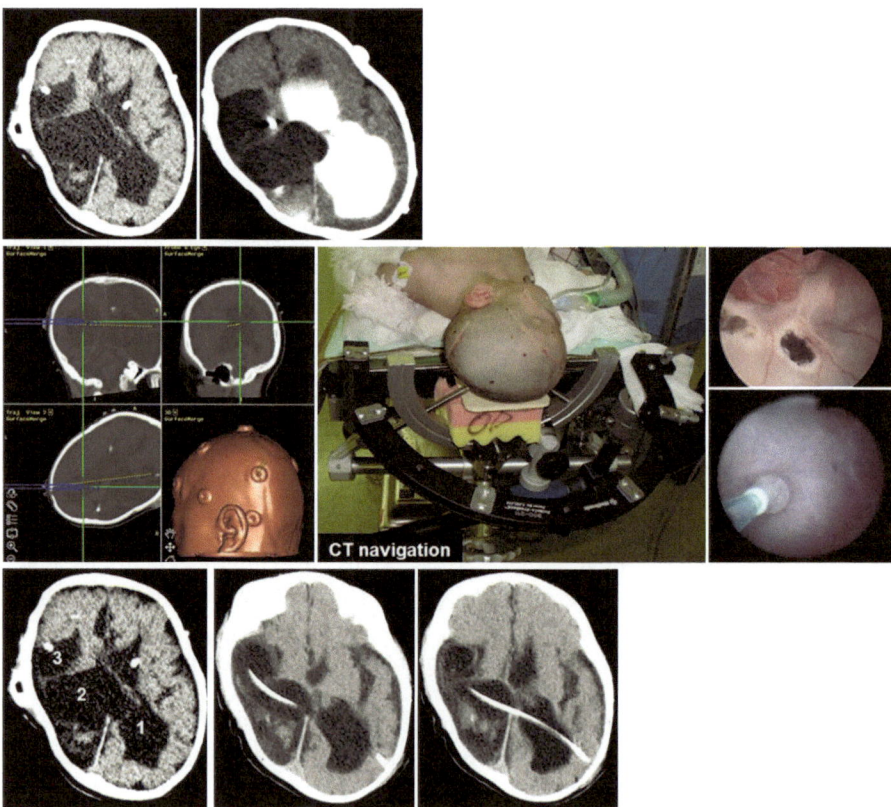

Fig. 20.5 A ventricular catheter with multiple side holes was passed from the enlarged ventricle to the cystic compartments following endoscopic fenestration under CT navigation. The catheter was then connected to a single VP shunt system

3. A burr hole was made and the first cyst (possibly, the rt.posterior horn) was fenestrated using a flexible endoscope.
4. A rigid endoscope was inserted and confirmed the second cyst (possibly, the rt.temporal horn) was located.
5. The flexible was again inserted and the wall of the second cyst was fenestrated.
6. A 12.5 Fr. endoscopic sheath was coaxially set to the rigid endoscope and was inserted to the second cyst. Then, the rigid endoscope was removed while the sheath was left as it was.
7. A ventricular catheter with multiple side holes up to 7 cm from the tip was passed through the sheath so that the tip was placed in the second cyst. The sheath was removed.
8. The ventricular catheter was connected to a new shunt valve (programmable one) and peritoneal catheter.
9. Previous shunts and CSF reservoirs were removed at the same time with the abovementioned procedure (single surgical setting).

Outcome: There was no complication after surgery, and a CT scan showed the ventricular catheter passing the all CSF cavities and connected them with the side holes (Fig. 20.5 lower). Several shunt revisions at the peritoneal side was required whenever she had abdominal surgeries. She also underwent functional posterior rhizotomy for severe spasticity 6 years later.

20.4.3 Hydrocephalus Associated with Intraventricular Cysts and Isolated Fourth Ventricle

Case 5
A 1-year-old boy whose congenital hydrocephalus is associated with multiple anomaly syndromes. The boy underwent a VP shunt soon after his birth at other hospital. Clinical course after the shunt was complicated with repeated shunt infection and resulted in multiple shunt surgeries when he was referred at the age of 3 months. After treatment of shunt infection, a new VP shunt with a programmable valve was installed. Since the fourth ventricle had been enlarged, a ventricular catheter was inserted to the fourth ventricle under endoscopic guidance. Recurrence of shunt infection was confirmed 3 months later.
Classification: Secondary complex hydrocephalus.
Treatment:

1. The infected shunt system was removed. At the same time, a CSF reservoir was placed instead of the shunt. In addition, a ventriculo-subgaleal shunt was made on the other side of the CSF reservoir.
2. After treatment of meningitis, a new shunt system with a programmable valve was installed.
3. One more shunt revision (peritoneal side) 2 months later, the patient became stable. However, follow-up CT scans revealed increased fourth ventricle and

cysts in the lateral ventricle. Ventriculography demonstrated cysts in the third and rt.lateral ventricles as well as isolated fourth ventricle (Fig. 20.6, upper and middle).

4. A small craniotomy was made on the rt.frontal bone adjacent to the anterior fontanelle (Video 2). The dura was opened and a flexible endoscope was inserted to the ventricle. Both cysts were fenestrated (Fig. 20.6, lower left). Then, the endoscope was advanced to the orifice of aqueduct, which had been obstructed by a membrane. It was penetrated by a biopsy forceps and enlarged by a 3 Fr. Fogarty catheter. A stylet was inserted in a ventricular catheter with multiple side holes. The catheter punctured the ventricle separately along the endoscope. The tip of the catheter was monitored under the endoscopic view and was guided to the aqueduct. The catheter passed into the fourth ventricle and the stylet was

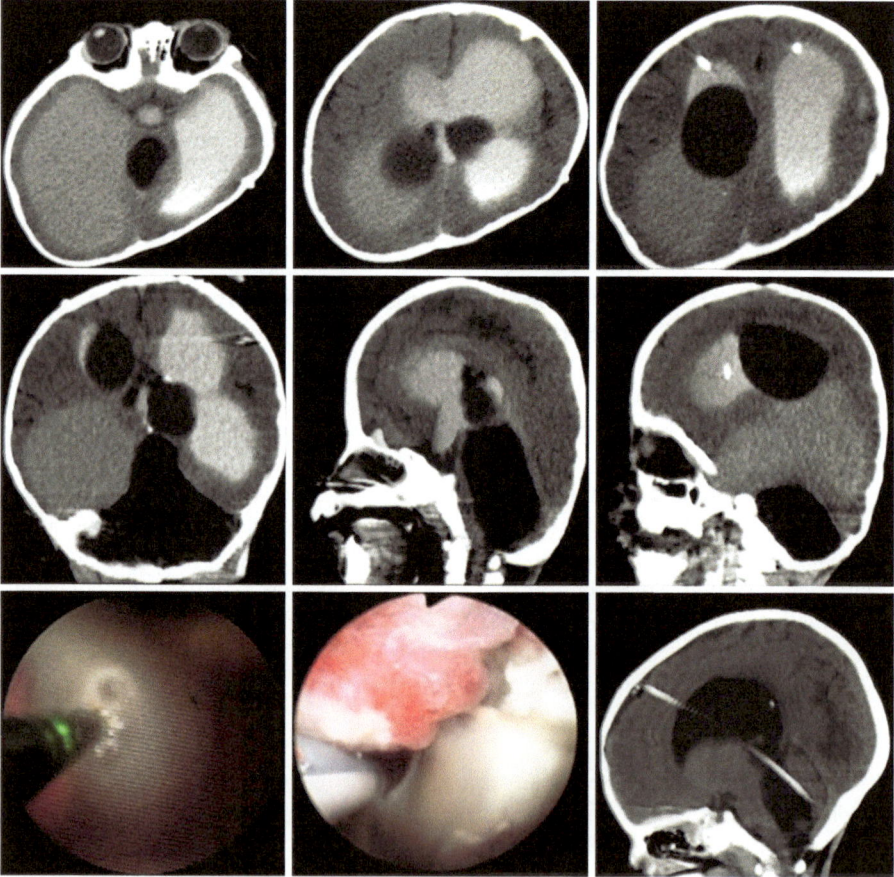

Fig. 20.6 Intraventricular cysts were endoscopically fenestrated first. Endoscopic aqueduct plasty was carried out, and a ventricular catheter was inserted into the fourth ventricle under direct observation. Since a VP shunt had already been installed on contralateral side, the catheter served as an internal shunt

removed (Fig. 20.6 lower center). The distal end of the catheter was ligated to the dura so that the catheter worked as an internal shunt between the fourth and lateral ventricles (Fig. 20.6 lower right).

Outcome: The fourth ventricle reduced in size after the surgery (Fig. 20.6 lower) and the respiratory function also improved. The shunt had been functioning until his death by a secondary complication due to facial anomaly a year later.

20.4.4 Hydrocephalus Associated with the CSF Overproduction

Case 6
A 5-month-old girl with congenital hydrocephalus associated with multiple anomaly syndrome (9p tetrasomy), post repair of congenital herniation of diaphragm.

The baby underwent a VP shunt when she was a month old and was discharged without any complication. However, she was readmitted 2 months later for ascites. Hyperplasia of the choroid plexus was suspected on CT (Fig. 20.7, upper left and center).

Classification: Primary complex hydrocephalus.
Treatment:

1. The peritoneal catheter was externalized from the chest wall and more than 500 ml CSF was drained daily (draining pressure: 20 cm H$_2$O).
2. Endoscopic choroid plexus coagulation (CPC) was performed 10 days later (Fig. 20.7 lower). The septum pellucidum was fenestrated to access the contralateral

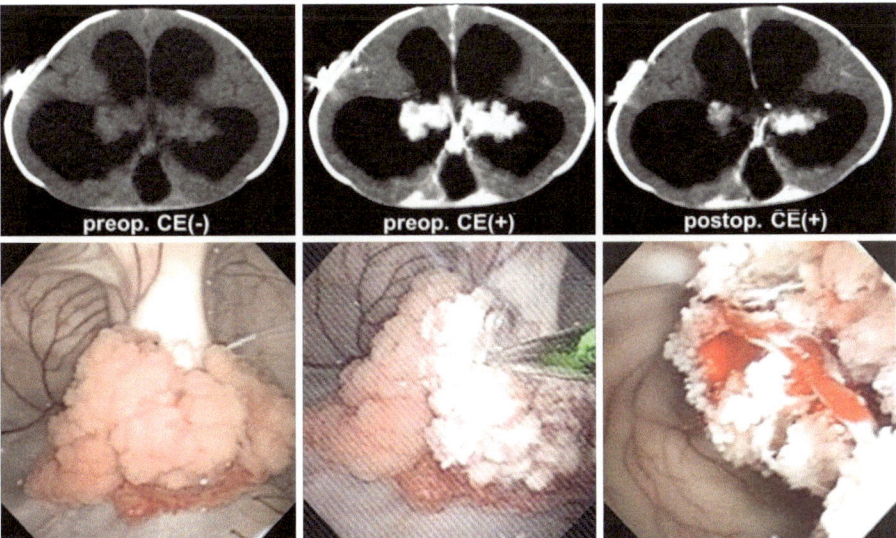

Fig. 20.7 Endoscopic coagulation of the choroid plexus is the choice of treatment before installing a VP shunt

choroid plexus. Postoperative-enhanced CT showed marked reduction of the enhancement of the choroid plexus (Fig. 20.7, upper right). Daily amount of the drained CSF reduced to around 200 ml/day soon after the surgery.

3. The externalized previous shunt system was removed and a VA shunt was newly installed a week later.

Outcome: The shunt remained stable; however, the baby had moderate neurodevelopmental delay.

20.4.5 Hydrocephalus Associated with an Intraventricular Tumor

Case 7

An 8-month-old girl with hydrocephalus due to third ventricular tumor. The previously healthy baby showed anorexia and was referred after a CT which revealed ventriculomegaly and a huge mass in the third ventricle. The anterior fontanelle was tense, suggestive of hydrocephalus. The MRI demonstrated the tumor locating in the hypothalamus to third ventricle up to the foramen of Monro (Fig. 20.8, left and center). Because of the tumor location, ETV was precluded for the treatment of hydrocephalus.

Classification: Primary complex hydrocephalus.
Treatment:

1. Endoscopic biopsy of the tumor was carried out from a burr hole placed on the right frontal.
2. After the biopsy, the baby was turned to prone position. A burr hole was made on the left parietal, and a ventricular catheter was inserted.

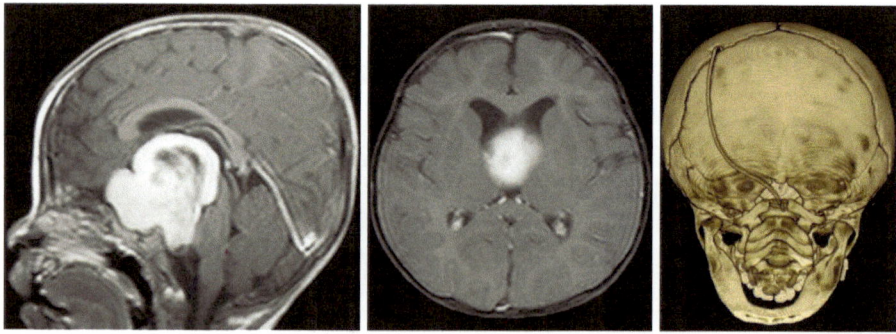

Fig. 20.8 Torkildsen shunt, an internal shunt which bridges the lateral ventricle to the cisterna magna, is applied for obstructive hydrocephalus when ETV is precluded from the indication

3. A small midline skin incision was made over the craniovertebral junction, and the midline portion of C1 cartilaginous lamina was removed. The ventricular catheter was passed through the muscle layer. Following dural incision at C1, the catheter tip was inserted in the cervical subarachnoid space about 1.5 cm in length. A Torkildsen shunt was established (Fig. 20.8, right).

Outcome: The anterior fontanelle became slack. The tumor was diagnosed as pilomyxoid astrocytoma and chemotherapy started.

20.4.6 Hydrocephalus in a Child Who Already Had a Major Abdominal Surgery

Case 8
A 3-month-old girl with hydrocephalus following intraventricular hemorrhage in neonate, postabdominal surgery (gastroenteritis). The baby was one of the twins who had undergone intrauterine treatment for twin-to-twin syndrome. She was born at 22 weeks' gestation with the body weight of 560 g, followed by intraventricular hemorrhage that formed intracerebral hematoma (Papile classification grade 4). Her postnatal course was complicated by severe necrotic enteritis which required surgical resection of necrotized intestinum. When hydrocephalus was diagnosed, VP shunt was precluded from the treatment option.
Classification: Primary complex hydrocephalus.
Treatment:

1. A CSF reservoir was placed on the right side and a ventriculo-subgaleal shunt was installed on the left side. The CSF was aspirated from the reservoir whenever the subgaleal pocket became tense.
2. Endoscopic third ventriculostomy (ETV) and bilateral CPC were performed when she was 6 months old. However, surgical outcome was suboptimal.
3. Retrograde ventriculo-SSS (superior sagittal sinus) shunt was added one and a half month later (Fig. 20.9, upper). The shunt was partially effective, but still aspiration of CSF was necessary from time to time because the subgaleal pocket became tense by accumulated CSF.
4. A ventriculoatrial (VA) shunt was installed 1 month after the ventriculo-SSS shunt. The right jugular vein was punctured under ultrasonographic guidance, and the atrial catheter was inserted by Seldinger procedure using a peel-off sheath (Fig. 20.9, lower). Total duration of treatment was 6 months.

Outcome: She had a shunt revision surgery (atrial side) 9 months later. Otherwise, the girl remained stable though multidisciplinary care at outpatient clinics was necessary.

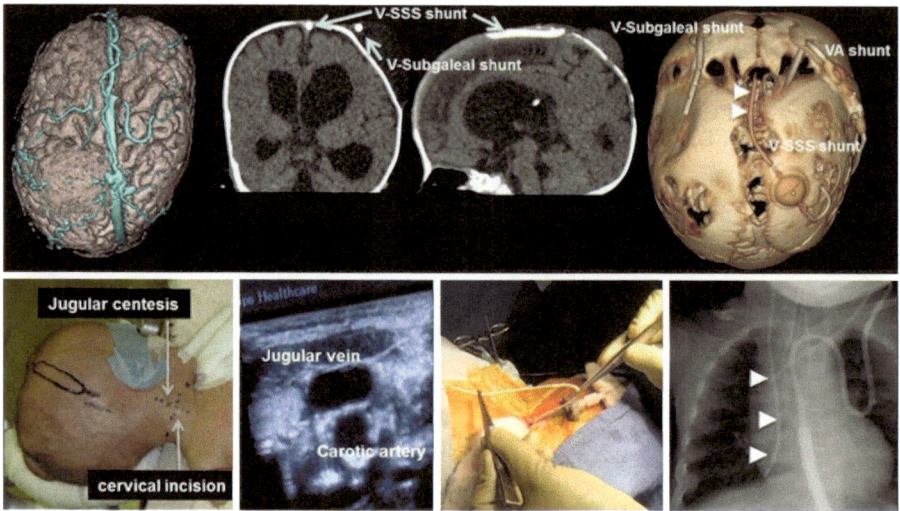

Fig. 20.9 When a VP shunt is impossible by abdominal problems, a VA or other CSF diversion procedures are considered for the alternative treatment for hydrocephalus. VA shunts are inserted to the jugular vein by a Seldinger method under ultrasonographic guidance. (*Upper right triangles*: catheter in the superior sagittal sinus. *Lower right triangles*: catheter in the *right* atrium)

20.4.7 Long-Lasting Meningitis due to Failed Management of Hydrocephalus

Case 9

A 10-month-old girl with hydrocephalus associated with myelomeningocele (MMC). The baby underwent repair of MMC at other hospital. Postoperative course was complicated by meningitis. An EVD was installed and a VP shunt followed after the CSF became clean. However, shunt infection recurred and an EVD was installed once more. Since then, she had received repeated EVDs. Finally, because hydrocephalus became more complex due to septation by long-lasting infection, she was referred for further treatment.

Classification: Secondary complex hydrocephalus.
Treatment:

1. Previous EVD which had subcutaneous tunnel only 10 cm from the wound and intermittent CSF leakage was removed, and a new one was placed with subcutaneous tunnel down to the chest wall. A CSF reservoir was inserted to the isolated right temporal horn which had been separated from the trigone by relatively thick adhesion (Fig. 20.10a, b upper row).
2. After sterilization of the CSF by intraventricular injection of vancomycin, a VP shunt was installed 2 weeks later. The ventricular catheter was intended to be inserted into the isolated right temporal horn using a flexible endoscope so that the shunt could be simplified. Color dye was injected into the temporal horn. It made the endoscopic localization of the temporal horn possible despite complex

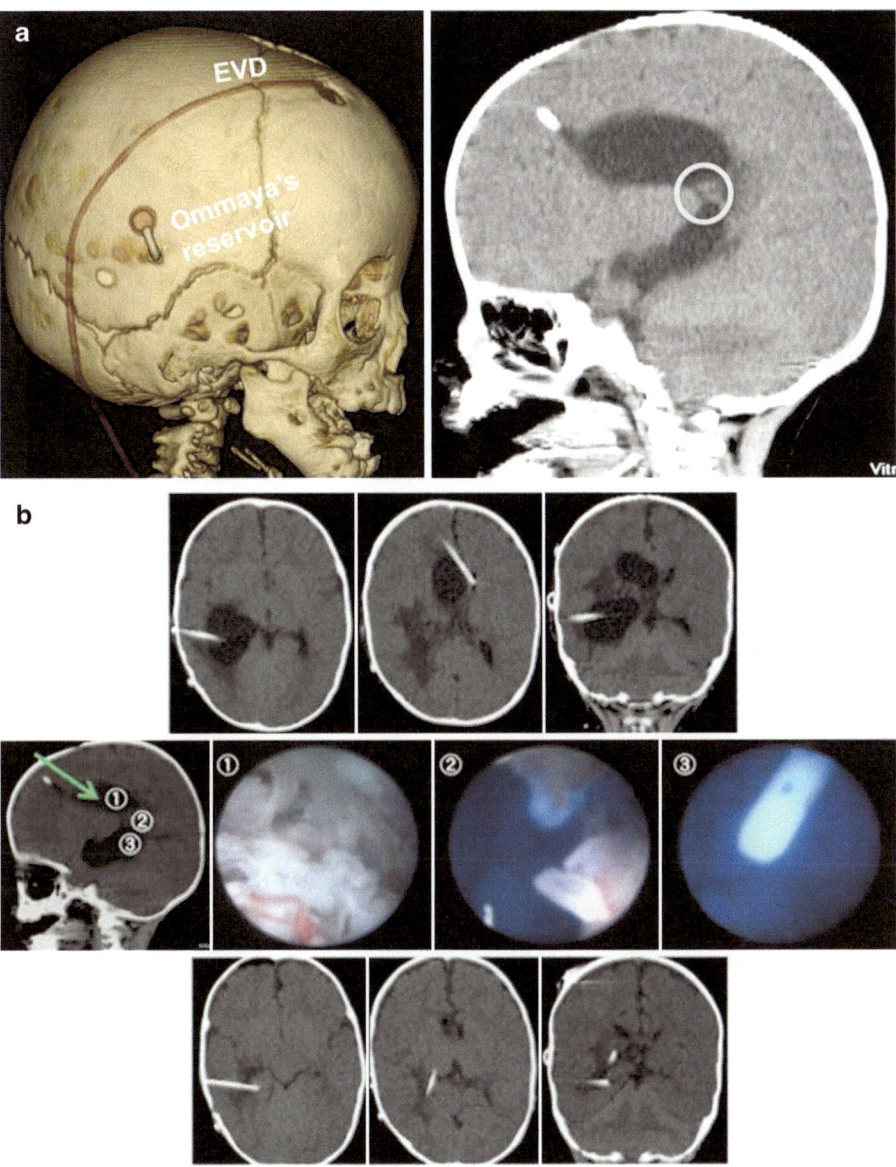

Fig. 20.10 (a) Poorly managed hydrocephalus with long-lasting meningitis. EVD was placed with a subcutaneous tunnel down to the right chest wall. A CSF reservoir is inserted to the right temporal horn which was enlarged by a thick septum (*circle*). (**b**) *Upper*: Meningitis was treated by EVD and a CSF reservoir. *Middle*: A VP shunt was installed. Color dye was injected through a CSF reservoir, first. Endoscopic fenestration and insertion of a ventricular catheter from the anterior to the temporal horn was carried out under the color dye guidance. (*1*) Viewed from the lateral ventricle to the septum. (*2*) The temporal horn filled with *blue* dye was observed through the fenestrated septum. (*3*) Inside the temporal horn, tip of the CSF reservoir was observed. *Lower*: CT scans after VP shunt. (**c**) The *left* lateral ventricle increased in size 2 weeks after the VP shunt. (**d**) An internal shunt with side holes was inserted stereotactically from the *right* frontal to pass the *right* anterior horn to the *left* anterior horn. (*1*) right temporal CSF reservoir. (*2*) Ventricular catheter of the VP shunt. (*3*) newly inserted internal shunt

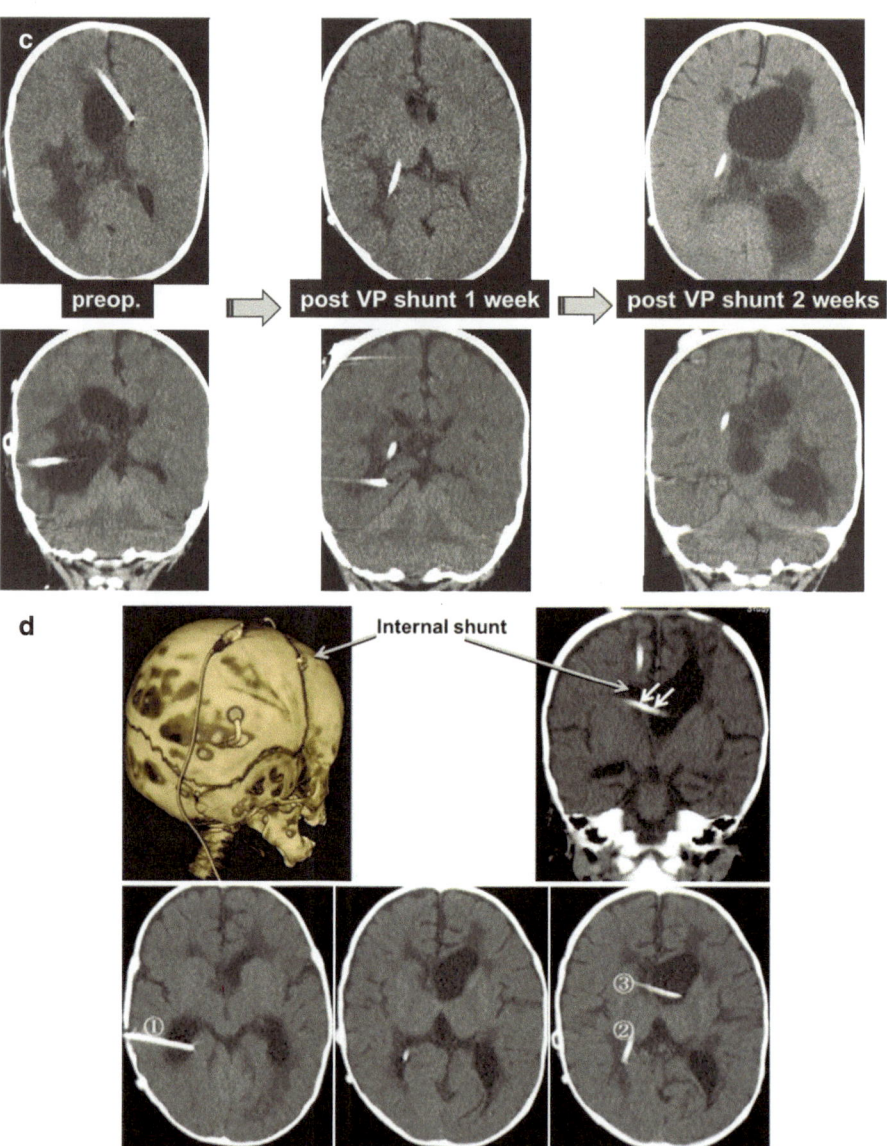

Fig. 20.10 (continued)

anatomy by adhesion (Fig. 20.10b-1) (Video 3). The blue-colored cystic cavity was fenestrated (Fig. 20.10b-2). The tip of the CSF reservoir was confirmed (Fig. 20.10b-3). Coaxially placed endoscopic sheath was slided into the cystic cavity of the temporal horn (Video 4). Then, a ventricular catheter with multiple side holes was passed into the sheath and inserted into the temporal horn while passing through the thick septum formed by adhesion. The catheter was connected to a shunt system with a programmable valve (Fig. 20.10b, lower).

3. Postoperative course was uneventful except when the left ventricle started enlarging (Fig. 20.10c). This time, surgery was done utilizing CT navigation because the right anterior horn became slit after the VP shunt. A 12.5 Fr. endoscopic sheath was used as a navigator. A small burr hole was opened on the right frontal bone, and the trajectory was decided to pass the right anterior horn to get into the left enlarged anterior horn. The sheath was placed in the left anterior horn, and a rigid endoscope was inserted to confirm that the sheath was there. A ventricular catheter with multiple side holes was inserted in turn and used as an internal shunt between the right and left anterior horn. The catheter was fixed on the periosteum of the burr hole edge (Fig. 20.10d). Total duration of the treatment was 1 and a half month.

Outcome:
The shunts have been functioning well, and she grows slowly with moderate neurodevelopmental delay.

20.5 Discussion

Complex hydrocephalus is often a product of poorly managed hydrocephalus. However, once hydrocephalus has turned into a complex one, an appropriate surgical strategy must be established prior to radical treatment.

20.5.1 Subdural Fluid Collection

Subdural fluid collection or hematoma after VP shunt has been regarded as the result of over drainage of CSF [4, 5]. Its incidence is estimated about 1–10%, despite the recent development of shunt valves [4, 6–9]. If the subdural fluid collection is large or symptomatic, and intractable for conservative treatment, further intervention is considered [5, 9].

Traditionally, subdural drainage or SP shunt is indicated for the treatment of subdural fluid collection like Case 1 [7, 10]. On the other hand, if the fluid collection is prominent, standard treatment often faced with difficulty. In such situation, it would be effective to infuse saline into the ventricle through a shunt valve while making a burr hole on the subdural fluid. The ventricle is easily re-expanded, and the elevated cortex pushes the subdural fluid out from the burr hole. Case 2 manifests immediate impact on treating the huge subdural fluid collection. The idea of saline infusion came from a punctured ball being re-expanded by air pump.

Recently, application of a programmable valve with antisiphon device is expected to reduce such a complication caused by over drainage of the CSF, though efficacy of programmable valve still remains controversial [11, 12]. It is also used as an alternative new shunt valve if the previous one was not the programmable or without antisiphon device.

20.5.2 Multiloculated Hydrocephalus

Most of complex hydrocephalus arise primarily or secondary to intraventricular hemorrhage or infection causing meningitis and/or ventriculitis. Multiloculated hydrocephalus is often the result of serious ventriculitis after hemorrhage or infection. It is composed of isolated ventricular compartments and/or periventricular cysts [1, 3].

Traditionally, it was treated by a cyst-peritoneal (CP) shunt from each fluid collection, which often ended up in multiple shunts [3]. Internal shunts which connect the fluid cavities by a single catheter, thus keep them in equal pressure, will help to make the shunt system simple. Case 3 is the good example to show how complex multiloculated hydrocephalus can be managed by a single VP shunt with the use of two internal shunts.

Recent advancement of neuroendoscopy enables to fenestrate each compartment, and the shunt could be simplified like Case 4 [13, 14]. The CT/MRI navigation system can help make fenestration on a deeply located cyst wall where endoscopic observation alone is not necessarily sufficient to identify the next cyst beneath the superficially located one [1, 3]. Without CT navigation, fenestration of the second cyst in Case 4 would have been more risky and tough to perform it.

20.5.3 Isolated Fourth Ventricle

Isolated fourth ventricle is the specific form of loculated ventricle [15, 16]. The most common pathology is the obstruction of the aqueduct due to hemorrhage or infection or the result of slit ventricle following over drained VP shunt [17]. It used to be treated by double shunts or double ventricular catheter connected to a single valve using a three-way connector with relatively high complication rate [3, 16].

Nowadays, endoscopic fenestration or cannulation to the fourth ventricle under endoscopic observation has emerged as the first procedure of choice [16, 18]. Since aqueduct plasty alone is prone to obstruction, it is recommended to cannulate the isolated fourth ventricle [16, 19]. Case 5 is the example of isolated fourth ventricle following repeated shunt infection. A ventricular catheter with multiple side holes can be safely inserted into the fourth ventricle, thus connecting the third and lateral ventricles to the fourth ventricle. The catheter is used as an ordinary ventricular catheter of a VP shunt or as an internal shunt if a VP shunt had already been installed.

20.5.4 Coagulation of the Choroid Plexus

Overproduction of the CSF by the hypertrophy of choroid plexus is a rare situation, but its clinical course would be more complicated. A simple VP shunt can result to the compartment syndrome of the abdominal cavity by the CSF fluid accumulation, causing expansion of abdomen to lead respiratory failure and other complications secondary to the increased abdominal pressure. Coagulation or resection of choroid plexus, thus, reduces the CSF production and is mandated to handle the situation in Case 6 before planning a VP shunt for the final solution. Again, endoscopic coagulation is the mainstream of treatment for hypertrophy of the choroid plexus, and its long-term efficacy has been established [20].

Coagulation of the choroid plexus is gaining a position of common procedure in accordance with ETV for the treatment of hydrocephalus [21, 22]. The technique and the range of coagulation would be the same for the hypertrophied choroid plexus. However, to make the coagulation more toward the tip of temporal horn, posterior approach through the trigone was selected in each side in Case 6 [23]. It allows more coagulation but needs bilateral procedures because fenestration of the ventricular septum does not suit in this approach.

20.5.5 Torkildsen Shunt

Torkildsen shunt, once regarded as a historical procedure to treat hydrocephalus more than 70 years ago, has been reevaluated its role in the time of modern neurosurgery [24, 25]. It is a kind of internal shunt which bridges the lateral ventricle to the cisterna magna. Torkildsen shunt is the best indicated when ETV is impossible due to a mass in the third ventricle or a lesion around the floor of the third ventricle [25]. Case 7 represents a typical case for the indication of Torkildsen shunt. Advantage of Torkildsen shunt is that it enables to avoid a VP shunt for obstructive hydrocephalus. When the both lateral ventricles are not communicating by obstruction of the foramen of Monro, endoscopic fenestration of the septum pellucidum will proceed Torkildsen shunt.

20.5.6 VA Shunt and Other CSF Diversion Procedures

VA shunt is no more a gold standard treatment for hydrocephalus but still has significant role in the management of complex hydrocephalus. VA shunt is currently applied only for limited cases with selected conditions since it has been replaced by VP shunt as the first selection of surgical procedure since the 1970s [26]. VA shunt would be best indicated when the abdominal condition precludes VP shunt placement like Case 8. When VA shunt is indicated, it is important to understand the VA shunt proper complications such as cardiopulmonary complications and shunt nephritis, in addition to standard complications common to VP shunt [27].

It is surely a great advantage to know a variety of CSF diversion procedures when a standard VP shunt failed repeatedly. Once the clinical course of complicated hydrocephalus seems to be heading to complex one, selection of one of other CSF diversion procedures can change the situation. Those procedures include ventriculo-subgaleal shunt (Wang), ventriculo-pleural shunt, [28] ventriculo-gallbladder shunt

[29], ventriculo-superior sagittal sinus shunt [30], ventriculo-iliac shunt [31], and so on. Case 8 illustrates the combination of those CSF diversion procedures before the hydrocephalus was finally stabilized by VA shunt.

20.5.7 Challenge to Complex Hydrocephalus

A real challenge for management of complex hydrocephalus was illustrated in Case 9. Lengthy maltreatment for infected hydrocephalus in the previous institute resulted in tough case of complex hydrocephalus. The first step of treatment was to control meningitis using an EVD and CSF reservoir. The longer the catheter for EVD, the fewer the infection rate. The distance of subcutaneous tunnel of the EVD has primary importance for preventing retrograde infection [32]. In addition, the use of antibiotic-impregnated or silver-bearing catheter would reduce the infection during EVD [33, 34]. The meningitis was successfully cured by EVD and radical intraventricular injection of VCM.

When inserting a new VP shunt is planned as the second step, it was intended to be simple with a single VP shunt system. An endoscopic fenestration to the isolated temporal horn was achieved using a color dye. It was used to guide endoscope to the isolated lesion beyond thicken septum through a nonlinear tract in the ventricle. It has been reported that color dye can be used safely for identifying a cystic lesion [35].

Development of a new pathology is not unusual like this complicated case. Unilateral ventriculomegaly is the result of pre-existing occlusion of the foramen of Monro. The slit ventricle in the shunted side precluded endoscopic fenestration. In this situation, stereotactic fenestration from the slit side was the choice of surgery. Fenestration was confirmed by an endoscope, and an internal shunt was placed to bridge the both sides of lateral ventricles. The case illustrates the best example of full use of multiple surgical managements including EVD, VP shunt, internal shunt, endoscopic fenestration, and other procedures to reach triumph over the complex hydrocephalus.

Conclusion
Most of complex hydrocephalus is the product of maltreated hydrocephalus. The best way to prevent complex hydrocephalus is to avoid complications during the initial treatment of hydrocephalus. The initial surgery is the key for preventing complex hydrocephalus, especially among the pediatric population.

Once the case resulted in complex hydrocephalus, every possible effort should be paid to avoid further complication. When a procedure went wrong several times in a series, it would be advised to change the treatment strategy and surgical procedures. If the endoscopic procedure is indicated, it should be applied first. Avoiding a VP/VA shunt placement is the best choice. An internal shunt with or without endoscopic procedure will help to install a shunt system simple, thus enables avoiding future complications related to the shunt system. For the treatment of complex hydrocephalus, full use of knowledge and experience as well as surgical procedures is tested.

Messages to Be Taken
1. Most of complex hydrocephalus is the result of failed previous shunt surgeries. Most of complex hydrocephalus is avoidable.
2. The CSF shunting procedures are regarded as a basic, standard one. However, treatment of complex hydrocephalus, especially those in pediatric cases, needs deep experience and consideration.
3. Every possible chance to apply endoscopic procedure should be encouraged to avoid a VP shunt.
4. If a VP shunt is unavoidable, the idea of "smart shunt" should be reminded. The simpler the shunt system, the better the clinical outcome.

References

1. Hassan SHA, Holekamp TF, Murphy TM, Mercer D, Leonard JR, Smyth MD, Park TS, Limbrick DD Jr. Surgical management of complex multiloculated hydrocephalus in infants and children. Childs Nerv Syst. 2015;31:243–9.
2. Simon T, Hall M, Riva-Cambrin J, Albert JE, Jeffries HE, LaFleur B, Dean JM, Kestle JRW, Hydrocephalus Clinical Research Network. Infection rates following initial cerebrospinal fluid shunt placement across pediatric hospitals in the United States. J Neurosurg Pediatr. 2009;4:156–65.
3. Spennato P, Cinalli G, Carannante G, Ruggiero C, DeCaro MLDB. Multiloculated hydrocephalus. In: Cinalli G, Maixner WJ, Sainte-Rose C, editors. Pediatric hydrocephalus. Milano: Springer; 2004. p. 219–44.
4. Cheok S, Chen J, Lazareff J. The truth and coherence behind the concept of overdrainage of cerebrospinal fluid in hydrocephalus patients. Childs Nerv Syst. 2014;30:599–606.
5. Hoppe-Hirsch E, Sainte-Rose C, REnier D, Hirsch JF. Pericerebral collections after shunting. Childs Nerv Syst. 1987;3:97–102.
6. Drake JM, Kestle JR, Milner R, Cinalli G, Boop F, Piatt J Jr, Haines S, Schiff SJ, Cochrane DD, Steinbok P, MacNeil N. Randomized trial of cerebrospinal fluid shunt valve design in pediatric hydrocephalus. Neurosurgery. 1998;43:294–303.
7. Gelabert-Gonzalez M, Aran-Echabe E, Serramito-Garcia R. Subdural collections: hygroma and haematoma. In: Di Rocco C, Turgut M, Jallo G, Martinez-Lage JF, editors. Complications of CSF shunting in hydrocephalus: Prevention, identification, and management. Heidelberg: Springer; 2015. p. 285–99.
8. Martínez-Lage JF, Pérez-Espejo MA, Almagro MJ, Ros de San Pedro J, López F, Piqueras C, Tortosa J. Syndromes of overdrainage of ventricular shunting in childhood hydrocephalus. Neurochirugia (Astur). 2005;16:124–33. (Spanish)
9. Pudentz RH, Foltz EL. Hydrocephalus: overdrainage by ventricular shunts. A review and recommendations. Surg Neurol. 1991;35:200–12.
10. Morota N, Sakamoto K, Kobayashi N, Kitazawa K, Kobayashi S. Infantile subdural fluid accumulation. Diagnosis and postoperative course. Childs Nerv Syst. 1995;11:459–66.
11. Gruber RW, Roehrig B. Prevention of ventricular catheter obstruction and slit ventricle syndrome by the prophylactic use of the Integra antisiphon device in shunt therapy for pediatric hypertensive hydrocephalus: a 25-year follow-up study. J Neurosurg Pediatr. 2010;5:4–16.
12. Hatlen TJ, Shurtleff DB, Loeser JD, Ojemann JG, Avellino AM, Ellenbogen RG. Nonprogrammable and programmable cerebrospinal fluid shunt valve: a 5-year study. J Neurosurg Pediatr. 2012;9:462–7.
13. Lewis AI, Keiper GL, Crone KR. Endoscopic treatment of multiloculated hydrocephalus. J Neurosurg. 1995;82:780–5.

14. Oka K, Ohta T, Kibe M, Tomonaga M. A new neurosurgical ventriculoscopy–technical note. Neuro Med Chir (Tokyo). 1990;30:77–9.
15. Ang BT, Steinbok P, Cochrane DD. Etiological differences between the isolated lateral ventricle and the isolated fourth ventricle. Childs Nerv Syst. 2006;22:1080–5.
16. Ogiwara H, Morota N. Endoscopic transaqueductal or intraventricular stent placement for the treatment of isolated fourth ventricle and pre-isolated fourth ventricle. Childs Nerv Syst. 2013;29:1563–7.
17. James HE. Spectrum of the syndrome of the isolated fourth ventricle in posthemorrhagic hydrocephalus of the premature infant. Pediatr Neurosurg. 1990–1991;16:305–8.
18. Fritsch MJ, Kienke S, Manwaring KH, Mehdom HM. Endoscopic aqueductoplasty and intraventriculostomy for the treatment of isolated fourth ventricle in children. Neurosurgery. 2004;55:372–7.
19. Schroeder C, Fleck S, Gaab MR, Schweim KH, Schroeder HWS. Why does endoscopic aqueductoplasty fail so frequently? Analysis of cerebrospinal fluid flow after endoscopic third ventriculostomy and aqueductoplasty using cine phase-contrast magnetic resonance imaging. J Neurosurg. 2012;117:141–9.
20. Ogiwara H, Uematsu K, Morota N. Obliteration of the choroid plexus after endoscopic coagulation. J Neurosurg Pediatr. 2014;14:230–3.
21. Kulkarni AV, Riva-Cambrin J, Browd SR, Drake JM, Holubkov R, Kestle JRW, Limbrick DD, Rozzelle CJ, Simon TD, Tamber MS, Wellons JC III, Whitehead WE. Endoscopic third ventriculostomy and choroid plexus cauterization in infants with hydrocephalus: a retrospective hydrocephalus clinical research network study. J Neurosurg Pediatr. 2014;14:224–9.
22. Warf BC. Comparison of endoscopic third ventriculostomy alone and combined with choroid plexus cauterization in infants younger than 1 year of age: a prospective study in 550 African children. J Neurosurg. 2005;103(6 Suppl):475–81.
23. Morota N, Fujiyama Y. Endoscopic coagulation of choroids plexus as treatment for hydrocephalus: indication and surgical outcome. Childs Nerv Syst. 2004;20:816–20.
24. Alp MS. What is a Torkildsen shunt? Surg Neurol. 1995;43:405–6.
25. Morota N, Ihara S, Araki T. Torkildsen shunt: re-evaluation of the historical procedure. Childs Nerv Syst. 2010;26:1705–10.
26. Keucher RT, Mealey J. Long-term results after ventriculoatrial and ventriculoperitoneal shunting for infantile hydrocephalus. J Neurosurg. 1979;50:179–86.
27. Massimi L, Di Rocco C. Complications specific to the type of CSF shunt: atrial shunt. In: Di Rocco C, Turgut M, Jallo G, Martinez-Lage JF, editors. Complications of CSF shunting in hydrocephalus: Prevention, identification, and management. Heidelberg: Springer; 2015. p. 177–85.
28. Richardson MD, Handler MH. Minimally invasive technique for insertion of ventriculopleural shunt catheters. J Neurosurg Pediatr. 2013;12:501–4.
29. Aldana PR, James HE, Postlethwatt RA. Ventriculogallbladder shunts in pediatric patients. J Neurosurg Pediatr. 2008;1:284–7.
30. Samadani U, Mattielo JA, Sutton LN. Ventriculo sinus shunt placement: technical case report. Neurosurgery. 2003;53:778–80.
31. Tubbs RS, Tubbs I, Loukas M, Cohen-Gadol AA. Ventriculoiliac shunt: a cadaveric feasibility study. J Neurosurg Pediatr. 2015;15:310–2.
32. Khanna RK, Rosenblum ML, Rock JP, Malik GM. Prolonged external ventricular drainage with percutaneous long-tunnel ventriculostomies. J Neurosurg. 1995;83:791–4.
33. Fichtner J, Güresir E, Seifert V, Raabe A. Efficacy of sliver-bearing external ventricular drainage catheter: a retrospective analysis. J Neurosurg. 2010;112:840–6.
34. Raffa G, Marseglia L, Gitto E, Germano A. Antibiotic-impregnated catheters reduce ventriculoperitoneal shunt infection rate in high-risk newborns and infants. Childs Nerv Syst. 2015;31:1129–38.
35. Yamaguchi S, Hida K, Takeda M, Mitsuhara T, Morishige M, Yamada N, Kurisu K. Visualization of regional cerebrospinal fluid flow with a dye injection technique in focal arachnoid pathologies. J Neurosurg Spine. 2015;22:554–7.

Posterior Fossa Anomalies and Hydrocephalus

21

Uppendra Chowdhary, Abdulraaq Al Ojan,
Faisal Al Matrafi, and Ahmed Ammar

21.1 Historical Aspects Related to Chiari Malformation

It was Cleland who described the first case of Chiari malformation in 1883. Some 8 years later, Hans Chiari (an Australian neuropathologist) described the Chiari I to III malformation. Chiari's co-worker Julius Arnold added certain further details to Chiari II malformation, and, hence, Chiari II malformation is also known as Arnold-Chiari malformation. Later on, a rare condition was described and added in the classification of Chiari malformation including Chiari III malformation, which is associated with encephalocele [1, 2]. Chiari himself published further studies in 1896 and expanded his initial classification by adding a fourth subgroup to the original scheme.

21.2 Epidemiology

There are several types of Chiari malformation as shown in Table 21.1. The most common type of Chiari malformation seen in clinical practice is Chiari II malformation, which is always associated with presence of spina bifida, and most of these spina bifida are associated with hydrocephalus and are of the open type, but in some cases of closed meningomyelocele or closed lipomeningomyelocele, hydrocephalus can still occur. If CT scan or MRI scan is done in children with open spina bifida, then up to 90% of these children will show structural hydrocephalus, but only 40% of these patients will develop symptoms related to hydrocephalus. For Chiari II

U. Chowdhary
The Onion Loft, The Fields, Lower Caldecote, Beds SG18 9BA, UK

A. Al Ojan • F. Al Matrafi • A. Ammar (✉)
Department of Neurosurgery, Kig Fahd University Hospital, Imam Abdulrahman Bin Faisal University, Al Khobar, Saudi Arabia
e-mail: ahmed@ahmedammar.com

© Springer International Publishing AG 2017
A. Ammar (ed.), *Hydrocephalus*, DOI 10.1007/978-3-319-61304-8_21

malformation, the prevalence rate was thought to be low in the pre-MRI scan era, but since the availability of MRI scan, the incidence of Chiari II malformation has increased. The prevalence rate is shown in literature to be between 0.1 and 0.5% of all babies born.

Table 21.1 Types of Chiari malformations

Type	Image	Features
0		It is seen in patient with syringohydromyelia. It presents as crowded posterior fossa and alteration of normal anatomy and tonsils never exceed the level of foramen magnum
1		Cerebellar tonsillar herniation >5 mm below foramen magnum
1.5		Cerebellar tonsillar herniation >5 mm not associated with distorted brainstem or fourth ventricle
2		These cases are usually associated with syringomyelia, cerebellar vermis, brainstem, and fourth ventricle herniation

Table 21.1   (continued)

Type	Image	Features
3		Type II + occipital or cervical encephalocele
4		Debated to be Dandy-Walker variant or part of Chiari malformation. We prefer to consider such pathology as Dandy-Walker variant

21.3   Aetiology: Chiari Malformations

There are several types of Chiari malformation, types 0, 1, 1.5, 2, 3, and 4 as summarised in Table 21.1. We hereby focus on Types I and II as they are the most common types and need special attention and frequently need surgical intervention.

Chiari I malformation is supposed to have a genetic basis, and there have been suggestions that the linkage is to chromosomes 9 and 15. It has been postulated that in Chiari I malformation, there is a disorder of para-axial mesoderm that leads to formation of a small posterior fossa. This leads to overcrowding of structure in posterior fossa and herniation of cerebellar tonsils below the foramen magnum. This, then, leads to disturbance of the CSF circulation and absorption, leading a small percentage of patients to develop hydrocephalus [3, 4]. In patients with Chiari I malformation, there is a higher incidence of hydrocephalus than associated incidence of syringomyelia. Some support to the hereditary abnormalities in the mesodermal development is the association of connective tissue disorder such as hypermobility syndromes (e.g. Ehlers-Danlos syndrome).

The development of Chiari II malformation in association with open spina bifida case is thought to be a leakage of CSF from the open spina bifida. There have been studies using intrauterine ultrasound and lately of in utero magnetic resonance imaging of the foetus in which a very large percentage of foetuses with open spina bifida have developed hydrocephalus in utero [5, 6]. Batty R et al., in their study, had shown this and also found out that 23% of the foetuses who had spina bifida had not developed Chiari II deformity. Hence, the aetiology of development of this congenital abnormality is still not fully established [7, 8].

21.4   Pathophysiology: Chiari Malformations

21.4.1 The Pathophysiology Chiari I Malformation

The development of hydrocephalus in a patient with Chiari I malformation leads to increased intracranial pressure, which initially becomes symptomatic with coughing or sneezing. Subsequently there is constant headache, and by this stage, various other associated abnormalities with Chiari I malformation become symptomatic such as symptoms and signs of brainstem, medulla, and upper spinal cord compression, cerebellar dysfunction due to compression which leads to imbalance and ataxia, and abnormality of CSF circulation at the craniovertebral junction [1, 9], which may lead to syringomyelia and central cord syndrome. Cardiac-gated phase-contrast magnetic resonance imaging has been used in trying to elucidate the abnormal circulation of CSF in such cases [2, 6, 10, 11].

Acquired Chiari I malformation also occurs in patients who had lumbo-peritoneal shunt. The exact mechanism is uncertain, but craniospinal pressure gradient has been implicated [2, 11–13].

21.4.2 The Pathophysiology of Chiari II Malformation

This is much more complex. Initially, it was thought that the prenatal formation of spina bifida leads to pulling of the cerebellar tonsils through the foramen magnum is the main mechanism of producing Chiari II malformation, but this was thought not to be the whole story as most of the patients with Chiari II malformation have multiple skull and brain abnormalities, which include hypoplasia of the cerebellum, partial or even total agenesis of the corpus callosum, microgyria, various other abnormalities of the thalamus, etc. Hence, in modern times, it is regarded that there is a combination of the tethered cord and continuous leakage of CSF at least until surgical repair has been done soon after birth combined with various abnormalities of the brain itself which may be the reason for the formation of Chiari II malformation [14, 15].

21.5   Clinical Presentation: Chiari Malformations

Chiari I malformation with hydrocephalus occurs in both children and adults. The initial presentation which can be attributable to the presence of hydrocephalus is headache and neck pain worsened by coughing or sneezing, which later on may become an almost constant severe headache [10, 16]. With such symptoms, there are associated symptoms that may have more debilitating features due to compression of the brainstem, medulla, upper cervical cord, and cerebellum, and in a significant percentage of patients, there is syringomyelia. In this case, it presents with ataxia, incoordination, dysarthria, dysphagia, nystagmus, weakness of the hands and dissociated suspended sensory loss, and even myelopathy.

In Chiari II malformation, the clinical presentation seen in infants and very young children presents with brainstem dysfunction of swallowing/feeding difficulties, weak cry, and difficulty in breathing, which may present as stridor or even periodic apnoea and weakness of the upper limbs. In these newly born infants, the diagnosis of associated hydrocephalus becomes clinically apparent due to the enlarged head and full or even tense anterior fontanels.

Diagnostic workup: For Chiari I malformation, when there is a clinical suspicion of hydrocephalus, an MRI scan of the head and of the craniovertebral junction should be done. This is the gold standard of investigation [3, 17]. The CT scan of the head and craniovertebral junction may also be useful if MRI scan is not available. This imaging modality will show the presence of dilated ventricles, which in the vast majority of cases involve all the four ventricles. At the craniovertebral junction, they will show the descent of the apex of the cerebellar tonsils below the foramen magnum. The measurement is taken from the line between inion and the basi-occiput. The consensus is that the tip of the cerebellar tonsil should be a minimum of 5 mm below this line. There is a minority opinion that there should be a range defining this pathology, and this should be between 3 and 5 mm descent of the apex of the cerebellar tonsils. For comprehensive anatomical diagnosis, it is important to get an MRI scan of the cervical and thoracic spine carried out to show presence or absence of syringomyelia and any evidence of thoracic kyphoscoliosis [12, 13].

Once the diagnosis of open spina bifida has been made in utero or at birth, then an MRI scan of the head should be done as soon as the newly born baby is stable. For practical purposes the MRI scan (or in some cases a CT scan of the head) may be postponed until the repair of the open meningomyelocele has been carried out and then the imaging is done within the next 1–2 days. Early craniocervical decompression as the definitive surgical procedure has also been advocated [17, 18]. The MRI scan of the head and brain would show the presence and degree of the hydrocephalus, the descent of the cerebellar tonsils and its extent, abnormality of the tectum, corpus callosum of the gyri of the cerebral hemispheres, and abnormality of the basal ganglia, cerebellum, etc. In a substantial number of such patients, the cerebellar tonsils are like a peg and may extend to a significant distance into the cervical spinal canal [13, 19].

In infants with open spina bifida, clinically, the measurement of the head circumference is likely to show enlarged head beyond the 50th percentile line of head circumference chart due to the presence of the hydrocephalus.

21.6   Management of Patients with Chiari I Malformation

The symptoms in Chiari malformation patients could arise from mid-teens to adulthood. Once the anatomical diagnosis has been established by either CT scan or MRI scan, then, there may be a period of observation in patients showing mild symptoms to observe them and see if any deterioration has happened over a period of a few months to a year. This is because in some patients in spite of anatomical confirmation of Chiari I malformation, there may not be progression of symptoms. The surgical indication in cases of Chiari malformation would be, in such cases, that there is a clinical deterioration or presence of complications such as hydrocephalus, syringomyelia, etc. Once the clinical signs of raised intracranial pressure to hydrocephalus have become established and are confirmed by imaging, then the surgical management should be both for correcting the craniocervical abnormality and insertion of a ventricular peritoneal shunt to control the hydrocephalus. There are some reports that once the craniovertebral abnormality has been dealt with by bony decompression and duraplasty to enlarge the cisterna magna, the hydrocephalus has disappeared. But in the majority of patients who have hydrocephalus with Chiari malformation, the hydrocephalus is likely to persist even after the surgical decompression of the craniovertebral malformation. Some advocate studying cerebrospinal fluid flow by magnetic resonance imaging to evaluate the post-operative state [15, 18].

21.7 Management of Patients with Chiari II Malformation

Surgical management of Chiari malformation is recommended for symptomatic patients who have cerebellar tonsil herniation and compression on the medulla and upper cervical spinal cord. There are several types of surgical techniques as summarised in Table 21.2.

In Chiari II malformation, the urgent and primary aim soon after birth is to close the open spina bifida so that the dural defect is repaired in a watertight fashion, and a full thickness skin cover is provided over the meningomyelocele. As the majority of such patients are likely to have hydrocephalus, imaging either CT scan or an MRI scan is necessary soon after the repair of the open spina bifida to diagnose the extent of the hydrocephalus. Following this, and within 24–72 h, insertion of a ventricular peritoneal shunt should be done. It is mandatory that just before the insertion of a ventriculo-peritoneal shunt, the ventricular CSF taken through the open fontanel should be sterile [20]. There are some surgeons who would put in a ventricular peritoneal shunt at the end of the repair of the open meningomyelocele, but the problem there is that the sterility of the CSF is not known, and once a ventricular peritoneal shunt becomes infected, then it becomes a major problem.

Table 21.2 Surgical techniques of Chiari decompression

Surgical techniques
Suboccipital craniectomy without duraplasty
Suboccipital craniectomy + duraplasty
Suboccipital craniectomy + duraplasty + tonsillectomy
Suboccipital craniectomy + duraplasty + C1 laminectomy
Suboccipital craniectomy + duraplasty + C1–C2 laminectomy

There have been sporadic attempts made to diagnose meningomyelocele in utero and then to repair the open meningomyelocele in utero [16, 21]. This is a complex and very specialised surgical procedure and has yielded variable results. A vast majority of the patients with open meningomyelocele are treated soon after birth both to close the meningomyelocele. In some selected cases, aspiration of ventricular CSF in utero is undertaken and the main aim being to reduce brain damage due to severe degree of hydrocephalus. Endoscopic third ventriculostomy is an alternative in selected cases for treating hydrocephalus [3, 4, 9, 11, 14, 17].

21.8   The Surgical Treatment of the Abnormalities at the Craniovertebral Junction in Chiari I Malformation

In patients with Chiari malformation with hydrocephalus, the first type of surgical procedure that should be done is the bony decompression of the foramen magnum with a laminectomy of C1 and C2 associated with a duraplasty [5, 8, 13, 22]. There are some controversies whether the dura should be incised and whether a watertight patch is put on or not. Most of the literatures are in favour of duraplasty with a pericranial patch insertion in a watertight manner. Those who do not advocate duraplasty sight the post-operative development of pseudo-meningocele and CSF leak, which can lead to meningitis and ventriculitis.

The surgical treatment of the hydrocephalus in patients with open spina bifida has already been alluded to, and once the hydrocephalus has been diagnosed by imaging, then the ventricular peritoneal shunt should be done in 24–72 h with a preoperative determination that the ventricular CSF is sterile. There are reports of endoscopic third ventriculostomy being performed for hydrocephalus in Chiari I malformation [4, 14].

In some of the patients with Chiari II malformation, there is advocacy that they should also have posterior fossa decompression with cervical laminectomy up to the tip of the descended cerebellar tonsil. This matter is controversial.

21.9   Basilar Invagination

Basilar invagination is a developmental abnormality. Here, the odontoid protrudes abnormally into the foramen magnum. The aetiology is multifaceted. Most of the cases have osteogenesis imperfecta or related osteochondrodysplasias [22, 23]. The clinical presentation is due to cervico-medullary compression or symptoms very similar to those due to Chiari I malformation. Posterior bony decompression is the method of treatment in most symptomatic cases, but in a minority, ventral bony decompression may be needed [13, 19, 22, 24].

21.10 Rare Syndromes Associated with Maldevelopment of Caudal Brainstem Area

At the beginning of this chapter, we stated that for the sake of completion, we are going to mention three rare syndromes, which are associated with congenital maldevelopment of the cerebellum and in some cases that of the brainstem and even of the midbrain. The reason to include these rare or very rare syndromes is because of association of hydrocephalus (or ventriculomegaly) in patients with these rare syndromes.

21.10.1 Joubert Syndrome

Joubert syndrome is a congenital dysgenesis or agenesis of the vermis so that the fourth ventricle communicates directly into the cisterna magna. The classical Joubert syndrome has been noted to be due to autosomal recessive or linked inheritance. The patients present from the neonatal age to infancy with hypotonia, cerebellar ataxia, global development delay, oculomotor apraxia, and breathing problems [25–27]. There are reports of variations of Joubert syndrome, and one of the more common variations of Joubert syndrome is tecto-cerebellar dysraphism with occipital encephalocele. The author has published a series of four such patients in 1989. None of these patients had either hydrocephalus or ventriculomegaly [23, 28]. There are some opinions in the literature (viz. Poretti [29]), where it has been postulated that the tecto-cerebellar dysraphism with occipital encephalocele is not a distinct disorder but part of the Joubert syndrome spectrum. The author, in his paper published in Surgical Neurology 1989, has stated that because of the presence of occipital encephalocele along with much more widespread dysraphism of the midbrain tectum and also because these patients have survived beyond infancy where most of the classical Joubert syndrome patients do not survive beyond infancy, it was regarded as a distinct entity. Again, we emphasise that the inclusion of Joubert syndrome in this chapter is related to the presence of occipital encephalocele in the author's series along with all cases of Joubert syndrome where there is a malformation just above the craniovertebral junction which is marked dysgenesis or agenesis of the vermis.

21.10.2 Wildervanck Syndrome

This is a very rare syndrome, which is also given the descriptive name of cervico-oculo-acoustic syndrome. This is characterised by Klippel-Feil anomaly, bilateral abducens palsy with retracted eye bulbs, and hearing loss. In a few cases, there have been associated abnormalities showing brainstem and cerebella diastematomyelia accompanied by vermian hypoplasia, tonsillar herniation resulting in tri-ventricular hydrocephalus in such children [25, 28]. The presence, in some patients, of hydrocephalus is the reason that we have included this in this chapter.

21.10.3 Goldenhar-Gorlin Syndrome

The descriptive name for this syndrome is oculoauriculovertebral dysplasia. This is a very rare syndrome, and the patient presents with numerous central nervous system anomalies which include occipital encephalocele, hydrocephalus, aqueduct stenosis, agenesis of corpus callosum, etc. [24, 27]. The reason to include this syndrome in this chapter is the presence of multiple hindbrain congenital abnormality associated with hydrocephalus.

21.11   Outcome and Prognosis of Chiari I Malformation

The outcome and prognosis is being discussed in relation to the common diseases of Chiari I and Chiari II malformation. The other syndromes that have been mentioned are rare, and because of variable structural and clinical presentation, it is not possible to give a definitive prognosis in some cases except to say that if the associated hydrocephalus is present in such patients with these rare syndromes and that becomes symptomatic and there is an associated serial enlargement of the ventricular size and of the head circumference, then insertion of a CSF diversion shunt may become necessary.

The outcome and prognosis in patients with Chiari I malformation is mostly dependent upon the surgical treatment of the primary lesion which is the small posterior fossa with tonsillar descent beyond the rim of the foramen magnum causing CSF obstruction and in some such cases production of hydrocephalus. If such patients do not have other complications than hydrocephalus such as severe symptomatic syringomyelia and/or a severe degree of upper thoracic kyphosis, then the outcome after the primary posterior fossa plus C1 and C2 decompression with or without duraplasty along with insertion of ventricular peritoneal shunt would arrest the clinical picture both those related to the cerebellar descent and that of hydrocephalus.

Regarding the patients with Chiari II malformation, the outcome again is much more related to the success of the repair of the open meningomyelocele especially prevention of infection in the CSF pathways primarily. In most of such cases, hydrocephalus is an active process and would need the ventriculo-peritoneal shunt within 42–72 h of repair of the meningomyelocele, but in some cases due to various reasons including CSF pathway infection, the ventricular peritoneal shunt insertion may have to wait until the CSF infection has been treated successfully. In this group of patients, there is now a growing tendency to do a third ventriculostomy rather than putting a ventricular peritoneal shunt and in selected groups of patients where a third ventriculostomy has proven to have much less complication and has been successful in controlling the hydrocephalus.

References

1. Battal MD, Kocaoglu M, Bulakbasi N, et al. Cerebrospinal fluid flow imaging by using phase-contrast MR technique. Br J Radiol. 2011;84(1004):758–65.
2. Isik N, Elmaci I, Silav G, Celik M, Kalelioglu M. Chiari malformation type III and results of surgery: a clinical study: report of eight surgically treated cases and review of the literature. Pediatr Neurosurg. 2009;45(1):19–28.
3. Di Rocco C, Frassanito P, Massimi L, Peraio S. Hydrocephalus and Chiari type I malformation. Childs Nerv Syst. 2011;27(10):1653–64.
4. Tubbs R, Shoja M, Ardalan M, Shokouhi G, Loukas M. Hindbrain herniation: a review of embryological theories. Ital J Anat Embryol. 2008;113(1):37–46.
5. Erbengi A, Oge HE. Congenital malformation of the craniovertebral junction: classification and surgical treatment. Acta Neurochir (Wien). 1994;127:180–5.
6. Gilbert J, Jones K, Rorke L, Chernoff G, James HE. Central nervous system anomalies associated with meningomyelocele, hydrocephalus, and the Arnold-Chiari malformation: reappraisal of theories regarding the pathogenesis of posterior neural tube closure defects. Neurosurgery. 1986;18(5):559–63.
7. Galarza M, Lòpez-Guerrero A, Martinez-Lage J. Posterior fossa arachnoid cysts and cerebellar tonsillar descent: short review. Neurosurg Rev. 2010;33(3):305–14.
8. Batty R, Vitta L, Whitby E, Griffiths P. Is there a causal relationship between open spinal dysraphism and Chiari II deformity? A study using in utero magnetic resonance imaging of the fetus. Neurosurgery. 2012;70(4):890–9.
9. Guillaume D. Minimally invasive neurosurgery for cerebrospinal fluid disorders. Neurosurg Clin N Am. 2010;21(4):653–72.
10. Mauer U, Gottschalk A, Mueller C, Weselek L, Kunz U, Schulz C. Standard and cardiac-gated phase-contrast magnetic resonance imaging in the clinical course of patients with Chiari malformation Type I. Neurosurg Focus. 2011;31(3):E5.
11. Yamada S, Tsuchiya K, Bradley W, Law M, Winkler M, Borzage MT, Miyazaki M, Kelly EJ, McComb JG. Current and emerging MRI imaging techniques for the diagnosis and management of CSF flow disorders: a review of phase-contrast and time-spatial labelling inversion pulse. AJNR Am J Neuroradiol. 2015;36(4):623–30.
12. Loukas M, Shayota B, Oelhafen K, Miller JH, Chem JJ, Shane Tubbs R, Oakes WJ. Associated disorder of Chiari Type I malformations. Neurosurg Focus. 2011;31(3):e3.
13. Payner T, Prenger E, Berger T, Crone K. Acquired Chiari malformations: incidence, diagnosis, and management. Neurosurgery. 1994;34(3):429–34.
14. Massimi L, Pravatà E, Tamburrini G, Gaudino S, Pettorini B, Novegeno F, Colosimo C, Di Rocco C. Endoscopic third ventriculostomy for the management of Chiari I and related hydrocephalus: outcome and pathogenetic implications. Neurosurgery. 2011;68(4):950–6.
15. Stevenson K. Chiari Type II malformation: past, present and future. Neurosurg Focus. 2004;16(2):E5.
16. Taylor F, Larkins M. Headache and Chiari I malformation: clinical presentation, diagnosis, and controversies in management. Curr Pain Headache Rep. 2002;6(4):331–7.
17. Pollack I, Kinnunen D, Albright L. The effect of early craniocervical decompression on functional outcome in neonates and young infants myelodysplasia and symptomatic Chiari II malformations: results from a prospective series. Neurosurgery. 1996;38(4):703–10.
18. Sivaramakrishnan A, Alperin N, Surapaneni S, Lichtor T. Evaluating the effect of decompression surgery on cerebrospinal fluid flow and intracranial compliance in patients with Chiari malformation with magnetic resonance imaging flow studies. Neurosurgery. 2004;55(6):1344–51.
19. Smith J, Shaffrey C, Abel M, Menezes A. Basilar invagination. Neurosurgery. 2010;66(3):A39–47.
20. Tamburrini G, Frassanito P, Iakovaki K, Pignotti F, Rendeli C, Murolo D, Di Rocco C. Myelomeningocele: the management of the associated hydrocephalus. Childs Nerv Syst. 2013;29(9):1569–79.

21. Adzick NS. Fetal surgery for spina bifida: past, present, future. Semin Pediatr Surg. 2013;22(1):10–7.
22. McLone D, Dias M. The Chiari II malformation: cause and impact. Childs Nerv Syst. 2003;19:540–50.
23. Kendall B, Kingsley D, Lambert S, Finn P. Joubert syndrome: a clinico-radiological study. Neuroradiology. 1990;31:502–6.
24. Di Lornezo N, Fortuna A, Guidetti B. Craniovertebral junction malformations. Clinical radiological findings, long-term results, and surgical indications in 63 cases. J Neurosurg. 1982;57(5):603–8.
25. Aleksic S, Budzilovich G, Greco M, McCarthy J, et al. Intracranial lipomas, hydrocephalus and other CNS anomalies in oculoauriculo-vertebral dysplasia (Goldenhar-Gorlin syndrome). Pediatr Neurosurg. 1984;11(5):285–97.
26. Balc S, Oguz K, Frat M, Boduroglu K. Cervical diastematomyelia in cervico-oculo-acoustic (Wildervanck) syndrome: MRI findings. Clin Dysmorphol. 2002;11(2):125–8.
27. Chowdhary UM, Ibrahim AW, Ammar AS, Dawoudu AH. Tecto-cerebellar dysraphism with occipital encephalocele. Surg Neurol. 1989;31(4):310–4.
28. Poretti A, Singhi S, Huisman TA, et al. Tecto-cerebellar dysraphism with occipital encephalocele: not a distinct disorder, but part of the Joubert syndrome spectrum? Neuropediatrics. 2011;42:170–4.
29. Sawin P, Menezes A. Basilar invagination in osteogenesis imperfecta and related osteochondrodysplasias: medical and surgical management. J Neurosurg. 1997;85(6):950–60.

Dandy-Walker Syndrome: A Challenging Problem

22

Ahmed Ammar and Abulrazaq Al Ojan

22.1 Introduction

Dandy-Walker syndrome was described for more than 100 years. Sutton in 1987 described the anatomical features of the disease. Dandy and Blackfan in 1914 described the disease clinically, and they made an association between hydrocephalus, posterior fossa cyst, and hypoplasia of cerebellar vermis. Dandy later in 1921 thought that the cause of this syndrome is either failure of foramen of Magendie and Luschka to develop in utero or due to obstruction of these foramina by the inflammatory process after birth. Walker and Taggart in 1941 described the cause of Dandy-Walker syndrome is the hypoplastic vermis with atresia of the foramina of Magendie and Luschka associated with hydrocephalus that always began during intrauterine life. Benda in 1954 was the one to name the syndrome as "Dandy-Walker syndrome (DWS)." He described it as a congenital malformation of cerebellar vermis that results in cyst formation and fourth ventricular hydrocephalus. Brodal and Hauglie-Hassen in 1959 suggested that anomalies of the fourth ventricle are developed at an earlier stage than the formation of foramina of Luschka and Magendie due to an increased intraventricular pressure [1, 2].

Dandy-Walker syndrome is a well-known etiology of congenital hydrocephalus in 70–90% of cases. It accounts approximately 1–4% of hydrocephalus cases. The incidence of DWS globally is about 1/30,000. Most cases are sporadic. Few cases runs in families, although the pattern of inheritance is not defined. During the first year of life, 80–90% of signs and symptoms of DWS may show up. Only 10–90% of signs and symptoms may appear late in childhood or adolescent period. Mental

A. Ammar, M.D., M.B.Ch.B., D.M.Sc. (✉) • A. Al Ojan
Department of Neurosurgery, King Fahd University Hospital, Imam Abdulrahman Bin Faisal University, Al Khobar, Saudi Arabia
e-mail: ahmed@ahmedammar.com

© Springer International Publishing AG 2017
A. Ammar (ed.), *Hydrocephalus*, DOI 10.1007/978-3-319-61304-8_22

retardation presents in 30–50% of the cases. The genetic causes of DWS have been the aim of several researchers. There are some impressive results. However, the responsible wide genome has not been determined. Some research findings are implicating that genetic deletions such as 3q23q25 might be the cause of Dandy-Walker syndrome in some cases [3].

DWS is presented as the complex spectrum of disorders described by several authors as (1) aplasia or hypoplasia of cerebellar vermis, (2) anterior-posterior enlargement of the posterior fossa, (3) upward displacement of torcula and transverse sinuses, and (4) cystic dilatation of fourth ventricle [4, 5]. However, we challenge this classification as there are some cases where the posterior fossa is not enlarged nor the torcula is elevated.

Dandy-Walker syndrome in some cases is associated with other malformation in the central nervous system and other systems in the cardiovascular system and genitourinary tract, extremity defects, abnormal facial features, PHACES Syndrome (posterior fossa malformation, hemangiomas, arterial anomalies, cardiac defects, eye abnormalities, sternal cleft, and supraumbilical raphe syndrome), and CNS malformation like holoprosencephaly, agenesis of corpus callosum, occipital encephalocele, and neural tube defect [1, 2, 6, 7].

The signs and symptoms of clinical presentation of DWS include increased intracranial pressure, signs and symptoms of cerebellar dysfunction, slow motor development, mental retardation, and cranial nerve dysfunction.

Prognosis of the disease depends on the extent of neurological impairment. Patients with marked neurological impairment have poor prognosis. On the other hand, those with less neurological impairment have a better prognosis depending on the presence of other developmental anomalies. Therefore, it is thought that identifying the chromosomal and genetic abnormalities is vital in predicting the outcome and progress the disease.

22.2   Toward a Better Understanding of DWS

DWS so far is raising several questions regarding the classification, the genetic causes, the radiological and clinical presentations, different methods of management, and the outcome. In order to answer some of the question, a research to study the impact of ICP gradient on the management and outcome was planned and preformed.

22.2.1 Research Material and Methods and Illustrative Cases

22.2.1.1   Materials and Methods
The period of the study was from June 1987 to June 2016. A number of cases diagnosed and treated at King Fahd University Hospital (KFHU), Al Khobar, Saudi Arabia, were 83 cases. Out of these 83 cases, only 21 cases are included in the study based on the inclusion and exclusion criteria.

22.2.1.2   Inclusion Criteria
(a) Newly diagnosed infants
(b) Follow-up for at least 2 years

22.2.1.3   Exclusion Criteria
(a) Cases lost to be followed up

22.2.1.4   Gender
The male to female ratio is 1:3.2 (male 5 and female 16). There were 19 cases delivered in our hospital diagnosed either prenatal or after birth and were referred soon after delivery to our service. Two cases were referred from other hospital; they were diagnosed but not treated there.

22.2.1.5   Management
VP shunt was inserted in all cases (21 cases): shunt revision in 12 cases replacing the shunt either as VP or VA or cystoperitoneal shunts, insertion of additional cystoperitoneal shunts in six cases, and shunt removal and insertion of cystoperitoneal shunt in three cases.

These cases were thoroughly studied. The radiological and clinical presentations of each case and the follow-up were correlated, and we came up with a conclusion that Dandy-Walker syndrome may be classified into four subgroups according to (1) upward position of the torcula and tentorium which may present high ICP, (2) Dandy-Walker cyst and associated CNS anomalies and chromosomal pattern, (3) the signs of compression on the brain stem, and (4) the skull deformity, which may indicate intrauterine high ICP. This new understanding has impact on the treatment.

Case 1
A 1-month-old infant was born prematurely at 32 GW. The child was diagnosed with intrauterine and at birth as having DWS and hydrocephalus. Medium-low pressure (40–70 mmH$_2$O) VP shunt was inserted from another hospital. Few weeks postoperatively, the infant became more sleepy most of the time and it was difficult to arouse him. Therefore, the patient was referred to KFHU. MRI and CT scan were performed and showed enlarged posterior fossa and signs of increased ICP in the posterior fossa (Fig. 22.1a, b), which are:

1. The cerebellum is markedly compressed.
2. High tentorium insertion and pushed-up torcula and transverse sinus.
3. The skull is bulging out as a result of the high ICP during intrauterine life (remolding of the occipital bones).
4. Brain stem was compressed and distorted.

TCD was performed and indicated clear differential pressure and ICP gradient as ICP in the posterior fossa was estimated to be 10 mmHg, where in the supratentorial is estimated to be 7 mmHg (Fig. 22.2).

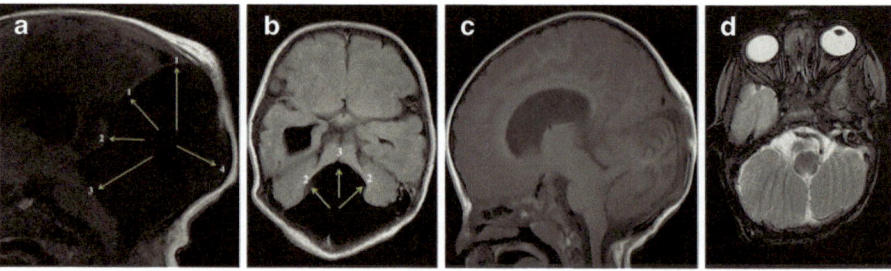

Fig. 22.1 A 1-month-old baby boy born prematurely at 32 GW. Preoperative MRI images of the case of high DWS cyst ICP. (**a**) MRI in sagittal view. *Arrows* indicate (*1*) upward positioning of the tentorium, transverse sinus, and torcula. (*2*) Compressed cerebellum. (*3*) Compression on the brain stem. (*4*) *Arrow* indicating out bulging skull deformity. (**b**) MRI axonal cuts show compression on the cerebellum and brain stem. *Arrow* indicates (*2*) points to the cerebellum, (*3*) points to the brain stem. (**c**) Eleven months later, postoperative follow-up MRI (sagittal cuts) showed remarkable changes, expanding the cerebellum and relieving the pressure in the brain stem, tentorium, and transverse sinus. (**d**) MRI axial cuts confirmed the findings in the sagittal cuts

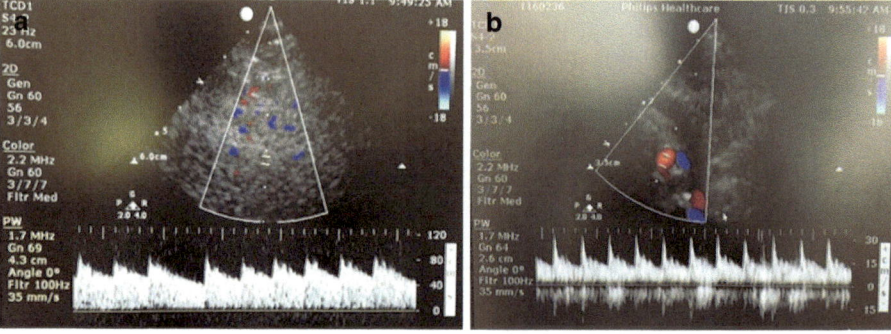

Fig. 22.2 (**a**) Estimated supratentorial ICP is 7 mmHG and (the calculation is MFV = 80 + (50 × 2)/3, Mfv = 60, PI = 80–50/60, PI = 0.5, ICP = (0.5 × 11.1) + 1.43 supratentorial ICP = 7), and (**b**) the infratentorial ICP is 10 mmHg (calculation is MFV = 30 + (15 × 2)/3, Mfv = 20, PI = 30–15/20, PI = 0.75, ICP = (0.75 × 11.1) + 1.43 infratentorial ICP = 10 mmHg)

Therefore, medium pressure cystoperitoneal shunt was inserted, and the VP shunt was kept not changed in place. The child made a remarkable recovery, woke up, and regained ability to suck. Eleven months later, follow-up MRI (Fig. 22.1c, d) showed marked changes:

1. The ICP in the posterior fossa was markedly reduced and became equal to the supratentorial ICP.
2. The cerebellum expanded and gained reasonable shape and size.
3. Brain stem restored its normal shape.
4. The tent in general is relaxed; of course the torcula remained the same.

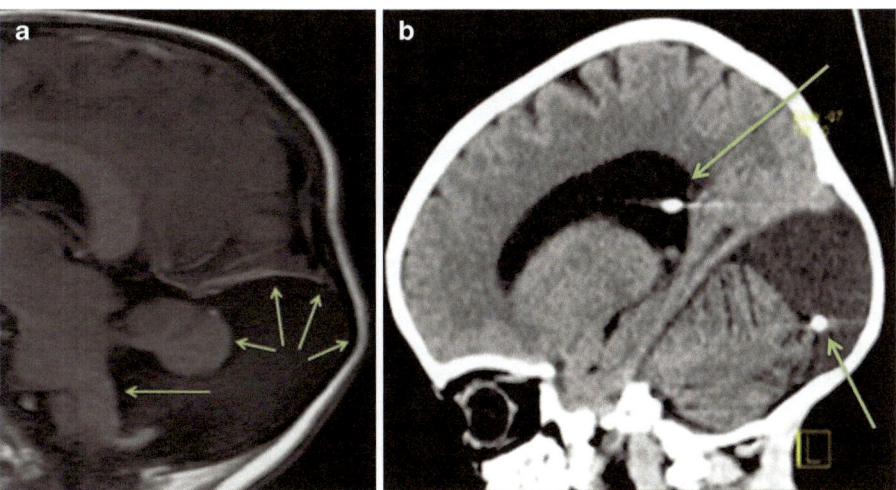

Fig. 22.3 (**a**) One-month-old full-term baby diagnosed as a case of DWS associated with hydro-cephalus. ICP was high but remained the same in supratentorial and infratentorial compartments. *Arrows* indicate that the cerebellum is compressed, brain stem is flattened, and the tentorium is slightly bulging upward. VP shunt and cystoperitoneal shunts of medium-low pressure were inserted. (**b**) Two years later, follow-up CT scan showed the tips of the two shunts in the lateral ventricles and in the DWS cyst in the posterior fossa. Cerebellum has remarkably expanded

A follow-up on the child after 28 months showed that the patient is doing very well with normal cognitive function and normal growth milestone.

Case 2
A 1-day-old, full-term, neonate boy was diagnosed as a case of DWS with a very small posterior fossa, and the position of the tentorium is very low with marked hydrocephalus.

TCD was performed and indicated high supratentorial ICP. VP shunt was inserted. The neonate tolerated the procedure and did not suffer any postoperative complication. Follow-up CT scan and MRI showed signs of decrease ICP in both supra- and infratentorial compartments. The child has been followed up for 5 years and is growing well with mild to moderate delay in the cognitive function.

Case 3
A 1-month-old, full-term neonate was received in the emergency room as he was drowsy, vomiting, and lethargic for few days after the insertion of high-pressure VP shunt. CT scan showed evidence of DWS; the tentorium is positioned in the standard level, and no indication of the differentiation of ICP in both, supratentorial or infratentorial compartments (Fig. 22.3a). The child was taken to the OR and the valve was changed to medium low, and another cystoperitoneal shunt was inserted (the same valve pressure in both shunts). On postoperative follow-ups, the child

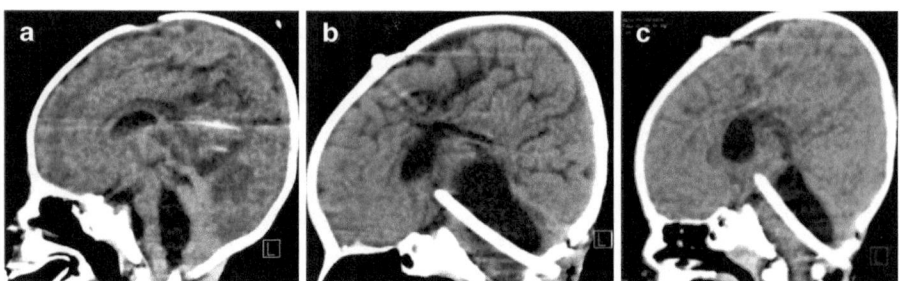

Fig. 22.4 (**a–c**) For illustrative Case 4. (**a**) The pressure in the posterior fossa was lower than supratentorial ICP. (**b**) VP shunt revision and insertion of third ventricle-DWS cyst and fourth ventricle-peritoneal shunt. (**c**) CT scan 3 days post-op showed the shunt in place and general signs of reduction of ICP

made good recovery. A 6-month follow-up showed the cerebellum is expanding, and child remained well (Fig. 22.3b). Two years later, the child is doing very well mentally but seems to have some cognitive functions and psychological problems as he is very emotional and has the tendency to be aggressive with his brothers. He is an intelligent child.

Case 4

A 10-month-old baby girl was brought by frustrated parents to the outpatient clinic. She was treated in a different hospital as a case of hydrocephalus and DWS. She was operated on for 18 times for insertion of EVDs, repeated insertions and revisions of VP shunts. At admission, she was febrile (temperature was 39.5 °C) and was unconscious, lethargic, and floppy. She was not able to raise her head nor suck to swallow. CT scan upon admission showed dilated lateral ventricles, the third ventricle is dilated too, the small posterior fossa and the cerebellum are pushed backward and downward, and the insertion of the tentorium and the torcula was not elevated. The brain stem is under pressure and anatomically distorted (Fig. 22.4a). The VP shunt was removed and EVD was inserted. CSF culture isolated *Actinobacteria* organism. The patient received full course of vancomycin. The patient responded well to the treatment. Four weeks later, VP shunt was inserted. The patient did not improve. Unfortunately, she deteriorated as she again lost the ability to suck and was sleepy most of the time.

We decided to insert third ventricle-cyst and fourth ventricle-peritoneal shunt. The ventricle catheter is designed inside the operation room as follows (Fig. 22.5):

1. Insert the part of the catheter which contains eyes (openings) in the third ventricles.
2. Measuring the length of the aqueduct of Sylvius and try to avoid that part of the catheter that comprises any opening.
3. Perform small and multiple openings/cuts in the part located in the fourth ventricle and inside the cyst.

New device

Third vent-fourth vent and stuning shunt

Fig. 22.5 Showing the third ventricular-fourth ventricular and DWS cyst catheter, which should be designed in the operating room according to the configuration and sizes of measurements of the length of third ventricle, aqueduct, and fourth ventricle

4. The catheter is connected to medium-low pressure valve subsequently to perito-neal catheter. Three days later, the child woke up and gradually regained sucking abilities and smiles (Fig. 22.4b). On follow-up, 12 months later, the patient is doing well and started to grow and sit. However, she may have significant delay in milestones and cognitive functions. CT scan showed that the shunt is in place and there are indications that the pressure is reduced inside the posterior fossa (Fig. 22.4b).

22.3   Analysis and New Proposed Concept of Treatment

DWS can be classified in different ways such as:

1. Genetics can be classified as syndromic and non-syndromic DWS.
2. Association with hydrocephalus or not.
3. Association with hydrocephalus and other CNS anomalies.
4. ICP gradient between supratentorial and infratentorial compartments (Table 22.1).

22.3.1 High-Pressure DWS Cyst

(a) ICP in the posterior fossa is higher than the ICP in the supratentorial compart-ment as estimated by TCD.
(b) High tentorial (torcula) insertion and upward displacement of the transverse sinus.
(c) Absent vermis. The cerebellum is compressed and placed either upward, medial, or in other locations in the posterior fossa.
(d) Distorting the typical configuration of the brain stem.
(e) Skull (occipital bone deformity).
(f) Enlarged posterior fossa.

Table 22.1 ICP gradient classification DWS

Group	ICP gradient differential pressure	High tentorium and torcula	Cerebellum compression	Brain stem distortion	Skull deformity	Size of the posterior fossa
(a) High-pressure DWS cyst	Post. fossa pressure > supratentorial ICP	Upward position	Significantly distorted and absent vermis	Compressed and distorted	Bulging deformity	Enlarged
(b) Balanced-pressure DWS cyst	Post. fossa Pressure = supratentorial ICP	Normal position	Atrophied, absent vermis	Normal or mildly compressed	Normal shape	Normal or slightly enlarged
(c) Low-pressure DWS cyst	Post. fossa pressure < supratentorial ICP	Lower position	Atrophied, absent vermis	Normal or distorted	Flattened occipital bone	Reduced in size
(d) DWS variants' cysts	Post. fossa pressure ≥ or ≤supratentorial ICP	Normal or slightly elevated	Atrophied, compressed, and absent vermis	Normal or distorted	Normal or bulging or flattened occipital bone	Normal or reduced or slightly enlarged

22.3.2 Balanced-Pressure DWS Cyst

(a) The ICP in the posterior fossa is equal to the ICP in the supratentorial compartment as estimated by TCD.
(b) Normal located tentorial (torcula) insertion and normal positioned transverse sinus.
(c) Absent vermis. The cerebellum is atrophied, either placed upward, medial, or in other locations in the posterior fossa.
(d) Distorting the normal configuration of the brain stem
(e) Skull is usually normal shaped.
(f) Normal size posterior fossa

22.3.3 Low-Pressure DWS Cyst

(a) The ICP in the posterior fossa is lower than the ICP in the supratentorial compartment as estimated by TCD.
(b) Downward tentorial (torcula) insertion and downward displacement of the transverse sinus.
(c) Absent vermis. The cerebellum is atrophied either placed upward, medial, or in other locations in the posterior fossa.
(d) Distorting the normal configuration of the brain stem.
(e) Skull is usually normal shaped or flattened.
(f) Normal or small size posterior fossa.

22.3.4 DWS Variant Cyst

(a) The ICP in the posterior fossa is higher or lower than the ICP in the supratentorial compartment. In some cases, the ICP is equal in both compartments as estimated by TCD.
(b) Downward, upward, or normal placed tentorial insertion, torcula, and transverse sinus
(c) Absent vermis. The cerebellum is atrophied either placed upward, medial, or in other locations, in the posterior fossa.
(d) Distorting the normal configuration of the brain stem.
(e) Skull is usually normal shaped or distorted.
(f) Normal size posterior fossa in most cases.

According to these findings, the treatment plan can be designed (Algorithm 1).

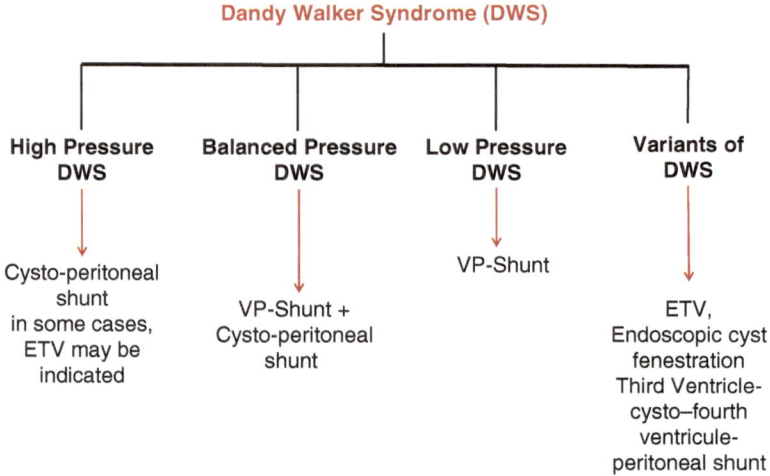

22.4   The Following Observations Have Been Obtained from the Literature Reviews and Our Experience

1. ICP varies between different cranial and brain compartments due to several factors. In some cases, supratentorial ICP is higher than the infratentorial ICP, which is demonstrated by hydrocephalus and low tentorial position and small posterior fossa and estimated by TCD. In other cases, the infratentorial ICP is higher than the supratentorial pressure and can be demonstrated by high tentorial level, bulging of occipital bones, and signs of compression on the brain stem, and the cerebellum is compressed and estimated by TCD as well. ICP sometimes is raised equally in both compartments. Changing of the level of the tentorium insertion may be caused by genetic abnormalities or due to ICP gradients between the supratentorial and infratentorial compartments during the period of molding the skull intrauterine.
2. In the management and insertion of shunts, the abovementioned facts should be taken into consideration so:
    (a) In case of elevated ICP within the DWS cysts indicated as high tent insertion (infratentorial ICP is higher than supratentorial ICP), cystoperitoneal shunt should be inserted and may or may not add VP shunt as well. The option of ETV in addition to cystoperitoneal shunt is very valid option.
    (b) In case of reduced ICP inside the DWS cyst indicated by low tent insertion (the supratentorial ICP is thought to be higher than the infratentorial ICP), there is no need for cystoperitoneal shunt; VP shunt is enough.
    (c) In cases that ICP is the same or nearly the same in both compartments, it is recommended to insert two shunts, VP shunt and cystoperitoneal shunt.
    (d) In cases of DWS variants, each case should be managed according to its merits. The cases of arachnoid cyst or isolated fourth ventricles or postinfec-

tion adhesions or arachnoid cysts or Blake's pouch cyst are included in this group. The ICP varies greatly in these cases. The recommendation is to communicate the ventricular systems to the fenestrated cysts and the subarachnoid space to a peritoneal shunt. Reconstruction of the aqueduct and insertion of third ventricle-fourth ventricle drain or third ventricle-fourth ventricle-peritoneal shunt can be inserted. The proximal catheter should be carefully prepared by keeping the eyes (halls) in the third ventricle, and the part of the catheter that passes through aqueduct of Sylvius must not have any holes or openings. However, additional holes (eyes) should be made in the part of the catheter in the cyst and in the fourth ventricle. This catheter should be connected to valve and peritoneal catheter.

3. ETV is a valid option for treatment especially in children after the age of 2 years.
4. Chromosomal and genetic studies should be performed on these patients in order to predict the long-term outcome. Patient with chromosomal and genetic abnormalities may have delayed milestones and cognitive function.
5. The cognitive functions and psychological developments of these children are affected. Therefore, they should be followed up by pediatric psychologist the earliest possible.

22.5 Discussion

Classification of posterior fossa midline cysts as DWS has been debated, updated, or revised in several documented literatures [1, 8–11]. Blake's pouch cyst had been recognized as a separate entity within DWS.

Klein O. et al. in 2006 summarized their knowledge about DWS (DWM), definition, types of treatment, and prognosis. They defined DWM as "a malformation associating hypoplasia of the vermis, pseudo cystic fourth ventricle, upward displacement of the tentorium, torcular and lateral sinuses and anterio-posterior enlargement of the posterior fossa. It is frequently associated with genetic anomalies, brain malformations (anomalies of gyration, grey matter heterotopias, meningoceles, corpus callosum agenesis) or systemic malformations (heart, orthopedic, intestinal, urogenital and facial anomalies). DWS is sometimes considered as part of many syndromes" [12].

Most of the cases of DWS are associated with hydrocephalus. On the debate if hydrocephalus should be considered as part of this syndrome or a result of this syndrome, we believe that both thoughts are correct. However, hydrocephalus should be treated not as an isolated problem. DWS cyst should be considered in the treatment, so an efficient treatment should consider hydrocephalus and its types and causes in combination with DWS cysts types [1]. ICP gradient in both supratentorial and infratentorial compartments should be considered as well. Treatment, when necessary, is still controversial. The main options remaining are cyst fenestration, ventriculo- and/or cystoperitoneal shunts, and more recently endoscopic third ventriculostomy.

22.6 Genetic and Chromosomal Abnormalities

Pascual-Castroviejo I. et al. in 1991 studied 38 cases of DWS and found out female to male ratio of 3:1. They reported several associated anomalies such as six angioma cases, and ophthalmic, cardiac, and limb anomalies had been reported as well. Associated CNS anomalies are agenesis of corpus callosum, occipital meningocele, and macrocephaly. Therefore, they concluded that it must be genetic causes or DWS [13]. Thirty-eight cases of Dandy-Walker malformation (DWM) are presented. A female predominance of 3:1 was found. Thirty-two cases (84%) were diagnosed within the first year of life. Of these, 17 cases (44.7%) were diagnosed at birth. Ten cases (26%) were delivered by cesarean section. Thirteen infants (34%) had a birth weight below 3000 g. Several associated malformations were observed, the most frequent being capillary angioma (six cases); cardiac malformations, ophthalmic anomalies, agenesis of the corpus callosum, malformed limbs, and occipital meningocele were also seen. These observations indicate that DWM represents a disorder of the midline central nervous system indicating marked genetic and etiologic heterogeneity with the possibility of showing clinical and pathological intra- and extracranial anomalies. Macrocephaly was the most frequent physical finding, appearing in 31 cases (82%). Seventeen (44.7%) patients died (11 before 6 months of age, 3 between 6 and 12 months, and 3 after 1 year). Postmortem studies were performed in 13 patients. Three cases were lost to follow-up. Mental retardation (IQ below 70) was found in 11 cases (58% of survivors), low intellect (IQ between 70 and 85) in 4 cases, and only 2 patients showed normal intellectual development (IQ more than 85). High incidence of malformations having genetic and environmental origins, as well as high early mortality of patients with DWM, indicates the complexity of this syndrome, which involves the midline developmental field structures. It is not an isolated malformation of posterior fossa in most cases [13–15].

New research implicated that genetic deletions such as 3q23q25 might be the cause of Dandy-Walker syndrome in some cases [3]. D'Antonio A. et al. in 2016 reported their meta-analysis of 22 studies that include 531 cases of fetus DWS. They found that 16.3% of isolated DWS have chromosomal abnormalities [2].

There are several variants of DWS such as Joupert syndrome, tecto cerebellum dysraphism [6] and arachnoid midline cyst. It has been proven that cerebellum plays a role in the development of different cognitive functions [16].

22.7 Prenatal Diagnosis

The prenatal diagnosis of DWS has been discussed in several literatures [7, 12, 16–19]. It is important to reach the diagnosis early in order to build a strategy to treat each case. The value of fetal MRI as the accurate diagnostic procedure has been debated. Klein O. et al., in 2003, stated that good MRI is necessary to diagnose fetal DWS [12]. However, Guibaud L. et al. (2012) stated that MRI may be not accurate enough [18, 20].

22.8   The Nature History

The nature history of DWS was debated as well [5, 21, 22]. The predication of the outcome, mortality, subnormal cognitive function and IQ, delayed milestones, varieties of neurological disorders, and nature history of DWS are very difficult to precisely estimate. The outcome is highly dependent on:

1. Genetic disorders and chromosomal abnormalities [21]
2. Associated CNS and other systems congenital anomalies
3. The association of hydrocephalus
4. The ICP gradient

The fact that most of the cases of DWS are diagnosed during hospital admissions [12], the prediction of the nature history cannot be accurate. The delay in milestone and cognitive function disorders may be caused by cerebellum developmental abnormalities. DWS among other cerebellum lesions provided evidences indicating that cerebellum has significant role in the development and regulation of the cognitive function. These observations encouraged several scientists to study these phenomena, and they found that cerebellum is connected with the higher function center in the cerebral cortex. Cognitive functions had been examined with functional neuroimaging and clearly showed that cerebellum is activated [14, 21]. Kozio LF. et al. in 2014 called the cerebellum as "supervised learning machine," as the cerebellum is thought to have roles in neurocognitive development, IQ, language function, learning abilities, working memory, and executive function. Understanding the relationship of the cerebellum and developing of cognitive function should help in the management of these patients and predict the outcomes of DWS [23, 24].

In DWS with cases of cognitive functions abnormalities, it is recommended to study the cerebellar vermis with MRI T2. The study of different types of cerebellum showed that the cerebellar vermis with one or only two identified lobe(s) is rather associated with other cerebellar anomalies, poor cognitive functions, and poor outcome. However, vermis containing three identified lobes and two deep fissures usually associated with very good outcome [12, 19].

22.9   Mortality

The mortality of DWS is not well determined. It is thought to be related to associated CNS anomalies or other congenital anomalies. Salihu HM. et al. in 2008 reported that the mortality in their studies that included 196 cases is 250/1000 cases of DWS. They confirmed the thought that DWS is a heterogeneous not a homogenous disease, as the survival is correlated to the existence of other congenital anomalies. They observed that the mortality increases six times if two or more organs are affected as well [25]. Salihu HM. et al. (2009) expanded their study in New York State, USA, and found distinct racial variation in DWS mortality rate. They found that there is no difference between non-Hispanic blacks and non-Hispanic white in nonmetal mortality; however, postnatal mortality increases eight times [26]. In another study by Pascual-Castroviejo et al. in 1991, they found that the rate of mortality is 44.7% (17 cases died out of 38 cases and most of them within the first 12 months of life) [13]. Only three cases died after first year. They reported as well that these patients were deficient in their cognitive functions such as reduction of IQ, IQ below 70 (58% of survived cases) and IQ between 70 and 85 in four cases, and only two patients enjoyed normal IQ.

22.10 The Proposed New DWS Classification and Guideline for Management

There are major disappointments between the neurosurgeons and patient's parent on the success of the treatment. It is so often that the methods of treatment especially VP shunts is revised once or more. There is no general consensus on the best method to treat DWS. The question is, should hydrocephalus be considered as part of DWS or not?

Management of hydrocephalus is challenging as shunt failures or shunt revision is the norm! It is possible that shunt's failure to drain enough CSF and reduce the ICP is due to the failure of understanding the pathophysiology and CSF dynamics in such patient impacting the insertion of the right shunt in the right place [27, 28]. There are clear diversities in the opinion and plans to treat DWS. This includes VP shunt, cystoperitoneal shunt, double shunts, and endoscopic third ventriculostomy (ETV) [12, 29]. We added the option of third ventricle-fourth ventricle DWS cystoperitoneal shunt.

The literature is full of different trials, either combined shunt or posterior fossa shunt or only VP shunt. ETV had been tried as well. Sikorski CW. in 2005 suggested the use of endoscope to frontally replace ventricular catheter to drain both the lateral ventricles and DWS to avoid the possible complications of double shunting.

Warf BC. in 2011 presented his successful experience by performing ETV in combination choroidal plexus cauterization. The general consensus is that there are different types of DWS and posterior fossa cysts. Therefore, the surgical treatment should be designed after carefully studying each cases [30]. We present our suggested guideline for the treatment of DWS, which is based on three factors. Hu CF. et al. (2011) reported a successful treatment of a 6-month-old girl who suffered DWS by ETV [31]. Therefore, every case of DWS or its variants should be studied thoroughly, and a management strategy should be built for every case, which include the chosen method of surgical management, follow-up, psychological therapy, and treatment of other anomalies.

The prognosis and long-term outcome of DWS is thought to be dependent on the severity of genetic and chromosomal anomalies. The midline cyst and the ICP gradient between the posterior and supratentorial fossa may play a role as well. There are several types of midline cysts. However, DWS should only be diagnosed based on these findings (as previously reported by Klein O. et al. 2003) as having these features: (a) a large cyst communicating with fourth ventricle, (b) a small rotated raised cerebellar vermis, (c) an upward displacement of the tentorium, (d) enlarged posterior fossa, (e) anterolateral displacement of normal cerebellar hemisphere, and (f) normal brain stem [7, 12, 20].

Conclusion

DWS is a complex syndrome, which has several variants in radiological, different clinical presentations, and may be of genetic profile. There is no general consensus on the best method to manage and treat DWS. The strategy and method of management and follow-up should be specially designed for each case according to the type and causes of the clinical signs and symptoms. We offer a modified clinical and management classification of DWS based on the pressure inside DWS in relation to the supratentorial pressure, association with hydrocephalus, and other CNS anomalies. This proposed new concept of classification and understanding may be used as guidelines for management of DWS. Children with DWS may grow up; however, in considerable number of cases, the psychological development and cognitive functions are affected and children need attention as early as possible in the policy of treatment.

References

1. Barkovich AJ, Kjos BO, Norman D, Edwards MS. Revised classification of posterior fossa cysts and cystlike malformations based on the results of multiplanar MR imaging. AJR Am J Roentgenol. 1989;153(6):1289–300.
2. D'Antonio F, Khalil A, Garel C, Pilu G, Rizzo G, Lerman-Sagie T, Bhide A, Thilaganathan B, Manzoli L, Papageorghiou AT. Systematic review and meta-analysis of isolated posterior fossa malformations on prenatal ultrasound imaging (part 1): nomenclature, diagnostic accuracy and associated anomalies. Ultrasound Obstet Gynecol. 2016;47(6):690–7.
3. Alanay Y, Aktaş D, Utine E, Talim B, Onderoğlu L, Cağlar M, Tunçbilek E. Is Dandy-Walker malformation associated with "distal 13q deletion syndrome"? Findings in a fetus supporting previous observations. Am J Med Genet A. 2005;136(3):265–8.
4. Hirsch JF, Kahn AP, Renier D, Sainte-Rose C, Hirsch EH. The Dandy-Walker malformation. A review of 40 cases. J Neurosurg. 1984;61:515–22.
5. McClelland S 3rd, Ukwuoma OI, Lunos S, Okuyemi KS. The natural history of Dandy-Walker syndrome in the United States: a population-based analysis. J Neurosci Rural Pract. 2015;6(1):23–6.
6. Chowdhary UM, Ibrahim AW, Ammar AS, Dawoudu AH. Tecto-cerebellar dysraphism with occipital encephalocoele. Surg Neurol. 1989;31(4):310–4.
7. Forzano F, Mansour S, Ierullo A, Homfray T, Thilaganathan B. Posterior fossa malformation in fetuses: a report of 56 further cases and a review of the literature. Prenat Diagn. 2007;27(6):495–501.
8. Calabrò F, Arcuri T, Jinkins JR. Blake's pouch cyst: an entity within the Dandy-Walker continuum. Neuroradiology. 2000;42(4):290–5.
9. Correa GG, Amaral LF, Vedolin LM. Neuroimaging of Dandy-Walker malformation: new concepts. Top Magn Reson Imaging. 2011;22(6):303–12.
10. Has R, Ermiş H, Yüksel A, Ibrahimoğlu L, Yildirim A, Sezer HD, Başaran S. Dandy-Walker malformation: a review of 78 cases diagnosed by prenatal sonography. Fetal Diagn Ther. 2004;19(4):342–7.

11. McClelland S 3rd, Ukwuoma OI, Lunos S, Okuyemi KS. Mortality of Dandy-Walker syndrome in the United States: analysis by race, gender, and insurance status. J Neurosci Rural Pract. 2015;6(2):182–5.
12. Klein O, Pierre-Kahn A, Boddaert N, Parisot D, Brunelle F. Dandy-Walker malformation: prenatal diagnosis and prognosis. Childs Nerv Syst. 2003;19(7–8):484–9.
13. Pascual-Castroviejo I, Velez A, Pascual-Pascual SI, Roche MC, Villarejo F. Dandy-Walker malformation: analysis of 38 cases. Childs Nerv Syst. 1991;7(2):88–97.
14. Baillieux H, De Smet HJ, Paquier PF, De Deyn PP, Mariën P. Cerebellar neurocognition: insights into the bottom of the brain. Clin Neurol Neurosurg. 2008;110(8):763–73.
15. De Smet HJ, Paquier P, Verhoeven J, Mariën P. The cerebellum: its role in language and related cognitive and affective functions. Brain Lang. 2013;127(3):334–42.
16. Aletebi FA, Fung KF. Neurodevelopmental outcome after antenatal diagnosis of posterior fossa abnormalities. J Ultrasound Med. 1999;18(10):683–9.
17. Gandolfi Colleoni G, Contro E, Carletti A, Ghi T, Campobasso G, Rembouskos G, Volpe G, Pilu G, Volpe P. Prenatal diagnosis and outcome of fetal posterior fossa fluid collections. Ultrasound Obstet Gynecol. 2012;39(6):625–31.
18. Guibaud L, Larroque A, Ville D, Sanlaville D, Till M, Gaucherand P, Pracros JP, des Portes V. Prenatal diagnosis of 'isolated' Dandy-Walker malformation: imaging findings and prenatal counselling. Prenat Diagn. 2012;32(2):185–93.
19. Klein O, Pierre-Kahn A. Focus on Dandy-Walker malformation. Neurochirurgie. 2006;52(4):347–56.
20. Spennato P, Mirone G, Nastro A, Buonocore MC, Ruggiero C, Trischitta V, Aliberti F, Cinalli G. Hydrocephalus in Dandy-Walker malformation. Childs Nerv Syst. 2011;27(10):1665–81.
21. Imataka G, Yamanouchi H, Arisaka O. Dandy-Walker syndrome and chromosomal abnormalities. Congenit Anom (Kyoto). 2007;47(4):113–8.
22. Kollias SS, Ball WS Jr, Prenger EC. Cystic malformations of the posterior fossa: differential diagnosis clarified through embryologic analysis. Radiographics. 1993;13(6):1211–31.
23. Koziol LF, Budding D, Andreasen N, D'Arrigo S, Bulgheroni S, Imamizu H, Ito M, Manto M, Marvel C, Parker K, Pezzulo G, Ramnani N, Riva D, Schmahmann J, Vandervert L, Yamazaki T. Consensus paper: the cerebellum's role in movement and cognition. Cerebellum. 2014;13(1):151–77.
24. Koziol LF, Budding DE, Chidekel D. Adaptation, expertise, and giftedness: towards an understanding of cortical, subcortical, and cerebellar network contributions. Cerebellum. 2010;9(4):499–529.
25. Salihu HM, Kornosky JL, Druschel CM. Dandy-Walker syndrome, associated anomalies and survival through infancy: a population-based study. Fetal Diagn Ther. 2008;24(2):155–60.
26. Salihu HM, Kornosky JL, Alio AP, Druschel CM. Racial disparities in mortality among infants with Dandy-Walker syndrome. J Natl Med Assoc. 2009;101(5):456–61.
27. Paladini D, Quarantelli M, Pastore G, Sorrentino M, Sglavo G, Nappi C. Abnormal or delayed development of the posterior membranous area of the brain: anatomy, ultrasound, diagnosis, natural history and outcome of Blake's pouch cyst in the fetus. Ultrasound Obstet Gynecol. 2012;39:279–87.
28. Sikorski CW, Curry DJ. Endoscopic, single-catheter treatment of Dandy-Walker syndrome hydrocephalus: technical case report and review of treatment options. Pediatr Neurosurg. 2005;41(5):264–8.
29. Ohaegbulam SC, Afifi H. Dandy-Walker syndrome: incidence in a defined population of Tabuk, Saudi Arabia. Neuroepidemiology. 2001;20(2):150–2.
30. Warf BC, Dewan M, Mugamba J. Management of Dandy-Walker complex-associated infant hydrocephalus by combined endoscopic third ventriculostomy and choroid plexus cauterization. J Neurosurg Pediatr. 2011;8(4):377–83.
31. Hu CF, Fan HC, Chang CF, Wang CC, Chen SJ. Successful treatment of Dandy-Walker syndrome by endoscopic third ventriculostomy in a 6-month-old girl with progressive hydrocephalus: a case report and literature review. Pediatr Neonatol. 2011;52(1):42–5.

Part VI

Fetal Hydrocephalus

Fetal Examination for Hydrocephalus

23

Arwa Sulaiman Al Shamekh, Noura Al Qahtani,
and Ahmed Ammar

23.1   Development of the Fetal Ventricular System

Although a comprehensive review of the development of the fetal central nervous system is vital to grasp the pathophysiology of hydrocephalus and its associated findings entirely, it is beyond the scope of this chapter. We will, instead, focus on and briefly describe the development of the ventricular system and mention a few significant central nervous system developmental milestones.

Fetal central nervous system (CNS) first appears as a neural plate in the third gestational week. The neural plate invaginates centrally, on the 18th post-conceptional day, thus forming a neural groove and bilateral neural folds. These neural folds converge until they fuse to form the neural tube, the process of neurulation, during the fourth week. The neural tube remains in connection with the amniotic fluid by its central canal-like structure, the neurocele. By the end of the fourth week, the rostral and caudal neuropores have closed [1]. Closure of the rostral neuropore results in the formation of three primary brain vesicles, the prosencephalon, the mesencephalon, and the rhombencephalon, and closure of the caudal neuropore obliterates the remaining connection between the neurocele and the amnion, thus giving rise to the primitive ventricular system. During the fifth gestational week (GW), five secondary brain vesicles develop: telencephalon and diencephalon, mesencephalon, metencephalon, and myelencephalon [2–6].

A.S. Al Shamekh (✉)
College of Medicine, Imam Abdulrahman Bin Faisal University, Al Khobar, Saudi Arabia
e-mail: arwa.shamekh@gmail.com

N. Al Qahtani
Department of Obstetrics and Gynecology, Imam Abdulrahman Bin Faisal University,
King Fahd University Hospital, Al Khobar, Saudi Arabia

A. Ammar
Department of Neurosurgery, King Fahd University Hospital, Imam Abdulrahman Bin Faisal
University, Al Khobar, Saudi Arabia
e-mail: ahmed@ahmedammar.com

© Springer International Publishing AG 2017   311
A. Ammar (ed.), *Hydrocephalus*, DOI 10.1007/978-3-319-61304-8_23

The sixth gestational week marks the formation of the future cerebral hemispheres and the future choroid plexus. The first choroid plexus is evident in the fourth ventricle at the seventh gestational week. The ventricles continue their differentiation in the eighth gestational week, with the frontal and posterior horns evident. By the end of the eighth week, the embryonal period proper has ended. Beginning the fetal period, the fetal brain increases in size, weight, and surface area with the concomitant increase in the size of the ventricles. Throughout the next weeks, the sulci and gyri appear, and the growth of the ventricles continues until the 20th gestational week, where the ventricular growth halts and the ventricular size remains constant throughout gestation [7, 8]. Table 23.1 demonstrates the gestational age of the fetus and the corresponding significant CNS developmental milestones and what can be visualized during US examination.

Table 23.1 Stages of the development of the fetal ventricular system with a few important general CNS developmental milestones

First trimester (embryonic)				
4th GW	5th GW	6th GW	7th GW	8th GW
• Three primary brain vesicles • Cephalic neuropore closes, followed by caudal neuropore (giving rise to primitive ventricular system) • Optic ventricle formed • Telencephalon cavity visible	• Five secondary brain vesicles • Lateral ventricles begin to form • Rhomboid fossa visualized • Lateral ventricles indented by medial and lateral ventricular eminences	• Future cerebral hemispheres evident • Interventricular foramina decrease in size • Cellular aggregations on the ventricular walls (future choroid plexus) • Epiphysis cerebri and olfactory tubercle	• Choroid plexus of fourth ventricle evident • Areae membranaceae rostralis and caudalis on the roof of the fourth ventricle • Dura mater is formed • Olfactory bulb and optic capsule formed	• Choroid plexus of lateral ventricles evident • Anterior and inferior horns of lateral ventricles visible • Optic tract reaches optic chiasm and cortical plate evident • Thalami are formed
First trimester (fetal)				
9th GW	10th GW	11th GW	12th GW	
• External capsule evident • Olivary nucleus with five components	• Commissure of the fornix evident	• First cerebral sulcus appears	• Corpus callosum evident • Crura cerebri visualized • Myelination of spinal cord	
Second trimester				
• Fetal cerebellum is similar to adult's • Ventricular expansion halts • Periinsular sulci evident • Brain cortical sulcation visualized • Cerebral closure of lateral sulcus • Adenohypophysis fully differentiated				
Third trimester				
• Primary sulci evident • Insular, cingular, and occipital secondary sulci evident • Fetal brain is similar to adult				

23.2 Ultrasonography (US)

The earliest recorded implementation of ultrasonography in identifying fetal anatomy was in 1958, when Donald et al. published his paper "The Investigation of abdominal masses by pulsed ultrasound" in *The Lancet*, containing the first printed ultrasound image of the fetal head and various other gynecological and obstetrical issues. It was not until the late 1960s and the early 1970s, so-called the ultrasonic boom, when the use of ultrasonography became widespread and standardized [9]. Currently, ultrasonography is the prenatal screening modality of choice. The particular examination of the fetal central nervous system is termed neurosonography.

Detection and sound interpretation of ultrasound findings stem from a well-founded understanding of the basic physics behind it. Ultrasound imaging utilizes generation of ultrasonic sound waves (over 20,000 Hz) employing a transducer and producing echoes, when met with an organ or tissue, are reflected back into the transducer, also known as the pulse-echo principle. Pulses are generated when an electric current is passed through piezoelectric crystals causing them to change their shape producing sound waves. Once these sound waves are reflected back into the piezoelectric crystals, an electric current is created corresponding with the intensity of the waves. The electric current is then "translated" by a computer into an image displayed on the screen. One pulse generates a scan line, and if multiple pulses are generated along different points, multiple scan lines are generated yielding a cross-sectional image of the scanned organ/tissue [10–12].

23.2.1 Modes

For the sake of brevity, we have included the most commonly used modes when performing a fetal neurosonogram.

(a) *B-Mode*: The brightness mode. It is the basic/2D mode that is used in fetal imaging. It displays a grayscale two-dimensional image of the fetal brain.
(b) *Doppler Mode*: Doppler ultrasound utilizes the changing frequency associated with blood flow to allow visualization of the vessels (color Doppler) and measurement of blood flow parameters (spectral Doppler).
(c) *Three-Dimensional Ultrasound (3D-US)*: 3D-US utilizes phased array probes to create multiple 2D images that are processed and then reconstructed as three-dimensional images. Volumetric assessment of the fetal brain and ventricles is possible in this mode [13]. Tomographic ultrasound imaging (TUI) is a subdivision that produces multiplanar high-definition images, similar to those of MRI.
(d) *Four-Dimensional Ultrasound (4D-US)*: With the fourth dimension being time, 4D-US allows real-time 3D imaging of the fetal brain.

23.2.2 Positioning and Planes

The International Society of Ultrasound in Obstetrics and Gynecology (ISUOG) guidelines state that the transabdominal route is the method of choice to perform the standard fetal CNS exam in low-risk pregnancies. However, the use of the transvaginal route has been recently advocated; some examples necessitating a transvaginal scan include examining a pregnancy less than 10 weeks of gestation and assessing fetal anatomy lying deeply in the maternal pelvis or when measuring nuchal translucency is not possible due to a vertical field position. The transabdominal exam usually yields three axial planes: the transventricular plane (allows visualization of the lateral ventricles and cavum septum pellucidum), the transcerebellar plane (allows visualization of the frontal horns of the lateral ventricles, CSP, thalami, cerebellum, and cisterna magna), and the transthalamic plane (allows visualization of the frontal horns of the lateral ventricles, CSP, thalami, and hippocampal gyri and is used to calculate the biparietal diameter and head circumference). It may be possible to produce two sagittal planes, midsagittal (allows visualization of the corpus callosum, CSP, brainstem, pons, and posterior fossa) and parasagittal (allows visualization of the entire lateral ventricles, choroid plexus, and periventricular tissue), and four coronal planes, transfrontal (allows visualizing the interhemispheric fissure, sphenoid bone, and orbits), transcaudate (allows visualizing the anterior horns of the lateral ventricles, genu of CC, and CSP), transthalamic (allows visualizing the thalami, interventricular foramina, third ventricle, and insulae), and transcerebellar (allows visualizing the occipital horns of lateral ventricles, interhemispheric fissure, and cerebellum) using the transabdominal route, but they are much more easily acquired using the transvaginal route. Table 23.2 summarizes the main differences between the transabdominal and transvaginal routes. To begin the exam, the mother is asked to lie supine and expose her abdomen. A conductive gel is then applied to the transducers head and then applied to the abdomen. The fetal head is located, and an adequate and symmetrical axial view is obtained. To obtain the correct atrial diameter, in the transventricular plane, the widest horizontal line between the medial and lateral walls is taken [14, 15].

Table 23.2 Comparison between the transabdominal and transvaginal route of the obstetric ultrasound

Transabdominal route	Transvaginal route
Basic screening exam	Dedicated neurosonogram (±3D)
Full bladder	Empty bladder
Supine position ± tilting	Lithotomy position
3.5–6 MHz	7–9 MHz
Greater penetration, somewhat poorer resolution	Less penetration, better resolution
Lower frequencies allow greater depth of field examination	High frequencies minimize depth of field
Movement of probe restricted by symphysis pubis and maternal ribs	Movement of probe restricted within the vagina
Bowel gas may pose some difficulties	Not affected by bowel gas
Types of probes: convex, linear, or phased array	Transvaginal probe (may or may not be phased array)
3 axial planes ±2 sagittal, 4 coronal planes (depending on fetal lie and presentation)	2 sagittal +4 coronal planes (if fetus is in cephalic presentation)

23.2.3 Normal Anatomy

During the first trimester (by the 11th GW), 2D-US can be used to measure the BPD and HC following the technique for the specific nomogram. It is possible to detect the lateral ventricles that are filled with the choroid plexus, the interhemispheric fissure, the falx, and the cranial bones that are beginning to ossify. 3D-US has enabled the sonographer to examine the fetal brain earlier than 11 GW, with more intricate detail depicting the primary brain vesicles with less procedural time. Volumetric assessment and reconstruction of the early brain vesicles are possible from the 7th to 12th GW using 3D-US inversion mode as demonstrated by Kim et al. (2008). During the basic CNS exam performed at 20 weeks' gestation, the following are noted: the shape of the head, lateral ventricle (atrial) size and choroid plexus, presence of CSP, thalami, cerebellum and intact vermis, cisterna magna depth (2–10 mm), and any spine abnormalities [16]. A patient undergoes dedicated neurosonography if the basic CNS exam is suspicious for abnormalities or if they are high risk for CNS anomalies. TUI provides a meticulous CNS exam; structures such as gyri are clearly visible, and 3D reconstruction angiography is easily obtained. Volume contrast imaging (VCI) can be used in any brain structure to provide volume measurements that are helpful in estimating gestational age; analysis of these measurements is further implemented to objectively and longitudinally assess ventriculomegaly or space-occupying lesions [10, 17, 18].

On Doppler US, fetal (GA 28.4 ± 4.5) middle cerebral artery flow velocity parameters were demonstrated by Malinger et al. (2011), as follows: peak systolic velocity (PSV) 35.1 ± 10.9 cm/s and pulsatility index (PI) 2.14 ± 0.7 [19–21].

23.2.4 Pathological Anatomy

Fetal ventriculomegaly is defined as when the atrial diameter of the lateral ventricle is >10 mm, regardless of gestational age, and is divided into three categories: mild ventriculomegaly (10–12 mm), moderate ventriculomegaly (12.1–15 mm), and severe ventriculomegaly (>15 mm). Fetal hydrocephalus occurs when ventriculomegaly occurs in the background of increased intracranial pressure. Differentiation between the two requires detecting signs of increased ICP, such as obliterated subarachnoid spaces, separation of the choroid plexus from the medial wall of the lateral ventricle >3 mm, giving the dangling choroid sign, and diminishing venous pulsation on Doppler [10, 22].

Ventriculomegaly without signs of raised ICP, on the other hand, is thought to occur as a result of a pathological process rendering the brain more compliant to the expanding size of the ventricular system. When ventriculomegaly is detected on US, a detailed systematic examination must be performed to detect or exclude any other underlying anomalies. Table 23.3 shows possible associated pathologies that might be detected in utero and a brief description of each condition. Fetal hydrocephalus can be broadly classified into three categories: simple hydrocephalus, primary/dysgenetic hydrocephalus, and secondary hydrocephalus. Simple hydrocephalus results

Table 23.3 Possible associated syndromes and anomalies with fetal hydrocephalus

Other CNS anomalies (>80%)			
Aqueductal stenosis	Chiari malformations [23, 24]	Dandy-Walker malformation (DWM) [25]	Neural Tube Defects (NTDs) [26, 27]
• Congenital or acquired • Unknown mechanism, may be inherited (Bickers-Adams-Edwards syndrome) • US detection usually at 20 GWs • US findings: severe ventriculomegaly, parenchymal thinning, dangling choroid sign, dilated third ventricle, intact posterior fossa, and ± macrocephaly • MRI findings: US findings + subarachnoid space loss + obliteration of cerebral aqueduct + absent flow-void signal at level of aqueduct	• Downward displacement of cerebellum and/or brainstem through the foramen magnum ± obstruction • US findings: severe ventriculomegaly, inferiorly displaced cerebellum and obliterated cisterna magna (banana sign), anterior angulation of frontal bones (lemon sign, not specific), and a lumbosacral myelomeningocele • MRI findings: US findings + small posterior fossa, low tentorial attachment, and a low torcula + beaked tectal plate causing aqueductal stenosis, + crowded foramen magnum + myelomeningocele + tethered cord	• Triad of vermian hypoplasia + posteriorly extending cystic dilatation of the fourth ventricle + large posterior fossa with torcular-lambdoid inversion • US findings: enlarged cisterna magna (≥10 mm) + vermian aplasia + trapezoid-shaped gap between cerebellar hemispheres. The vermis is not fully formed before the 18th GW, thus decreasing specificity [28] • MRI findings: vermian hypoplasia + cystic dilatation of the fourth ventricle + enlarged posterior fossa with torcular-lambdoid inversion	• Failure of closure of the neural tube on the 27th day of conception • Encompasses myelomeningocele (50%), anencephaly (40%), encephalocele (5%), craniorachischisis, exencephaly, tethered cord, and iniencephaly (1%) • Myelomeningocele US findings: open neural tube defect with splayed/divergent posterior element • Myelomeningocele MRI findings: laminar eversion, absence of posterior elements, and presence of the neural placode

Posterior fossa cysts [25]	Polymicrogyria [29]	Holoprosencephaly
• Blake pouch cyst: absent communication between subarachnoid space and fourth ventricle due to failure of fenestration of Blake pouch • Blake pouch cyst imaging findings: retro-/inferocerebellar cystic lesion + anterosuperior displacement of choroid plexus + cyst communicating with enlarged fourth ventricle • Arachnoid cyst: duplicated arachnoid producing fluid-filled cyst • Arachnoid cyst findings: well-circumscribed cystic lesion + absent communication with fourth ventricle	• Cortical development malformation resulting in abnormal arrangement and folding of cortex ± fusion of gyri • Occurs after 20 GWs • May be detected BY ultrasound, findings include: overfolding of the cortical ribbon + cortical hyperechogenicity [30] • Better diagnosed by MRI, findings: focal cortical thickening ± signal changes	• Failure of or incomplete separation of the midline intracranial structures (during the fourth to sixth GW) • Alobar (most severe), semilobar, and lobar (least severe) and middle interhemispheric variant/syntelencephaly • Alobar HPE US findings: monoventricle + fused thalami + absent CC + absent interhemispheric fissure + absent CSP + absent falx + absent third ventricle + dorsal cyst + ACA and MCA replaced by tangled branches of ICA and BA + fused cortex (pancake, cup, or ball shape) + facial anomalies (proboscis, cyclopia, mononostril, hypotelorism, cebocephaly) • Alobar HPE MRI findings: US findings + absent olfactory tract ± abnormal optic nerves • Semilobar HPE US findings: absent SP + monoventricle with occipital and temporal horns + rudimentary falx + incomplete interhemispheric fissure + partially/completely fused thalami + hypoplasia/agenesis of CC + incomplete hippocampus ± mild facial anomalies (hypotelorism, cleft lip) • Semilobar HPE MRI findings: US findings + absent olfactory tracts and bulbs • Lobar HPE US findings: fused frontal horns of lateral ventricles + wide communication with third ventricle + forniceal fusion + absent SP + normal/hypoplastic SP + displaced ACA (snake under the skull sign) + normal thalami + interhemispheric fissure + falx • Lobar HPE MRI findings: US findings • MIH findings: absent CSP + absent CC + absent falx + prominent cisterns + abnormal fusion of frontal and parietal lobes

(continued)

Table 23.3 (continued)

Non-CNS anomalies

Craniofacial anomalies	Cardiovascular anomalies	Genitourinary anomalies
• Cleft lip ± palate: US findings include a vertical hypoechoic lesion on the upper lip of the fetus through an angled coronal plane. The palate can be visualized in the axial plane. 3D + 4D US aids in further assessment • Low-set ears: lower position of the pinna ≥2SD below population average • Bilateral optic atrophy • Facial bone anomalies • Acrocephalosyndactylies	• Ventricular or atrial septal defect • Congenital pulmonary stenosis or patent ductus arteriosus • Transposition of the great vessels • Left hypoplastic heart syndrome These are only examples; much more anomalies have been described	• Renal agenesis/hypoplasia, cystic dysplasia, renal fusion, ureteral duplication, and other anomalies
Gastrointestinal anomalies	Skeletal anomalies	
• Omphalocele, gastroschisis, esophageal and anorectal anomalies, and CDH are some of the reported associated anomalies	• Congenital talipes equinovarus/club foot most commonly detected foot anomaly	

Syndromic

Acrocephalosyndactylies	Acrocephalopolysyndactylies	Achondroplasia	Fetal alcohol syndrome [31]
• Calvarial anomalies + digit anomalies (syndactyly only) • Apert syndrome: autosomal dominant; US findings include abnormal cranial shape, midfacial hypoplasia, and bilateral syndactyly. Other anomalies may be detected by US and MRI. • Pfeiffer syndrome: autosomal dominant, three subtypes (type 2, most severe), can be detected antenatally, US and MRI findings of cloverleaf skull, hypertelorism, and proptosis, and broad digits are suggestive of PS	• Calvarial anomalies + digit anomalies (syndactyly + polydactyly) • Types I–IV • Type I/Noack syndrome: now considered a variant of PS • Type II/Carpenter syndrome • Type III/Sakati-Nyhan syndrome • Type IV/Goodman syndrome • Possible other anomalies detected on US and MRI: cloverleaf skull + cardiac anomalies + cryptorchidism + multiple limb anomalies	• Rhizomelic dwarfism • Sporadic or autosomal dominant (homozygousity lethal) • FGFR3 mutation • Usually detected in third trimester • US findings (2D + 3D) and MRI findings: short femur length (<5th percentile), trident hand, frontal bossing, facial dysmorphism, large cranial vault with small skull base, narrow foramen magnum, cervicomedullary kink, brainstem elevation, posterior vertebral scalloping, gibbus, short pedicle canal stenosis, laminar thickening, intervertebral disc widening, increased lumbosacral angle, anterior rib flaring, horizontal acetabular roof, tombstone iliac wings, champagne glass type pelvic inlet, metaphyseal flaring	• Result of prenatal maternal alcohol consumption • US and MRI findings: short palpebral fissures, epicanthus, smooth philtrum, midfacial hypoplasia, long thin upper lip, microcephaly, agenesis of CC, focal cortical thickening, congenital scoliosis, radioulnar synostoses, ASD/VSD, ptosis, optic nerve hypoplasia, renal dysplasia, genital hypoplasia, IUGR
Lissencephaly-pachygyria spectrum [32]	Osteopetrosis	Sotos syndrome [33, 34]	X-linked hydrocephalus [35]

(continued)

Table 23.3 (continued)

• Congenital cortical malformations causing absent/minimal sulci • Type I: classic lissencephaly/subcortical band heterotopia spectrum. US and MRI findings: few shallow sulci + shallow sylvian fissure (hourglass/figure-8 appearance), thick cortex (12–20 mm), hypoplastic anterior CC, CSP + CV. • Type II: cobblestone lissencephaly. US and MRI findings: abnormal sulci (small sylvian fissure + hourglass/figure-8 appearance), multinodular cortical surface, posterior cephalocele, brainstem abnormalities, cerebellar abnormalities, and globe abnormalities	• Abnormal osteoclasts producing thick sclerotic bone that is also weak • Autosomal recessive (infantile, lethal type) • US findings: increased bone density, multiple fractures, shortened long bones, evidence of IUGR	• Cerebral gigantism • Autosomal dominant • US and MRI findings: fetal overgrowth, macrocephaly (inverted pear-shaped head), polyhydramnios, increased nuchal translucency, CNS anomalies, and renal anomalies	• AKA MASA syndrome/L1 syndrome/CRASH syndrome (no longer used) • Xq28 mutation • Variable mental retardation, hydrocephalus, spasticity, flexion deformity of thumbs • US findings: ventriculomegaly, agenesis of CC, flexed adducted thumbs • MRI findings: US findings + bilateral absence of pyramidal tracts
Fryns syndrome [33]	Goldenhar syndrome [33]	Hydrolethalus	Metachromatic leukodystrophy [36]

Syndrome	Findings
Mucopolysaccharidoses	• Multiple congenital anomaly syndrome • Autosomal recessive • US and MRI findings: craniofacial anomalies (dysmorphia, hypertelorism, microphthalmia, long philtrum, flat/broad nasal bridge, macrostomia, cleft lip ± palate, micrognathia, low-set ears, hydrocephalus, CC agenesis, neuronal heterotopias), thoracic anomalies (CDH/eventration, pulmonary hypoplasia), and limb anomalies
Multiple pterygium syndrome	• Complex set of anomalies of the ears, eyes, and vertebrae • US and MRI findings (not specific): preauricular tags, hemifacial microsomia, unilateral microphthalmia/anophthalmia, transverse facial clefts, vertebral segmentation errors, abnormal amniotic fluid volume (AFV)
Neu-Laxova syndrome [33, 37]	• Lethal set of multiple anomalies • Autosomal recessive • US and MRI findings: cleft lip ± palate, hydrocephalus, CC agenesis, DWM, midline brain structures absent, micrognathia, polydactyly, and club feet
Meckel-Gruber syndrome [33]	• Most common hereditary leukodystrophy • Autosomal recessive • MRI findings: periventricular deep WM signal changes, sparing subcortical U fibers (butterfly pattern) + T1 hypointensity + T2 hyperintensity (tigroid pattern); MRS shows decreased NAA, increased myo-ins and lactate

(continued)

Table 23.3 (continued)

• Excessive accumulation of mucopolysaccharides due to deficient lysosomal enzymes • Type IH and type IV may display hydrocephalus • Type IH: Hurler syndrome, most severe, autosomal recessive, death in the first decade of life (usually from cardiac disease) • Type IV: Morquio syndrome, 30–40-year life expectancy (cause of death is usually C2 myelopathy) • Type I US and MRI findings: macrocephaly, narrow foramen magnum, shortening and widening of long bones, paddle-ribs, gibbus, cardiac anomalies • Type IV US and MRI findings: platyspondyly, coxa valga, hypertelorism, dolichocephaly, and multiple distal musculoskeletal anomalies	• Multiple disorders characterized by soft tissue webs over the neck and joints • Lethal and nonlethal subtypes • Lethal type: autosomal/x-linked recessive, incompatible with life • Lethal type US and MRI findings: cutaneous webs across joints, cystic hygroma, hydrops fetalis, craniofacial anomalies, polyhydramnios, and fetal hypokinesia	• Lethal multiple malformation syndrome • Autosomal recessive • US and MRI findings: hypokinesia, IUGR, short neck, CNS anomalies (microcephaly, lissencephaly, and cerebellar hypoplasia), limb anomalies, pulmonary hypoplasia, fetal anasarca, facial anomalies (proptosis, hypertelorism, micrognathia, flattened nose, and optic anomalies)	• Triad of renal cystic dysplasia, holoprosencephaly/occipital encephalocele, and postaxial polydactyly [38] • Pseudotrisomy 13 • US and MRI findings: occipital encephalocele, enlarged echogenic kidneys, polydactyly, oligo-/anhydramnios, microcephaly
VACTERL–H association	Down syndrome [33, 39]	Triploidy	Skeletal dysplasias [40]

• Vertebral anomalies • Anorectal anomalies • Cardiac anomalies • TEF (tracheo-esophageal fistula) and EA (esophageal atresia) • Renal anomalies • Limb anomalies • Hydrocephalus	• Trisomy 21 • Maternal nondisjunction (95%), paternal Robertsonian translocation (2%), mosaic trisomy (3%) • US findings: nuchal translucency >3 mm, nuchal fold thickness >6 mm, hypoplastic nasal bone, echogenic bowel, cardiac anomalies, gastrointestinal anomalies, ventriculomegaly, craniofacial anomalies	• Lethal aneuploidy due to extra chromosomal set • US and MRI findings are nonspecific and include: IUGR, body asymmetry, macrocephaly, hydrocephalus, oligohydramnios, placental abnormalities, syndactyly	• Wide spectrum of abnormal bone formation pathologies • Types that display hydrocephalus: thanatophoric, campomelic, and chondrodysplasia punctata • Thanatophoric dysplasia: two subtypes, death within first hours of extrauterine life • TD US and MRI findings: narrow thoracic cavity, short thick bowed tubular bones, macrocephaly with frontal bossing, cloverleaf skull (type II) [41] • Campomelic dysplasia: 97% mortality within the first year of life • CD US and MRI findings: lower extremity bowing, narrow thoracic cavity, hypoplastic scapulae, hydronephrosis, cardiac anomalies • Chondrodysplasia punctata (CDP): rhizomelic (lethal) and non-rhizomelic form. Epiphyseal calcified deposits • CDP US and MRI findings: polyhydramnios, facial anomalies, skeletal stippling

Walker–Warburg syndrome [33]

- HARDE syndrome
- Lethal congenital muscular dystrophy
- Autosomal recessive
- US and MRI findings: hydrocephalus, neuronal migration anomalies (lissencephaly), mesencephalic-pontine junction kink, Dandy–Walker continuum, encephalocele, CC agenesis, microphthalmia, microtia, genital anomalies, cleft lip ± palate

when an abnormal CSF circulation pathway is present, such as aqueductal stenosis, foramen of Monro atresia, or maldeveloped arachnoid granulation. Dysgenetic hydrocephalus is a product of abnormal cerebral development. Secondary hydrocephalus happens as a result of intracranial tumors, intraventricular hemorrhage, and infectious processes [42]. Table 23.4 demonstrates the possible pathologies under each classification and what can be visualized in each abnormality. Figures 23.1a, b, 23.2, and 23.3a–c are images of fetuses that had ventriculomegaly on routine ultrasound associated with other anomalies.

The progression of hydrocephalus can be documented in seven stages: (1) ventriculomegaly, (2) high ICP, (3) dangling choroid, (4) obliterated subarachnoid space, (5) extended dura and superior sagittal sinus, (6) diminished venous pulsation, and (7) skull enlargement.

As mentioned previously, TUI creates intricate images that compete with MRI. Detailed structural assessment of the hydrocephalic brain is possible in this mode. VCI is a novel method that allows volumetric assessment of the lateral ventricles in hydrocephalic fetuses. Extraction and analysis of lateral ventricle volume and correlation with intracranial cavity volume (ventricle occupying rate) can be of help during follow-up of ventriculomegaly cases [48, 49]. Due to various pathological conditions resulting in hydrocephalus, no one pattern of flow velocity changes is expected. Some studies have shown that isolated mild ventriculomegaly is associated with normal flow velocity patterns; others have shown that in the context of IVH, PI increases significantly. To record useful and accurate flow-related parameters in hydrocephalic fetuses, a distinct classification, such as the one mentioned above, should be used to categorize these parameters. Algorithm 1 shows a possibly sound approach to the diagnosis of fetal hydrocephalus. Although serial US examinations are recommended, there is no clear consensus how often they should be.

Algorithm 1 Assessment of ventriculomegaly

23.2.5 Advantages

1. Sound waves are nonionizing, making it a valuable tool in evaluating fetal tissue.
2. More readily available.
3. Cost-effective.
4. Hardly any contraindications.
5. Real-time imaging.
6. Not affected by metal.
7. Easy to access other organs during the exam.
8. Doppler imaging allows in-depth evaluation of physiologic parameters.

23.2.6 Limitations

1. Operator dependent.

Table 23.4 Classification of hydrocephalus

Classification		Grading/subclassification	Findings	Prognosis
Simple		Aqueductal stenosis	• Ventriculomegaly • Normal ventricular orientation • Smooth ventricular walls • Intact cortex	Variable prognosis [43] Possible improvement with in utero surgery? [44]
Primary/dysgenetic		Syndromic	Discussed in Table 23.3	The more severe the anomaly, the worse the prognosis
		Associated with other anomalies		
Secondary	Post-hemorrhagic (IVH)	Grade I	Hemorrhage limited to subependymal matrix	Grades I and II are usually associated with a good prognosis
		Grade II	Hemorrhage spilling into ventricles <50% filling, ventriculomegaly <15 mm	Grades III and IV are usually associated with a poor prognosis [45]
		Grade III	Ventricular spilling >50%, accompanied by ventriculomegaly, no parenchymal injury	
		Grade IV	Grades I–III hemorrhage + periventricular parenchymal hemorrhage	
	Tumors	Teratomas (most common)	• Usually supratentorial • Teratomas display mixed solid and cystic components • Calcification • Hypervascularization • Intratumoral hemorrhage	Poor prognosis (except choroid plexus papillomas)
		Non-teratomatous		
		Different origins		
	Infectious	Toxoplasmosis	• Ventriculomegaly • Subependymal germinolysis/cysts • Periventricular/cortical calcification • Cerebellar hypoplasia • Thickened lenticulostriate vessels (candlestick sign) • Polymicrogyria	Severe CMV infection: 30% mortality HSV (disseminated): % mortality if not treated (25% if treated) Varicella zoster: 30% mortality [46]
		Other (syphilis, VZV, parvovirus B19)		
		Rubella		
		Cytomegalovirus (CMV) (most common)		
		Herpes simplex virus (HSV) [47]		

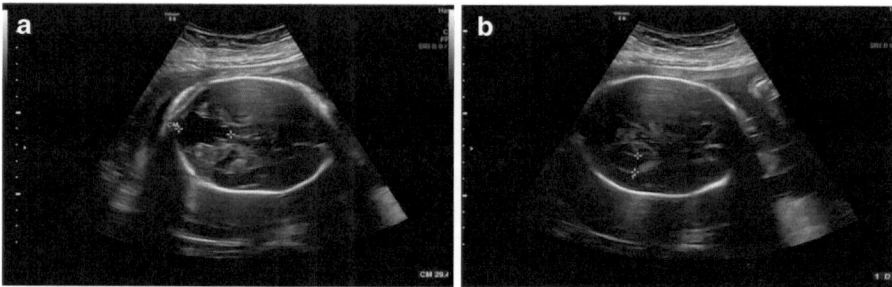

Fig. 23.1 (**a**) 27-week-old fetus with enlarged cisterna magna, (**b**) with mild ventriculomegaly (10 mm)

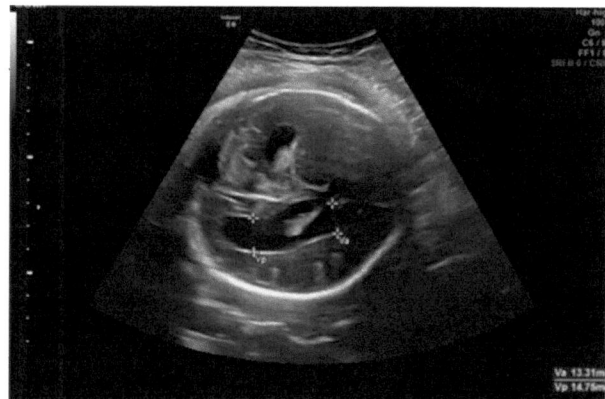

Fig. 23.2 A 33-week-old fetus with moderate ventriculomegaly (atrial diameter = 14.75 mm)

2. Large body habitus may pose some limitations.
3. Artifacts.
4. Possible risk of thermal heating or mechanical injuries to the fetus from high frequencies.
5. Decreased image quality in organs with high acoustical impedance (e.g., ossified calvaria).

23.2.7 Future

In the coming years, fetal ultrasound will not only be the golden standard in fetal screening, but it will also be relied on as a distinct diagnostic tool. Due to the recent advances in dedicated neurosonography (3D, 4D, Doppler imaging) that have allowed visualization of not only the complex anatomical structures in the growing

Fig. 23.3 A 31-week-old fetus with (**a**) moderate ventriculomegaly, (**b**) an enlarged cisterna magna (46.84 mm) and an occipital defect (encephalocele), and (**c**) polydactyly

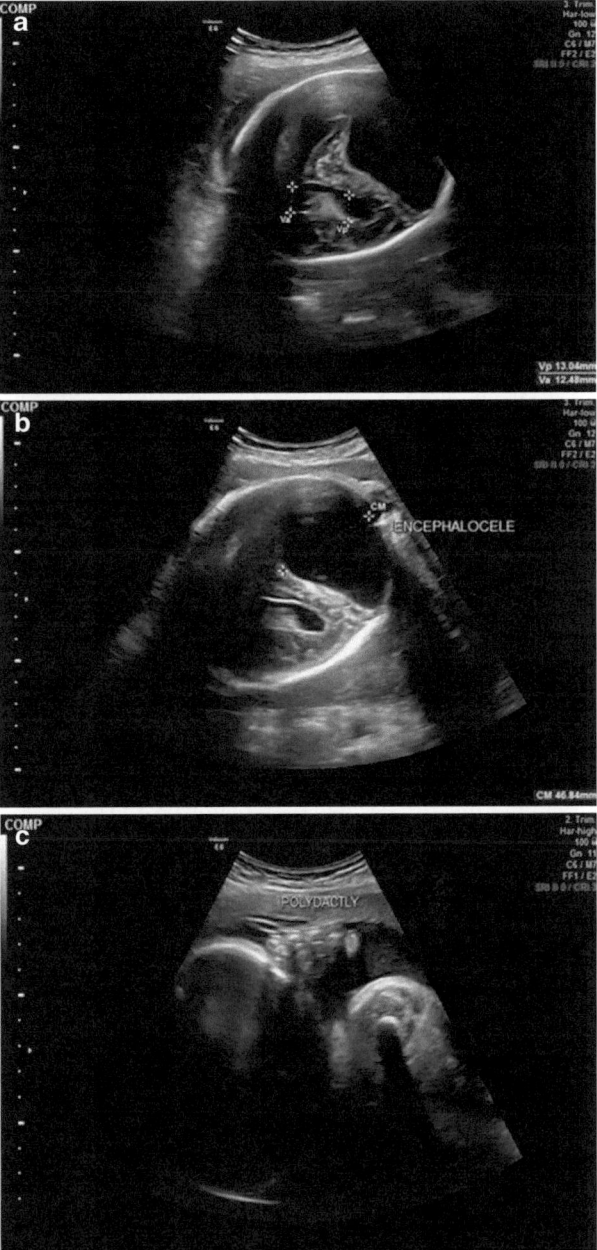

fetus, it has allowed an assessment of the neurobehavioral status in various fetal situations.

Rizzo et al. (2016) demonstrated satisfactory volumetric evaluation of the fetal brain when they applied 5D CNS+ software to simultaneously visualize the fetal biparietal diameter, occipitofrontal diameter, head circumference, posterior fossa, cerebellum, and atrial lateral ventricle diameter using three planes (axial, sagittal, and coronal). One drawback to this technique is the need for an experienced sonographer who is well equipped in interpreting fetal CNS anatomy [50].

An in utero neurobehavioral assessment was demonstrated using 4D US by Kurjak et al. (2012). They were the first to devise a score, the Kurjak's antenatal neurodevelopmental test (KANET), to detect, classify, and follow up fetuses that show signs of abnormal development. A significant drawback to this test is that it is time-consuming [51]. Another example is the Fetal Observable Movement System (FOMS), which is used to score facial movements [52]. Hydrocephalic fetuses have yet to be included in these studies, but that may change in the coming years.

Two other US modalities that might be able to detect microstructural changes in the fetal brain are US elastography/sonoelastography and molecular ultrasonography. Molecular ultrasonography is yet to be implemented clinically; sonoelastography has been used in a range of clinical applications but not in evaluating the brain. Both are novel applications and only time will tell if they are implemented as distinct diagnostic tools.

23.3   Magnetic Resonance Imaging (MRI)

Although the development of ultrafast T2-weighted MRI sequences occurred in the early 1980s, the era of fetal MRI began when these sequences were used to evaluate the fetus in the late 1990s and early 2000s. These sequences provided, as the name implies, extremely short acquisition times that proved to be particularly beneficial when imaging a moving fetus. The T2-weighted sequences allowed better delineation of brain surfaces and CSF pathways [53].

The physics behind MRI is a complicated subject; nevertheless, we have tried to make it as intelligible as possible. The MR system contains a magnet with a specific field strength (1.5 T or 3 T); liquid helium-immersed wires in the shape of loops surround this magnet. When these loops are subjected to an electric current, with the flow of electrons, a magnetic field is formed perpendicularly to the loops. When this large magnetic field is applied to the human body, the hydrogen protons in our body readjust themselves, so some will be aligned in the direction of the magnetic field and others in the opposite direction. Net magnetization is produced from the small remainder of protons aligned with the main magnetic field, thus giving the source for the MR signal. The rapid changing of magnetic and electric fields produces radiofrequency energy that is transmitted by an RF coil as an RF pulse. A 90° RF pulse rotates the net magnetization to the transverse plane; a 180° RF pulse rotates the net magnetization 180° in the $-z$ direction. Relaxation and dephasing times of the net magnetization are the backbone of acquiring basic MR sequences [54].

23.3.1 Indications and Contraindications

Indications:

1. Increased BMI of the mother, obscuring image quality of ultrasound

2. Oligohydramnios/polyhydramnios impairing ultrasound image quality
3. Need for visualizing the fetus as a whole, usually in the second and third trimester
4. Further evaluation of US-detected CNS malformations
5. Metabolic assessment in suspected hypoxic-ischemic injury
6. Screening for associated anomalies and syndromic diseases
7. Evaluating underlying cause for intrauterine growth restriction (IUGR)
8. Screening in suspected familial genetic syndromes
9. Maternal disease with a possible interference of fetal development
10. Assessing fetus after trauma to the maternal abdomen

Contraindications:

1. *Absolute*: Same as for any MRI exam, such as pacemakers.
2. *Relative*: Claustrophobia; maternal sedation can be considered.

23.3.2 Sequences

The most commonly used sequence in imaging the fetal brain is single-shot fast spin echo (SSFSE), which is a T2-weighted sequence. Balanced fast field echo (b-FFE) is another sequence that is preferred in examining the white matter tracts. T1-weighted images can be taken using two-dimensional gradient-echo (2D GRE) or turbo fast low-angle shot (FLASH). T1-weighted sequences can detect hemorrhage, calcification, and fat. The drawback to acquiring T1-weighted sequences is the long acquisition time (1–2 min), which might prompt maternal sedation. To summarize, T2-weighted sequences are the sequences of choice to initially evaluate the fetal brain; once a more detailed study is necessary, T1-weighted sequences are employed [55].

23.3.3 Positioning and Planes

The mother is asked to lie supine, with the head of the fetus centered within the coil. If the mother cannot tolerate the supine position, left lateral decubitus can be an option. Images of the three planes (axial, sagittal, and coronal) are taken simultaneously. Measuring the atrial diameter is done through the axial plane [56].

23.3.4 Anatomy

23.3.4.1 Normal Anatomy
Starting at 20–21 GW, fetal anatomy can be studied by MRI [57]. The normally visualized anatomy by gestational age can be seen in Table 23.2.

23.3.4.2   Pathological Anatomy

Once ventriculomegaly is detected, the following should be evaluated: presence of frontal horn dilatation, asymmetrical dilatation (>2–2.4 mm), ventricular wall abnormalities, contents of the ventricles (choroid plexus or septation), corpus callosum and CSP, cerebral parenchyma abnormalities, posterior fossa abnormalities, derangements of biometric values (BPD, HC, FOD, and craniocerebral index), and extracranial anomalies. These can be all detected by US but are better delineated by MRI (if dedicated neurosonography is not present) [58–61]. Refer to Tables 23.3 and 23.4 for a more detailed description of the possible findings.

23.3.5   Advantages and Limitations

23.3.5.1   Advantages
1. Nonionizing
2. Allows assessment of the whole fetus at once
3. Excellent delineation of CNS anatomy
4. Operator independent

23.3.5.2   Limitations
1. Cannot be used in fetuses <17 GW
2. Maternal weight limitations
3. Not as cost-effective as US
4. Motion artifacts, which may necessitate sedation
5. Detection of multiple skeletal anomalies
6. Theoretical risk of tissue heating

23.3.6   Future of Fetal MRI

The future of fetal MRI is already here. We are able to evaluate the fetal brain from a myriad of scopes (functional, structural, volumetric, etc.). These methods, when applied to the hydrocephalic fetus, will allow proper counseling, monitoring, and stratification of those who might benefit from therapeutic interventions.

23.3.7   Other Modalities

23.3.7.1   Diffusion-Weighted Imaging (DWI)

DWI utilizes Brownian motion, the random movement of water protons, in detecting minute structural changes in the cerebral parenchyma. Normal brain tissue allows limited random movement of water molecules, but when tissue insult occurs, such as stroke or infection, the normal barriers are disrupted, thus "allowing" increased random movement of these water molecules. The apparent diffusion coefficient (ADC), the net diffusion of water molecules, is calculated to differentiate true restriction from T2 shine through. True diffusion restrictions appear as hyperintense signals on DWI and hypointense signals on ADC [62].

The use of DWI in fetuses, however, is different. Not only does it help in detecting various fetal pathologies, but it also exhibits the normal process of maturation in the continuously developing fetal CNS. DWI allows quantitative assessment of normal and abnormal fetal CNS by calculation of ADC values in the various structures of the fetal brain. Multiple studies (Schneider et al. (2009), Raghini et al. (2003), Boyer et al. (2013)) have documented the normal ADC values in the developing fetal brain at various gestational ages [63–65].

Microstructural changes and physiologic derangements due to hydrocephalus can be quantified using ADC values and comparing them to the ADC values of normal fetal brains. Erdem et al. (2007) demonstrated the calculated ADC values in 12 fetuses with documented ventriculomegaly on ultrasound, which showed lower ADC values at frontal and occipital white matter and basal ganglia. The mechanisms behind the decreased diffusion in long-standing hydrocephalus are proposed to be due to the direct effects of increased intracranial pressure and reduced blood flow, causing ischemic injury to fetal brain tissue [66].

Although no long-term studies have been done to document outcome in relation to reduced ADC values in hydrocephalic fetuses to justify predicting prognoses of such patients, it can be said that DWI offers an in-depth analysis of the fetal CNS and a possible window for intervention (if possible) before irreversible damage occurs.

23.3.7.2 Magnetic Resonance Spectroscopy (MRS)

MRS depends on chemical shift, the changes in resonant frequency of the different chemical substances found within the anatomical structures. In other words, the metabolite peak seen in MRS correlates to the number of protons present in the tissue.

The various metabolites found on MRS, demonstrated by Story et al. (2011), in normal fetal brain tissue are summarized in Table 23.5. These metabolites are expressed when the water signal is suppressed. The levels and ratios of these metabolites changes with the changing gestational age of the fetus and if any pathologies are present [67].

N-Acetylaspartate (NAA) levels are initially low during early development with high levels of myoinositol and choline; with increasing gestational age, however,

Table 23.5 MRS neuronal markers (Story et al. 2011)

Metabolite	Association	Peak levels (ppm)
Myoinositol	Marker for: osmoregulation, nutrition, and detoxification of cells and an astrocytic marker	3.5
Choline	Marker for: myelination and membrane lipid metabolism	3.2
Creatine	Marker for: energy metabolism	3.0
N-Acetylaspartate	Marker for: neural cells and oligodendrocytes	2.0
Lactate	Marker for: anaerobic glycolysis	1.3

NAA levels begin to peak and myo-ins and Cho levels decrease. Although the presence of lactate in adult brain tissue is known to be pathological in origin, indicating the presence of anaerobic glycolysis in response to hypoxia, it is now postulated that in the rapid process of fetal brain maturation and differentiation, some tissues may resort to anaerobic metabolism, thus increasing lactate levels slightly [68–72].

In fetal hydrocephalus, Kok et al. (2003) found the myoinositol levels to be decreased, indicating a possible hypo-osmolar state. If the mechanism of injury is assumed to be hypoxic in nature, several findings are to be noted: decreased NAA levels, decreased Cr levels, increased myo-ins levels, and increased lactate levels which may all indicate poor outcome [73].

Loss in signal-to-noise ratio (SNR) in MRS is a significant drawback that is largely attributed to fetal movement; MRS imaging takes approximately 5 min, which may be improved/corrected using new novel techniques such as constructive averaging [74] or simply decreasing image acquisition times.

23.3.7.3   Diffusion Tensor Imaging (DTI) and Fiber Tractography (FT)

DTI, being highly sensitive to water molecules, allows visualization of white matter tracts in the brain and cerebral wall structures by means of mean diffusivity (MD), the molecular diffusion rate, and fractional anisotropy (FA), the directional preference of diffusion.

Three images may be derived from the DTI sequence: FA map, ADC map, and the color-coded map. The FA map depicts the shape of the diffusion tensor, with FA values ranging from 0 to 1 with higher values indicating more elongation of the diffusion ellipsoid [75]. The ADC map delineates the size of the diffusion ellipsoid, with an ADC value in the unit $10–3$ mm^2/s. The colormap, as the name suggests, assigns colors to the tracts, with red indicating left-right orientation, green indicating anterior-posterior orientation, and blue indicating superior-inferior orientation. Reconstruction of these maps yields a three-dimensional tractogram. Currently, neonatal and pediatric DTI atlases, much like growth percentile charts, demonstrating the anatomical tracts through various ages, are available online [74, 76, 77]. In pediatric hydrocephalus cases, compared with controls, postsurgical DTI showed signs of randomized network features with less small worldness, indicating abnormalities in integration and segregation of brain networks [78].

In the developing fetus, with the high water content of the CNS and the ongoing myelination process, any deviation in DTI parameters such as anisotropy may indicate WM development impairment [79–83].

In fetal hydrocephalus, evaluating the WM tracts and the cerebral wall is of importance in the diagnostic workup and determining possible prognosis, which can be interpreted from the normal parameters found in DTI of normal fetuses.

Regarding WM tract organization in fetal hydrocephalus, a question of whether in utero surgery might be of benefit arises. If detected and treated earlier, would there be less random reorganization of tracts leading to a much better neurocognitive outcome?

23.3.7.4 Functional Magnetic Resonance Imaging (fMRI)

When a stimulus produces an increase in local neuronal activity, thus increasing local blood flow and venous blood oxygenation, a blood-oxygen-level dependent (BOLD) effect is found. fMRI utilizes this effect and detects the local increase in signal (BOLD contrast) to demonstrate activation of certain areas of the brain.

Fetal fMRI has been used [84] to evaluate response to vibroacoustic and visual stimuli with some success, demonstrating a temporal and frontal lobe activation, respectively [85, 86]. Evaluation of resting-state network (RSN) connectivity in normal and pathological conditions in the fetus may help in judging the functional framework and severity of detected abnormalities [87].

Studies have yet to include and evaluate hydrocephalic fetuses using fMRI, but with the improving techniques and acquisition times, fMRI may play a significant role in the understanding of fetal brain development and function in the background of hydrocephalus.

23.3.7.5 Cerebrospinal Fluid Flow in Fetal Hydrocephalus

Assessing cerebrospinal fluid flow dynamics and physiology is becoming an area of interest in evaluating pathological conditions such as normal-pressure hydrocephalus (NPH) and Chiari malformations and determining response to certain treatments or interventions. CSF flow can be measured and visualized using phase-contrast magnetic resonance imaging (PC-MRI) and time-spatial labeling inversion pulse magnetic resonance imaging (time-SLIP MRI). Table 23.6 highlights the major differences between PC-MRI and time-SLIP MRI. Cine phase-contrast MRI in conjunction with computational fluid dynamics (CFD) can be used to contrive a three-dimensional, physiological model of the CNS to quantify CSF flow dynamics [88–103].

No studies have observed nor documented the CSF flow in fetal hydrocephalus using the abovementioned modalities, but it would be interesting to see in the coming years how they can be incorporated in the workup of such patients and

Table 23.6 Comparison of the methods used to assess CSF flow

PC-MRI	Time-SLIP MRI
• Quantitative and qualitative (midline sagittal acquisition) • Acquired by subtracting datasets collected from signal contrast between flowing and stationary spins • Utilizes velocity encoding value (V_{enc}) • Measures CSF pulsatility and stroke volume • Unreliable in gauging response to treatment due to variability • Requires large number of cardiac cycles to acquire • Can be used with CFD	• Qualitative • Acquired by tagging the CSF with radiofrequency pulses and using the CSF as a tracer to generate images. Stationary signals are suppressed • Utilizes the T1 interval • Determines if CSF flow is typical or atypical or no movement is visualized • Reliable tool in determining response to intervention by visualizing CSF flow pre- and post-intervention • CSF movement can be seen up to 5 s • Noninvasive

determining the response and outcome to correcting these abnormalities, whether post- or prenatally.

23.4 The Role of the Obstetric Physical Examination

The obstetric physical exam in the case fetal hydrocephalus has largely been over-shadowed by the routine use of ultrasound. It could be possible to detect an abnormality during the late second and third trimester. After taking a detailed history from the mother, the obstetric exam starts with a general abdominal examination and assessment of the fundal height. The fundal height may give a large for gestational age measurement, due to macrocephaly. After the fundal height is taken, Leopold's maneuvers are used to determine the fetus lie and presentation. The first maneuver, the fundal grip, establishes the presentation. Hydrocephalic fetuses are frequently observed to be in the breech presentation that is when the fetal head occupies the fundus. The second maneuver, the lateral grip, is used to determine the fetal lie (relationship of the long axis of the fetus to the long axis of the mother). Fetal lie can be longitudinal, transverse, or oblique. The third maneuver, the 1st pelvic grip, is an adjunct to the fundal grip and confirms the fetal presentation. The fourth maneuver, the 2nd pelvic grip, helps in assessing fetal engagement. It is important to note that a normal obstetric exam does not exclude fetal hydrocephalus, while an abnormal obstetric exam may prelude the diagnosis. After the physical exam is concluded, an ultrasound must be used to confirm the findings.

23.5 The Genetics Behind Fetal Hydrocephalus

The methods used in prenatal diagnosis of genetic abnormalities are classified into two groups: invasive methods and noninvasive methods. The invasive methods are amniocentesis, chorionic villus sampling, and percutaneous umbilical cord blood sampling. Noninvasive prenatal screening (NIPS) utilizes the principles of cell-free fetal DNA, the process of detection of fetal DNA in the mother's blood. Only four gene mutations have been associated with pure isolated ventriculomegaly: L1CAM (L1 cell adhesion molecule), AP1S2 (adaptor-related protein complex 1, sigma-2 subunit), MPDZ (multiple PDZ domain protein), and CCDC88C (coiled-coil domain-containing protein 88C) [104]. The genetic abnormalities seen in isolated and syndromic hydrocephalus are discussed in another chapter. As a general rule, genetic screening is usually done using any of the NIPS methods first and can be escalated to invasive methods if an increased risk of genetic abnormality is detected. To conclude, hydrocephalus is believed to be the manifestation of abnormal genetic mutations, their interaction, or both.

Conclusion

The case of the hydrocephalic fetus is a complex one. The recent advances in imaging technologies have allowed us to not only observe the fetal development in intricate detail, but we are also privileged to assess the fetal brain from a structural, perfusional, and functional standpoint. Integration of these modalities in the workup of fetal hydrocephalus will get us closer than ever to fully understanding the pathological process and helping us determine who might benefit from therapy the most. This integration is also a reminder that these methods are meant to complement each other and not compete.

The case of the hydrocephalic fetus is a complex one, but with the technology we have today, a little less so.

References

1. Moore KL, Persaud TVN, Torchia MG. The developing human: clinically oriented embryology. Amsterdam: Elsevier Health Sciences; 2015.
2. Müller F, O'Rahilly R. The human brain at stage 16, including the initial evagination of the neurohypophysis. Anat Embryol. 1989;179(6):551–69.
3. Müller F, O'Rahilly R. The human brain at stage 17, including the appearance of the future olfactory bulb and the first amygdaloid nuclei. Anat Embryol. 1989;180(4):353–69.
4. Müller F, O'Rahilly R. The development of the human brain from a closed neural tube at stage 13. Anat Embryol. 1988;177(3):203–24.
5. Müller F, O'Rahilly R. The first appearance of the future cerebral hemispheres in the human embryo at stage 14. Anat Embryol. 1988;177(6):495–511.
6. O'Rahilly R, Müller F. The development of the neural crest in the human. J Anat. 2007;211(3):335–51.
7. O'Rahilly R, Müller F. Ventricular system and choroid plexuses of the human brain during the embryonic period proper. Am J Anat. 1990;189(4):285–302.
8. Hans J, Lammens M, Hori A. Clinical neuroembryology: development and developmental disorders of the human central nervous system. New York: Springer; 2014.
9. Campbell S. A short history of sonography in obstetrics and gynaecology. Facts Views Vis Obgyn. 2013;5(3):213.
10. Kurjak A, Chervenak FA. Donald School textbook of ultrasound in obstetrics and gynecology. New Delhi: Jaypee Brothers Publishers; 2011.
11. Bushberg JT, Boone JM. The essential physics of medical imaging. Philadelphia: Lippincott Williams & Wilkins; 2011.
12. Kremkau FW, Forsberg F. Sonography principles and instruments. Amsterdam: Elsevier Health Sciences; 2015.
13. Kim MS, Jeanty P, Turner C, Benoit B. Three-dimensional sonographic evaluations of embryonic brain development. J Ultrasound Med. 2008;27(1):119–24.
14. International Society of Ultrasound in Obstetrics & Gynecology Education Committee. Sonographic examination of the fetal central nervous system: guidelines for performing the 'basic examination' and the 'fetal neurosonogram'. Ultrasound Obstet Gynecol. 2007;29(1):109.
15. Malinger G, Lev D, Lerman-Sagie T. Normal and abnormal fetal brain development during the third trimester as demonstrated by neurosonography. Eur J Radiol. 2006;57(2):226–32.
16. Chudleigh T, Smith A, Cumming S. Obstetric & gynaecological ultrasound: how, why and when. Amsterdam: Elsevier Health Sciences; 2016.
17. Salomon LJ, Alfirevic Z, Bilardo CM, Chalouhi GE, Ghi T, Kagan KO, Lau TK, Papageorghiou AT, Raine-Fenning NJ, Stirnemann J, Suresh S. ISUOG practice guidelines: performance of first-trimester fetal ultrasound scan. Ultrasound Obstet Gynecol. 2013;41(1):102.
18. Endres LK, Cohen L. Reliability and validity of three-dimensional fetal brain volumes. J Ultrasound Med. 2001;20(12):1265–9.

19. Malinger G, Svirsky R, Ben-Haroush A, Golan A, Bar J. Doppler-flow velocity indices in fetal middle cerebral artery in unilateral and bilateral mild ventriculomegaly. J Matern Fetal Neonatal Med. 2011;24(3):506–10.
20. Zalel Y, Almog B, Seidman DS, Achiron R, Lidor A, Gamzu R. The resistance index in the fetal middle cerebral artery by gestational age and ventricle size in a normal population. Obstet Gynecol. 2002;100(6):1203–7.
21. Degani S. Evaluation of fetal cerebrovascular circulation and brain development: the role of ultrasound and Doppler. Semin Perinatol. 2009;33(4):259–69. WB Saunders.
22. D'addario V, Rossi AC. Neuroimaging of ventriculomegaly in the fetal period. Semin Fetal Neonatal Med. 2012;17(6):310–8. WB Saunders.
23. Curnes JT, Oakes WJ, Boyko OB. MR imaging of hindbrain deformity in Chiari II patients with and without symptoms of brainstem compression. Am J Neuroradiol. 1989;10(2):293–302.
24. Hadley DM. The Chiari malformations. J Neurol Neurosurg Psychiatry. 2002;72(2):ii38–40.
25. Bosemani T, Orman G, Boltshauser E, Tekes A, Huisman TA, Poretti A. Congenital abnormalities of the posterior fossa. Radiographics. 2015;35(1):200–20.
26. Mirsky DM, Schwartz ES, Zarnow DM. Diagnostic features of myelomeningocele: the role of ultrafast fetal MRI. Fetal Diagn Ther. 2014;37(3):219–25.
27. Shaer CM, Chescheir N, Schulkin J. Myelomeningocele: a review of the epidemiology, genetics, risk factors for conception, prenatal diagnosis, and prognosis for affected individuals. Obstet Gynecol Surv. 2007;62(7):471–9.
28. Pilu G, Visentin A, Valeri B. The Dandy–Walker complex and fetal sonography. Ultrasound Obstet Gynecol. 2000;16(2):115–7.
29. Mavili E, Coskun A, Per H, Donmez H, Kumandas S, Yikilmaz A. Polymicrogyria: correlation of magnetic resonance imaging and clinical findings. Childs Nerv Syst. 2012;28(6):905–9.
30. Dhombres F, Nahama-Allouche C, Gelot A, Jouannic JM, Billette de Villemeur T, Saint-Frison MH, Ducou le Pointe H, Garel C. Prenatal ultrasonographic diagnosis of polymicrogyria. Ultrasound Obstet Gynecol. 2008;32(7):51–4.
31. Kfir M, Yevtushok L, Onishchenko S, Wertelecki W, Bakhireva L, Chambers CD, Jones KL, Hull AD. Can prenatal ultrasound detect the effects of in-utero alcohol exposure? A pilot study. Ultrasound Obstet Gynecol. 2009;33(6):683–9.
32. Barkovich AJ, Koch TK, Carrol CL. The spectrum of lissencephaly: report of ten patients analyzed by magnetic resonance imaging. Ann Neurol. 1991;30(2):139–46.
33. Benacerraf BR. Ultrasound of fetal syndromes. Amsterdam: Elsevier Health Sciences; 2008.
34. Chen CP. Prenatal findings and the genetic diagnosis of fetal overgrowth disorders: Simpson-Golabi-Behmel syndrome, Sotos syndrome, and Beckwith-Wiedemann syndrome. Taiwan J Obstet Gynecol. 2012;51(2):186–91.
35. Marín R, Ley-Martos M, Gutiérrez G, Rodríguez-Sánchez F, Arroyo D, Mora-López F. Three cases with L1 syndrome and two novel mutations in the L1CAM gene. Eur J Pediatr. 2015;174(11):1541–4.
36. Cheon JE, Kim IO, Hwang YS, Kim KJ, Wang KC, Cho BK, Chi JG, Kim CJ, Kim WS, Yeon KM. Leukodystrophy in children: a pictorial review of MR imaging features. Radiographics. 2002;22(3):461–76.
37. Rode ME, Mennuti MT, Giardine RM, Zackai EH, Driscoll DA. Early ultrasound diagnosis of Neu–Laxova syndrome. Prenat Diagn. 2001;21(7):575–80.
38. Ickowicz V, Eurin D, Maugey-Laulom B, Didier F, Garel C, Gubler MC, Laquerriere A, Avni EF. Meckel-Grüber syndrome: sonography and pathology. Ultrasound Obstet Gynecol. 2006;27(3):296–300.
39. Bethune M. Literature review and suggested protocol for managing ultrasound soft markers for Down syndrome: thickened nuchal fold, echogenic bowel, shortened femur, shortened humerus, pyelectasis and absent or hypoplastic nasal bone. Australas Radiol. 2007;51(3):218–25.
40. Benaicha A, Dommergues M, Jouannic JM, Jacquette A, Alexandre M, Le Merrer M, Ducou Le Pointe H, Garel C. Prenatal diagnosis of brachytelephalangic chondrodysplasia punctata: case report. Ultrasound Obstet Gynecol. 2009;34(6):724–6.

41. Burrows PE, Stannard MW, Pearrow J, Sutterfield S, Baker ML. Early antenatal sonographic recognition of thanatophoric dysplasia with cloverleaf skull deformity. Am J Roentgenol. 1984;143(4):841–3.
42. Ritner JA, Frates MC. Fetal CNS: a systematic approach. Radiol Clin N Am. 2014;52(6):1253–64.
43. Levitsky DB, Mack LA, Nyberg DA, Shurtleff DB, Shields LA, Nghiem HV, Cyr DR. Fetal aqueductal stenosis diagnosed sonographically: how grave is the prognosis? AJR Am J Roentgenol. 1995;164(3):725–30.
44. Emery SP, Greene S, Hogge WA. Fetal therapy for isolated aqueductal stenosis. Fetal Diagn Ther. 2015;38(2):81–5.
45. Elchalal U, Yagel S, Gomori JM, Porat S, Beni-Adani L, Yanai N, Nadjari M. Fetal intracranial hemorrhage (fetal stroke): does grade matter? Ultrasound Obstet Gynecol. 2005;26(3):233–43.
46. Sauerbrei A, Wutzler P. The congenital varicella syndrome. J Perinatol. 2000;20(8):548.
47. Brown ZA, Selke S, Zeh J, Kopelman J, Maslow A, Ashley RL, Watts DH, Berry S, Herd M, Corey L. The acquisition of herpes simplex virus during pregnancy. N Engl J Med. 1997;337(8):509–16.
48. Haratz KK, Oliveira PS, Rolo LC, Nardozza LM, Milani HF, Barreto EQ, Araujo Júnior E, Ajzen SA, Moron AF. Fetal cerebral ventricle volumetry: comparison between 3D ultrasound and magnetic resonance imaging in fetuses with ventriculomegaly. J Matern Fetal Neonatal Med. 2011;24(11):1384–91.
49. Hata T, Mori N, Tenkumo C, Hanaoka U, Kanenishi K, Tanaka H. Three-dimensional volume-rendered imaging of normal and abnormal fetal fluid-filled structures using inversion mode. J Obstet Gynaecol Res. 2011;37(11):1748–54.
50. Rizzo G, Capponi A, Persico N, Ghi T, Nazzaro G, Boito S, Pietrolucci ME, Arduini D. 5D CNS+ software for automatically imaging axial, sagittal, and coronal planes of normal and abnormal second-trimester fetal brains. J Ultrasound Med. 2016;35(10):2263–72.
51. Kurjak A, Stanojević M, Predojević M, Laušin I, Salihagić-Kadić A. Neurobehavior in fetal life. Semin Fetal Neonatal Med. 2012;17(6):319–23.
52. Hata T. Current status of fetal neurodevelopmental assessment: four-dimensional ultrasound study. J Obstet Gynaecol Res. 2016;42(10):1211–21.
53. Clouchoux C, Limperopoulos C. Novel applications of quantitative MRI for the fetal brain. Pediatr Radiol. 2012;42(S1):24–32.
54. Pooley RA. Fundamental physics of MR imaging. Radiographics. 2005;25(4):1087–99.
55. Saleem SN. Fetal MRI: an approach to practice: a review. J Adv Res. 2014;5(5):507–23.
56. Prayer D. Fetal MRI. 1st ed. Berlin: Springer; 2011.
57. Levine D, Hatabu H, Gaa J, Atkinson MW, Edelman RR. Fetal anatomy revealed with fast MR sequences. AJR Am J Roentgenol. 1996;167(4):905–8.
58. Brown JS, Levine D. MR volumetry of brain and CSF in fetuses referred for ventriculomegaly. AJR Am J Roentgenol. 2007;189(1):145–51.
59. Glenn OA, Barkovich AJ. Magnetic resonance imaging of the fetal brain and spine: an increasingly important tool in prenatal diagnosis, part 1. Am J Neuroradiol. 2006;27(8):1604–11.
60. Mailath-Pokorny M, Kasprian G, Mitter C, Schöpf V, Nemec U, Prayer D. Magnetic resonance methods in fetal neurology. Semin Fetal Neonatal Med. 2012;17(5):278–84.
61. Morris JE, Rickard S, Paley MNJ, Griffiths PD, Rigby A, Whitby EH. The value of in-utero magnetic resonance imaging in ultrasound diagnosed foetal isolated cerebral ventriculomegaly. Clin Radiol. 2007;62(2):140–4.
62. Bammer R. Basic principles of diffusion-weighted imaging. Eur J Radiol. 2003;45(3):169–84.
63. Boyer AC, GonÇalves LF, Lee W, Shetty A, Holman A, Yeo L, Romero R. Magnetic resonance diffusion-weighted imaging: reproducibility of regional apparent diffusion coefficients for the normal fetal brain. Ultrasound Obstet Gynecol. 2013;41(2):190–7.
64. Righini A, Bianchini E, Parazzini C, Gementi P, Ramenghi L, Baldoli C, Nicolini U, Mosca F, Triulzi F. Apparent diffusion coefficient determination in normal fetal brain: a prenatal MR imaging study. Am J Neuroradiol. 2003;24(5):799.

65. Schneider JF, Confort-Gouny S, Le Fur Y, Viout P, Bennathan M, Chapon F, Fogliarini C, Cozzone P, Girard N. Diffusion-weighted imaging in normal fetal brain maturation. Eur Radiol. 2007;17(9):2422–9.

66. Erdem G, Celik O, Hascalik S, Karakas HM, Alkan A, Firat AK. Diffusion-weighted imaging evaluation of subtle cerebral microstructural changes in intrauterine fetal hydrocephalus. Magn Reson Imaging. 2007;25(10):1417–22.

67. Story L, Damodaram MS, Allsop JM, McGuinness A, Wylezinska M, Kumar S, Rutherford MA. Proton magnetic resonance spectroscopy in the fetus. Eur J Obstet Gynecol Reprod Biol. 2011;158(1):3–8.

68. Azpurua H, Alvarado A, Mayobre F, Salom T, Copel JA, Guevara-Zuloaga F. Metabolic assessment of the brain using proton magnetic resonance spectroscopy in a growth-restricted human fetus: case report. Am J Perinatol. 2008;25(05):305–9.

69. Berger-Kulemann V, Brugger PC, Pugash D, Krssak M, Weber M, Wielandner A, Prayer D. MR spectroscopy of the fetal brain: is it possible without sedation? Am J Neuroradiol. 2013;34(2):424–31.

70. Brighina E, Bresolin N, Pardi G, Rango M. Human fetal brain chemistry as detected by proton magnetic resonance spectroscopy. Pediatr Neurol. 2009;40(5):327–42.

71. Evangelou IE, Du Plessis AJ, Vezina G, Noeske R, Limperopoulos C. Elucidating metabolic maturation in the healthy fetal brain using 1H-MR spectroscopy. Am J Neuroradiol. 2016;37(2):360–6.

72. Pugash D, Krssak M, Kulemann V, Prayer D. Magnetic resonance spectroscopy of the fetal brain. Prenat Diagn. 2009;29(4):434–41.

73. Kok RD, Steegers-Theunissen RP, Eskes TK, Heerschap A, van den Berg PP. Decreased relative brain tissue levels of inositol in fetal hydrocephalus. Am J Obstet Gynecol. 2003;188(4):978–80.

74. Shetty AN, Gabr RE, Rendon DA, Cassady CI, Mehollin-Ray AR, Lee W. Improving spectral quality in fetal brain magnetic resonance spectroscopy using constructive averaging. Prenat Diagn. 2015;35(13):1294–300.

75. Assaf Y, Ben-Sira L, Constantini S, Chang LC, Beni-Adani L. Diffusion tensor imaging in hydrocephalus: initial experience. Am J Neuroradiol. 2006;27(8):1717–24.

76. Deshpande R, Chang L, Oishi K. Construction and application of human neonatal DTI atlases. Front Neuroanat. 2015;9:138.

77. Shi F, Yap PT, Fan Y, Gilmore JH, Lin W, Shen D. Construction of multi-region-multi-reference atlases for neonatal brain MRI segmentation. Neuroimage. 2010;51(2):684–93.

78. Yuan W, Meller A, Shimony JS, Nash T, Jones BV, Holland SK, Altaye M, Barnard H, Phillips J, Powell S, McKinstry RC. Left hemisphere structural connectivity abnormality in pediatric hydrocephalus patients following surgery. Neuroimage Clin. 2016;12:631–9.

79. Bui T, Daire J, Chalard F, Zaccaria I, Alberti C, Elmaleh M, Garel C, Luton D, Blanc N, Sebag G. Microstructural development of human brain assessed in utero by diffusion tensor imaging. Pediatr Radiol. 2006;36(11):1133–40.

80. Gupta RK, Hasan KM, Trivedi R, Pradhan M, Das V, Parikh NA, Narayana PA. Diffusion tensor imaging of the developing human cerebrum. J Neurosci Res. 2005;81(2):172–8.

81. Huang H, Xue R, Zhang J, Ren T, Richards LJ, Yarowsky P, Miller MI, Mori S. Anatomical characterization of human fetal brain development with diffusion tensor magnetic resonance imaging. J Neurosci. 2009;29(13):4263–73.

82. Hüppi PS, Dubois J. Diffusion tensor imaging of brain development. Semin Fetal Neonatal Med. 2006;11(6):489–97.

83. Kasprian G, Brugger PC, Weber M, Krssák M, Krampl E, Herold C, Prayer D. In utero tractography of fetal white matter development. Neuroimage. 2008;43(2):213–24.

84. Gowland P, Fulford J. Initial experiences of performing fetal fMRI. Exp Neurol. 2004;190:22–7.

85. Jardri R, Houfflin-Debarge V, Delion P, Pruvo JP, Thomas P, Pins D. Assessing fetal response to maternal speech using a noninvasive functional brain imaging technique. Int J Dev Neurosci. 2012;30(2):159–61.

86. Garel C. New advances in fetal MR neuroimaging. Pediatr Radiol. 2006;36(7):621–5.
87. Schöpf V, Kasprian G, Brugger PC, Prayer D. Watching the fetal brain at 'rest'. Int J Dev Neurosci. 2012;30(1):11–7.
88. Battal B, Kocaoglu M, Bulakbasi N, Husmen G, Sanal HT, Tayfun C. Cerebrospinal fluid flow imaging by using phase-contrast MR technique. Br J Radiol. 2014;84(1004):758–65.
89. Chen G, Zheng J, Xiao Q, Liu Y. Application of phase-contrast cine magnetic resonance imaging in endoscopic aqueductoplasty. Exp Ther Med. 2013;5(6):1643–8.
90. Enzmann DR, Pelc N. Normal flow patterns of intracranial and spinal cerebrospinal fluid defined with phase-contrast cine MR imaging. Radiology. 1991;178(2):467–74.
91. Greitz D. Cerebrospinal fluid circulation and associated intracranial dynamics. A radiologic investigation using MR imaging and radionuclide cisternography. Acta Radiol Suppl. 1992;386:1–23.
92. Hentschel S, Mardal KA, Løvgren AE, Linge S, Haughton V. Characterization of cyclic CSF flow in the foramen magnum and upper cervical spinal canal with MR flow imaging and computational fluid dynamics. Am J Neuroradiol. 2010;31(6):997–1002.
93. Linninger AA, Xenos M, Zhu DC, Somayaji MR, Kondapalli S, Penn RD. Cerebrospinal fluid flow in the normal and hydrocephalic human brain. IEEE Trans Biomed Eng. 2007;54(2):291–302.
94. Matsumae M, Hirayama A, Atsumi H, Yatsushiro S, Kuroda K. Velocity and pressure gradients of cerebrospinal fluid assessed with magnetic resonance imaging: clinical article. J Neurosurg. 2014;120(1):218–27.
95. Sherman JL, Citrin CM. Magnetic resonance demonstration of normal CSF flow. Am J Neuroradiol. 1986;7(1):3–6.
96. Stoquart-El Sankari S, Lehmann P, Gondry-Jouet C, Fichten A, Godefroy O, Meyer ME, Baledent O. Phase-contrast MR imaging support for the diagnosis of aqueductal stenosis. Am J Neuroradiol. 2009;30(1):209–14.
97. Sweetman B, Linninger AA. Cerebrospinal fluid flow dynamics in the central nervous system. Ann Biomed Eng. 2011;39(1):484–96.
98. Velardi F, Hoffman HJ, Ash JM, Hendrick EB, Humphreys RP. The value of CSF flow studies in infants with communicating hydrocephalus. Childs Nerv Syst. 1986;2(3):139–43.
99. Wentland AL, Wieben O, Korosec FR, Haughton VM. Accuracy and reproducibility of phase-contrast MR imaging measurements for CSF flow. Am J Neuroradiol. 2010;31(7):1331–6.
100. Yamada S, Tsuchiya K, Bradley WG, Law M, Winkler ML, Borzage MT, Miyazaki M, Kelly EJ, McComb JG. Current and emerging MR imaging techniques for the diagnosis and management of CSF flow disorders: a review of phase-contrast and time–spatial labeling inversion pulse. Am J Neuroradiol. 2015;36(4):623–30.
101. Yamada S, Miyazaki M, Kanazawa H, Higashi M, Morohoshi Y, Bluml S, McComb JG. Visualization of cerebrospinal fluid movement with spin labeling at MR imaging: preliminary results in normal and pathophysiologic conditions. Radiology. 2008;249(2):644–52.
102. Öztürk M, Sığırcı A, Ünlü S. Evaluation of aqueductal cerebrospinal fluid flow dynamics with phase-contrast cine magnetic resonance imaging in normal pediatric cases. Clin Imaging. 2016;40(6):1286–90.
103. Quencer RM. Intracranial CSF flow in pediatric hydrocephalus: evaluation with cine-MR imaging. Am J Neuroradiol. 1992;13(2):601–8.
104. Kousi M, Katsanis N. The genetic basis of hydrocephalus. Annu Rev Neurosci. 2016;39:409–35.

The Dilemma of Prenatal Hydrocephalus: Grading and Classification of Fetal Hydrocephalus

24

Ahmed Ammar

24.1   Introduction

Currently, there is no global agreement nor consensus on how to deal with cases of hydrocephalus diagnosed in utero. The reason is not due to the unavailability of surgical tools and medical equipment but mainly due to uncertainty on the outcome of such procedures [1–6]. The ethical and religious values of both parents and their families, the society, and customs that they belong to are factors to consider and may influence the decision for prenatal intervention [7, 8]. Each option on dealing with prenatal hydrocephalus has several precautions and certain factors that should be considered before taking any decision. The trial of surgical repairs of fetal cardiac or urologic or diaphragmatic hernia showed some success [9–12]. These have brought encouragements to neurosurgeons in spina bifida repair and draining dilated ventricles during prenatal periods. Spina bifida repair was performed more often than the trials to treat hydrocephalus for several obvious reasons related to the understanding of the pathophysiology of hydrocephalus. Neurosurgeons in favor of prenatal intervention based their management on studies showing that the progression of hydrocephalus may have serious effect on the growth of the fetus, and subsequently the child's mental, neurologic, and cognitive functions may be affected too [2–5, 13–17]. Therefore, there is an assumption that an early reduction of ICP and reduced accumulation of CSF in dilated ventricles may alter the progress, and consequences of prenatal hydrocephalus subsequently improve the outcome. Other imposing teams believed that the problem is more serious and complicated than the dilated ventricles itself. It includes genetic profile of the fetus and the function of neural cells and tissues. Exposure of the pregnant mothers to unnecessary risks without enough evidence that it will be beneficial for the babies on a long term raised a very serious ethical issue.

A. Ammar, M.D., M.B.Ch.B., D.M.Sc.
Department of Neurosurgery, King Fahd University Hospital, Imam Abdulrahman Bin Faisal University, Al Khobar, Saudi Arabia
e-mail: ahmed@ahmedammar.com

© Springer International Publishing AG 2017   341
A. Ammar (ed.), *Hydrocephalus*, DOI 10.1007/978-3-319-61304-8_24

24.2   Illustrative Case

This case is seen frequently in different centers; it demonstrates the dilemma of prenatal hydrocephalus.

An obstetrician during routine examination of pregnant woman discovered that her 28 GW fetus suffers hydrocephalus. So, the obstetrician sought neurosurgical opinion. The mother is 32 years old young and healthy and that is her first pregnancy. US examination showed that her 28 GW fetus suffers progressive ventricular dilatation over a period of 4 weeks. No other congenital anomalies have been detected. A council meeting with the parents was arranged and organized. The medical team includes obstetrician, neonatologist, neurosurgeon, and genetic specialist. A social worker attended the meeting as well. The stressed parents asked several questions including:

(a) What is the future of the fetus? What are the possibilities for the fetus to survive?
(b) What are the causes of hydrocephalus?
(c) What are the options of treatment? The availability and safety of these methods!
(d) What are the risks accompanying this pregnancy on the mother and the fetus?
(e) What are the risks of each methods of intervention for the mother?
(f) What are the risks of each methods of intervention for the fetus?
(g) Would these methods of interventions be painful for the fetus?
(h) What are the risks or chances that future pregnancy may produce another hydrocephalic fetus?

The answers to these questions are explained to the parents in a comprehensible, scientific, and professional way. This case clearly illustrates the dilemma in which the mother and treating medical team faced in order to make decision to properly manage such a problem. So far, we do not have an absolute, proven, and evidence-based medicine answer for most of these questions.

24.2.1 Professional Medical Prospect

The medical/professional answers and action are and should be based on solid scientific facts without risking the health of the mother for any reason. The pathophysiology, development, and progression of hydrocephalus are very complicated processes and are not fully understood [6, 12, 18, 19]. Therefore, the prediction of the outcome and the future of the fetus with hydrocephalus cannot be predicted in most of the cases unless there are significant genetic abnormalities. In cases with multiple CNS anomalies, huge dilated ventricles and a very thin cortex, or the absence of certain parts of brain tissues and associated genetic alterations [4, 7, 19–21], the prediction of unfavorable long-term outcome is likely to occur. In other types of isolated hydrocephalus with mild or moderate ventricle dilatation

[6, 11, 22, 23], prenatal intervention may be considered and discussed. There are few and scattered reliable studies showing different experiences and outcome for intrauterine intervention of detected congenital anomalies such as spina bifida or hydrocephalus but failed to produce strong evidence on its benefits [1, 19, 24–28]. Ammar A. and Al-Jama F. et al. in 1996 [25] presented their limited experience in decompressing the dilated ventricles prenatally by using prolonged intrauterine transabdominal ventricular external drainage. In this case, the head circumference was reduced significantly to allow successful normal vaginal delivery. However, the long-term follow-up for such hydrocephalic fetus and her mother was not mentioned. This is exactly the case in most of the literatures presenting cases of long-term prenatal intervention with few exceptions such as Cavalcant DP. and Salomao MA. et al. in 2003 [29].

24.2.2 Ethical Prospect

The professional and ethical approach to the cases of prenatal hydrocephalus should be based on:

1. The mother's health status, opinion, and decision. The pregnant ladies have the right to know all the facts regarding their fetus and hydrocephalus. Different options of treatment should be discussed to her, and both parents should share in making the decision. The parents should be partners in dealing with such condition. At the end, mothers have the absolute right to make the final decision.
2. The fetal status and medical condition. Cases of fetus with isolated ventriculomegaly with normal chromosomal and genetic profile should be identified and differentiated from other cases of prenatal hydrocephalus with other congenital anomalies.
3. Availability of methods of investigations, surgical techniques, and postnatal care in the center where the fetus will be treated. In cases of any shortcoming in prenatal or postnatal care, the pregnant mother and the fetus should be referred to a center where the service is available.
4. The ethics, values, culture, and religious beliefs of the parents, their family, and their society should be considered as well.
5. The social and economic prospects of the health-care system, educational system, and social aids are factors to be considered.

24.3 Experience with Fetal Hydrocephalus

The demand and availability of the free prenatal examination for more than 30 years lead to an early diagnosis of different types of CNS congenital anomalies. We have been challenged, as other neurosurgeons, to use these important informations to improve the outcomes of hydrocephalus detected in utero [13, 30–35]. This continuous effort took several stages and studies. In brief, this experience can be summarized in these research studies which are:

(a) Study the size of the problem
(b) Option for early delivery, admission to NICU, and treatment
(c) Design neonate shunt (Ammar shunt)
(d) Fetal surgery
(e) Audit the experience
(f) Producing "Prenatal Hydrocephalus Management Guideline (PHMG)"

24.3.1 Study the Size of the Problem

Awary B. and Lardi A. et al. in 1997 [36] showed that the frequency of hydrocephalus is 1/1000 live births in Saudi Arabia. It also showed the frequency of CNS anomalies in aborted fetus. There are several studies that showed the prevalence of hydrocephalus in different geographical places in the world but may be differing by racial data too [13, 30–35]. In general, the incidence of hydrocephalus globally varies between 1 and 3/1000 LB, and the incidence of hydrocephalus in aborted fetus may be higher.

24.3.2 Early Delivery and Shunting as Option to Treat Hydrocephalic Fetus

The option of induced early delivery and insertion of VP shunt in the first few days after delivery was practiced in few centers in the world [6, 18, 19, 37, 38]. The scientific foundation of this approach was based on well-known facts, which are (a) most of the development of the brain and neural tissues and cells occurs intrauterine and (b) increased intracranial pressure and enlarged ventricles may alter the normal development and cause permanent neurological defects and cognitive function disorders. A designed protocol was made to induce early delivery and shunting of the neonate.

The inclusion criteria are:
1. The parents, especially the mother's acceptance and signed informed consent for such procedures.
2. Hydrocephalus is the only detected anomaly.
3. Hydrocephalus is a progressing type proven by regular ultrasonography examination every 2 weeks (two to three examinations).
4. The mother's general condition is good.
5. The age of the fetus between 28 and 34 GW.

Protocol, methods, and results:
1. Strict selection of the fetus according to the abovementioned criteria
2. Postnatal immediate and complete neonatology examination after birth to confirm the absence of other congenital anomalies

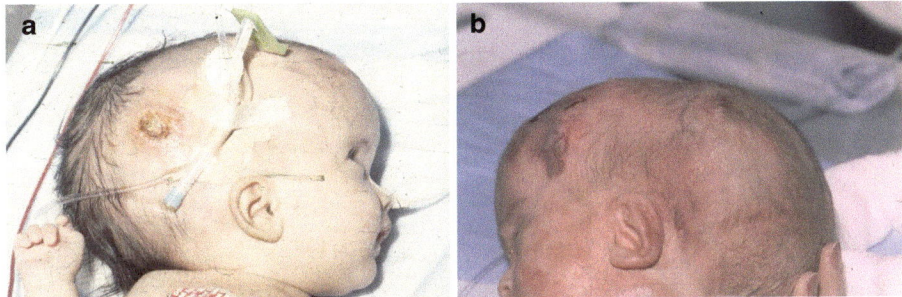

Fig. 24.1 (**a**) Shows pressure ulcer and skin breaking down over the bur hole valve which was inserted in an infant. (**b**) Shows areas of scalp ischemia and sloughing over the shunt course and under the head bandages, in premature baby

3. Admission to neonate intensive care unit (NICU)
4. Brain CT scan in Day 1
5. Insertion of VP shunt in Day 2 or 3
6. Follow-up CT scan after 1 week
7. Discharge in 10–15 days after removal of suture
8. Closed follow-up

There were 18 neonates treated by this protocol on the period between 1992 and 1994. It was noticed that the neonates delivered before or at 28 GW suffered CSF malabsorption in the abdomen and skin changes, sloughings, and necrosis over the shunt [25] course as shown in Fig. 24.1a, b. These skin changes and necrosis may be due to relatively large shunt tubes and valves, and these eventually lead to shunt infection, meningitis, and occasionally septicemia. Wrapping the head should be avoided as well.

The general outcome was acceptable. Most of these neonates showed good tolerance of the procedure.

During that time, in the late 1980s and early 1990s, we looked for smaller, softer, and low-profile shunts to avoid such preventable complications but failed to find any shunt with these specifications. Therefore, we designed a new shunt for premature and neonates having the features that we are looking for [24].

24.3.3 Designing a New Premature/Neonate Shunt

The shunt is a uni-shunt made of soft silicon with reservoir (biconcave to avoid any pressure on the skin) as shown in Fig. 24.2a. A major shunt manufacturing company agreed to produce this shunt [39]. The CSF flow and pressure were controlled through a flow control using the Poiseuille's equation, which controls the flow of the fluids inside a small narrowed tube. The equation is $Q = \dfrac{P\pi D^4}{128L\mu}$.

Fig. 24.2 (**a**) A uni-shunt
made of soft silicon with
reservoir. (**b**) Shows the
CSF flow and pressure
curve

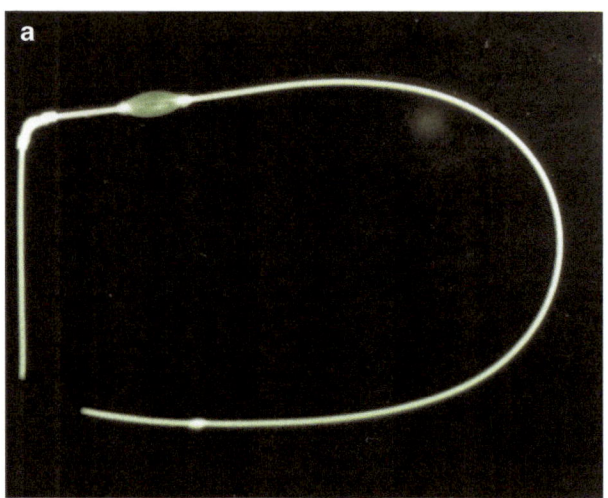

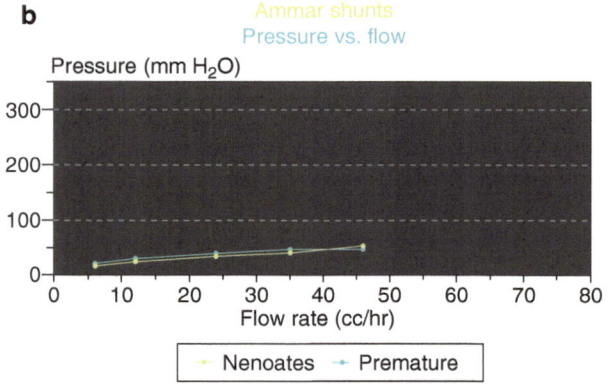

The flow and pressure of this shunt were studied and measured by Dr. Adolf
Ashraf in Heidelberg Institute. Figure 24.2b shows the CSF flow/pressure curve.
Ammar A. in 1995 [24] reported the experience of using two shunts. Several other
neonate shunts with smaller configurations and different types of valves have been
produced since then. In general, the results were encouraging.

24.3.4 Fetal Surgery

Several trials have been reported for successful intrauterine intervention [5, 18, 26,
28, 39–43]. These trials had been thoroughly studied, and we decided to explore the
possibility of such treatment. The inclusion criteria were:

1. The parents especially the mother must accept and sign informed consent for
   such procedures.

2. Enough time was given to discuss the procedure, indication, complication, and consequences with the mother.
3. Hydrocephalus is the only detected anomaly.
4. The dilatation of the ventricles is increasing by time and detected by repeated ultrasonography.
5. The mother's general condition is good.
6. The age of the fetus between 28 and 32 GW.

Prolonged intrauterine transabdominal ventricular external drainage [25].

The prolonged intrauterine transabdominal ventricular external drainage is the method that has been chosen to treat six hydrocephalus fetuses.

The aims of this procedure are:

(a) To reduce the intracranial pressure (ICP).
(b) To decrease the head circumference subsequently permitting vaginal delivery. Ammar A. et al. in 1997 reported a successful case of fetal hydrocephalus treated by this method as demonstrated by Fig. 24.3. However, the second case had a different outcome and the baby was born dead.

A formed committee studied thoroughly all the cases and reviewed all protocols and the outcomes and came up with the following recommendations with the following results and conclusions:

1. There is no clear evidence that prenatal intervention provides better chance for the development of these fetuses and newborns [3, 4, 43, 44].
2. The psychological stress of the mother is a serious issue and should be considered.
3. There is an ethical question on performing such procedures [7].
4. The complication of early delivery could be serious, and the shunt infection and malfunction are high.

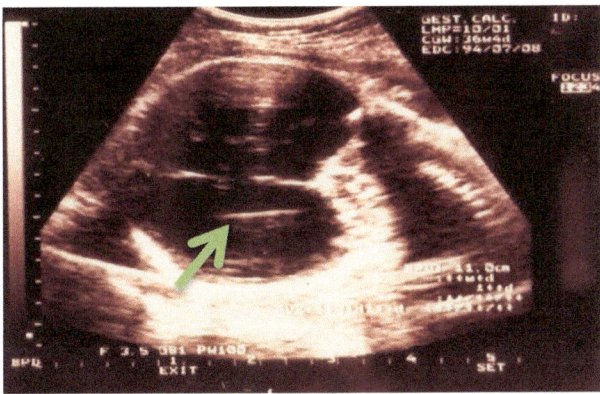

Fig. 24.3 Ultrasonography of the brain of 32 GW fetus shows dilated ventricles. The case is confirmed and diagnosed as hydrocephalus. *Arrow* indicates ventricular catheter inside the lateral ventricle to drain CSF

5. The pathophysiology of development of fetal hydrocephalus is not so far clear. The ventriculomegaly and prenatal hydrocephalus may be due to several causes; some are errors in the development such as aqueductal stenosis or as part of syndrome caused by genetic alteration. Further studies are badly needed [6, 14, 18, 38].

Therefore, further research studies should be performed including chromosomal and genetic studies. However, based on these experiences, the following guideline was produced.

24.3.5 Prenatal Hydrocephalus Management Guideline (PHMG)

1. Confirmation that the case is a definite case of hydrocephalus. Once the ventricular dilatation is detected, ultrasonography examination or MRI should be performed every 2 weeks. Hydrocephalus can be diagnosed with great certainty as there is gradual increase and progress in the size of the ventricles. Other CNS and general congenital anomalies should be looked for. If the size of the ventricles does not increase in size, most likely the case is a case of ventriculomegaly, caused by brain atrophy.
2. The confirmed hydrocephalus cases should go for chromosomal analysis by taking amniotic fluid samples.
3. If there are clearly chromosomal abnormalities and the case is thought to be syndromic, no intrauterine intervention should be performed. According to the severity of the chromosomal anomalies, US image findings, and the association of other anomalies, abortion may be considered.
4. In case of the absence of marked chromosomal anomalies and absences of any other CNS or general congenital anomalies, prenatal intervention may be considered in one of these forms:
   (a) Early delivery via CS at 32 GW. Careful neurological and general examination must be done immediately after birth. Brain CT scan should be done in Day 1 to confirm the diagnosis, followed by VP shunt or VA shunt. Complete genetic examination of the fetus.
   (b) Intrauterine intervention (shunts or draining) with endoscopic assistance.
   (c) Follow-up and regular fetal examination should be carefully conducted. The fetus should be delivered by CS. Careful general and neurological examination soon after delivery. CT scan should be performed at Day 1, followed by VP shunt or VA shunt. Complete genetic examination of the fetus.

24.4 Discussion

Improvement of the antenatal care challenges physicians and surgeons. Dolk H. and Loane M. et al. in 2010, the congenital anomalies with European Surveillance of Congenital Anomalies (EUROCAT), which has registrations of 1.5 million LB/year in 22 countries, found that 23.9/1000 births have major congenital anomalies in the period of 2003–2007. Eighty percent were live births, but 2.5% of live births with

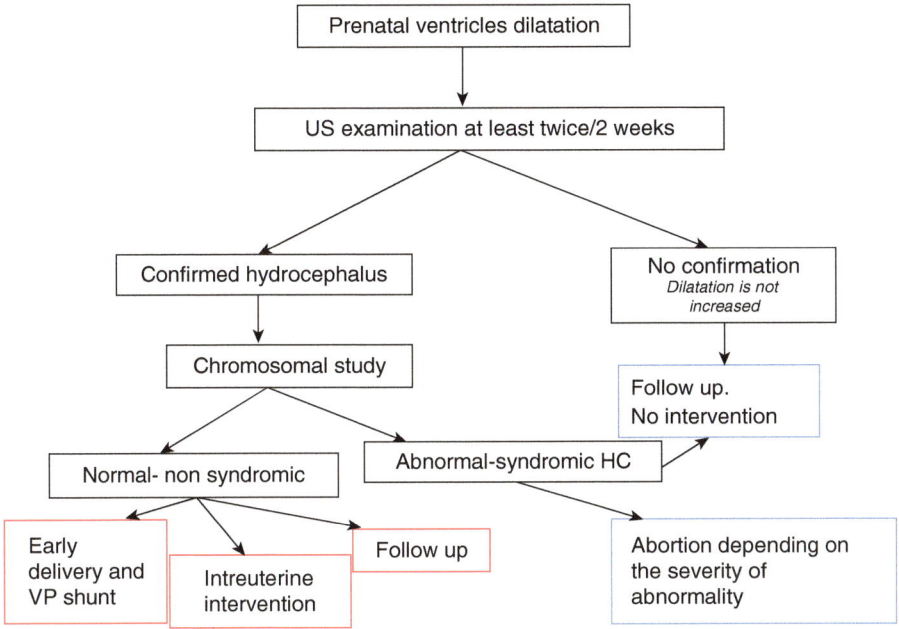

Algorithm 1 Summarizes the proposed Prenatal Hydrocephalus Management Guideline (PHMG).

congenital anomalies died in the first week of life. Two percent were stillbirths or fetal deaths from 20 weeks of gestation. 17.6% of all cases were terminations of pregnancy following prenatal diagnosis (TOPFA) [4]. They found out that the prevalence of hydrocephalus in Europe (22 countries) is 4.65/1000 live births. They also found out that 9% of hydrocephalus cases are from multiple pregnancies.

In China, a major study by Dai L. et al. (2006) between 1998 and 2004 included 4,282,536 live births in the period between 1998 and 2004. They identified 3648 cases of prenatal hydrocephalus. They found out the prevalence of congenital hydrocephalus is 11.5/10,000 births. Ninety-six percent were live births. They found out significant gender difference as males made 72% of these cases (total cases of hydrocephalus included in that study were 3648 cases). It was very interesting that they found regional differences as the major cases of isolated hydrocephalus were in north of China, while the syndromic hydrocephalus was found more in the south of China. According to their study the maternal age plays a role as risk factor for fetal hydrocephalus. The highest incidence of prenatal hydrocephalus was seen among ages less than 20 years old (11.42/10,000) [31]. It seems that the incidence of hydrocephalus in China and South America is higher than the rate of hydrocephalus in Europe and the USA.

Similar studies as that of China showed different results. Xie D. et al. in 2016, in their study performed in the period between 2005 and 2014, showed that the prevalence of congenital anomalies is 191.84/10,000 (perinatal). They found out as well that the prevalence of congenital hydrocephalus and neural tube defects remarkably reduced to 5.29/10,000 from 11.8. The prevalence of NTDCs has decreased as well from 7.87 to 1 [45].

In Brazil, Cavalcanti DP and Salomão in 2003 found out that the rate of identifying prenatal hydrocephalus has increased to 3.16/1000 births. The annual increasing rates of hydrocephalus since 1998 were significant ($p = 0.001$) [29].

There are several reports that demonstrate the good outcome for prenatal intervention in surgery, urology, cardiac surgery, and diaphragmatic hernia [2, 8, 12, 22]. The case is not the same in neurosurgery; there are several reports on intrauterine interventions in cases of spina bifida and hydrocephalus [9, 10, 26, 33, 34, 46].

24.5 The Dilemma of Fetal Hydrocephalus

Fetal hydrocephalus has remained challenging to neurosurgeons everywhere. The questions are does prenatal intervention help the fetus grow normally or at least improve the outcome of hydrocephalus? or does surgical intervention not alter the outcome and expose the fetus to new problems such as hemorrhage or infection which may worsen the case? The intervention may expose the pregnant ladies to serious hazards.

Holzgreve W. et al. in 1993 raised the concern about the prognosis and outcome of fetal hydrocephalus. In their experience, which included 118 fetuses, they had only 28 cases of isolated hydrocephalus, 64 cases suffered different anomalies, 26 cases had spina bifida associated with hydrocephalus plus hydrocephalus, 6 cases proved to have chromosomal abnormalities, and 6 cases had fetal infection. They found out the outcome of hydrocephalus diagnosed in utero and treated accordingly had worse outcome than the hydrocephalus diagnosed and treated after birth. They thought the reason is due to selection bias for the postnatal cases of hydrocephalus [19]. Nevertheless, this fact should be known to the parents especially the mothers before taking any decision. They recommended that intrauterine intervention to treat hydrocephalus should aim preventing fetal death. Gupta JK and Bruce FC et al. in 1994 came to the conclusion that there is no benefit of fetal surgery as treatment of fetal hydrocephalus. They also confirmed the previous observation that the outcome of fetal hydrocephalus is worse than the postnatal diagnosed hydrocephalus in spite of the accuracy of US diagnosis of prenatal hydrocephalus [19]. Vintzileos AM et al. in 1987 reported their series consisting of 20 cases of fetal hydrocephalus, 14 cases of them (70%) were associated with other anomalies, and only 30% had isolated hydrocephalus. They managed their cases prenatally by frequent ultrasonography, and every week fetal biophysical profile and as soon as the lung is matured, they deliver the fetus via Cesarean section. The results of this protocol were 45% (nine patients) alive, 20% (four patients) still death, 10% (two patients) postnatal death, and 20% (four cases) aborted [21].

24.6 The Progress of Fetal Surgery

Several experimental and clinical trials were performed early in the 1980s and early 1990s, which motivated several neurosurgeons to try such techniques, especially that techniques are not technically difficult and the diagnostic and surgical equipments and instruments are available.

Glick PL et al. in 1984 performed animal experiments to induce hydrocephalus and treated it by ventriculo-amniotic shunt in 28 fetal lambs and 21 intrauterine shunt (7 control and 17 fetal monkeys [8 shunts and 9 control]). They found out that the overall survival improved after the shunt and the ventricular sizes reduced and the head was decompressed. However, they detected several complications such as subdural hematomas, subdural hygromas, shunt infection, shunt obstruction, and the most important the histopathological cerebral damage due to hydrocephalus which did not improve. They recommended to continue research on animals before the use in human [10]. Mozik E. in 1985 seconded the suggestions of the previous authors and stressed that the selection of the cases may benefit the intrauterine intervention [6].

Drugan A. et al. in 1989 studied the natural history of hydrocephalus detected in utero and did not receive any intrauterine treatment. They analyzed the outcome of 43 cases and concluded that the outcome is poor in cases of ventriculomegaly associated with other CNS anomalies (severe hydrocephalus) and good for isolated nonprogressive hydrocephalus and variable for the progressive ones [5]. There is a debate on the value of intrauterine treatment or even postnatal treatment for cases of isolated nonprogressive ventriculomegaly, which could be a normal variant as indicated by von Koch CS. and Gupta N. et al. in 2003 [35].

Oi S. and Matsumoto S. et al. (1990), and Oi S. and Yamada H. et al. (1990), studied the pathophysiology and postnatal outcome of fetal hydrocephalus as well as the factors affecting the prognosis of intrauterine hydrocephalus and the controversies of intrauterine surgery. They operated on four cases by transabdominal transvaginal cephalocentesis and one transabdominal cephalocentesis. They found that intracranial pressure of such cases was high as well as the intracranial pressure with postnatal cases as they were operated postnatally for VP shunt. The factors they studied were (1) type of hydrocephalus, (2) underlying condition, (3) associated anomalies, (4) time of diagnosis and delivery, (5) fetal period after delivery, (6) head circumference and the degree of hydrocephalus immediately after delivery, and (7) the age of the fetus or the newborn at the time the treatment started. They found that the only significant factor to have an impact on the outcome is the fetal period after diagnosis of hydrocephalus until the treatment started. They called for a precise pathophysiological evaluation and emphasized upon the need to develop less invasive and reliable intrauterine surgical techniques [15, 23]. Furthermore, Davis GH in 2003 reviewed the results of fetal surgery and suggested that intrauterine intervention may be of some benefit in every selected cases of isolated hydrocephalus without chromosomal abnormalities. He agreed that the natural history of hydrocephalus has not been fully understood. He highlighted the fact that fetal hydrocephalus is a dynamic process, so mild hydrocephalus may progress to moderate or severe hydrocephalus, and moderate hydrocephalus may change to mild or even normalize the size of the ventricles [46]. Therefore, frequent and accurate examination of the ventricular size should be calculated and measured every time of examination.

Cavalheiro S. and Moron A.E. et al. in 2003 presented their experience with intrauterine surgery performed in the period between 1986 and 2001. They operated in 39 fetus aged between 24 and 32 GW, 20 cases treated by repeated cephalocentesis, 18 received ventriculo-amniotic shunt, and one case treated by endoscopic third ventriculostomy (ETV). Thirty-eight patients received VP shunt immediately after birth. The cases had been followed for median 5 years (1–14 years). Follow-ups showed that 26 were considered as normal, 6 have mild to moderate disability, and 7 were severely disabled. They concluded that the prenatal intervention could be useful in non-infected cases, with normal chromosomal pattern and with gestational age between 24 and 32 weeks [44]. This result could be debated as well. Is this good or normal outcome could be due to postnatal VP shunt especially in the absence of randomized double-blind well-controlled study? von Koch CS et al. in 2003 addressed the dilemma of hydrocephalus detected in utero. Again, they stressed the fact that prolonged intrauterine hydrocephalus has a detrimental effect on the cognitive function of these patients. The diagnostic procedures such as US and MRI allow diagnosis of hydrocephalus as early as 18 weeks. On the other hand, the surgical procedures and methods of interventions have not sufficiently progressed. Therefore, the results of intrauterine intervention in general are poor [35].

24.7   Classification of Fetal Hydrocephalus

Cavalcanti D.P. et al. in 2003 classified hydrocephalus in four subgroups (i.e., isolated hydrocephaly, hydrocephaly associated with congenital infection, syndromic hydrocephaly, and hydrocephaly associated with multiple defects). They thought that only cases of isolated hydrocephaly cases had been significantly benefited from the prenatal intervention [29]. It is important to modify this classification by adding cases of hydrocephalus + IVH group as this is a special group which needs special care, and the outcome is promising. The pathophysiology, types of hydrocephalus, associated anomalies, and chromosomal and genetic defects have great impact on the outcome of the hydrocephalus diagnosed in utero; therefore, it should be considered in any decision for prenatal intervention.

In general, fetal hydrocephalus was classified in different ways by several authors. We studied some of these classifications and came up with this grading system: (1) to provide standard informations for research and comparing the different results of intervention and management, (2) to help the neurosurgeons to make decision regarding cases of fetal hydrocephalus, and (3) to provide hydrocephalus fetus parent with simple grading system and classification, so they can grasp the facts and share in making decisions.

Fetal Hydrocephalus Grading System (FHGS)
Grade 1. Pre-ventriculomegaly (gray zone) when the ventricular atrium is 10 mm or less.
Grade 2. Mild when the ventricular atrium is between 10 and 15 mm.
Grade 3. Moderate when the ventricular atrium is more than 15 mm, and there is detected valuable cortex.

Grade 4. Very large ventricles, cortex is very thin and associated with other CNS anomalies.

Grade 5. IVH is detected within the dilated ventricles. The cases of hydrocephalus due to IVH group may have promising outcome.

Each type of hydrocephalus should be examined to classify its type, if it is syndromic with definite chromosomal and genetic anomalies, isolated hydrocephalus, and associated with other CNS anomalies like spina bifida and/or other brain anomalies.

Pagani G. and Thilaganathan B. et al. in 2014 studied the neurodevelopmental outcome of isolated hydrocephalus and found out that nearly 7.9% have abnormal development [17]. However, Scala C. et al. in 2016 found that the incidence of abnormal neurodevelopment in cases of isolated hydrocephalus is only 5.9% [47]. Chiu T.H. et al. in 2014 found that in Taiwan, the cases of mild ventriculomegaly without other anomalies have good prognosis (without intrauterine intervention) and grow without neurological deficit [14].

The physical and psychological impact on pregnant women with hydrocephalus is tremendous. The exposure of these ladies to different methods of intervention should be carefully studied and considered. Unfortunately, there are no enough studies on the short-term and long-term consequences of uterine scar caused by the surgical intervention.

Conclusion

We believe there is a room for prenatal intervention. However, there is a great need for more studies to understand the intrauterine CNS development, CSF dynamic, the ICP, and enlarged ventricles in cases of normal brain and in abnormal brains. The precaution and indication for selecting cases should be standardized. There are different methods of interventions; however, there is no clear evidence which method is superior to the others.

References

1. Cambria S, Gambardella G, Cardia E, Cambria M. Experimental fetal hydrocephalus. Ventriculo-amniotic shunt. Neurochirurgie. 1986;32(4):339–42.
2. Chaveeva P, Stratieva V, Shivachev H, Aktash S, Panova M, Shterev A. Fetal therapy: intrauterine thoraco-amniotic shunting in macrocystic type cystic adenomatoid malformation of the lung: review of the literature and case report. Akush Ginekol (Sofiia). 2016;55(Suppl 1 Pt 2):15–9.
3. Depp R, Sabbagha RE, Brown JT, Tamura RK, Reedy NJ. Fetal surgery for hydrocephalus: successful in utero ventriculoamniotic shunt for Dandy-Walker syndrome. Obstet Gynecol. 1983;61(6):710–4.
4. Dolk H, Loane M, Garne E. The prevalence of congenital anomalies in Europe. Adv Exp Med Biol. 2010;686:349–64.
5. Drugan A, Krause B, Canady A, Zador IE, Sacks AJ, Evans MI. The natural history of prenatally diagnosed cerebral ventriculomegaly. JAMA. 1989;261(12):1785–8.
6. Mrozik E. Problems of and possibilities in fetal surgery. Geburtshilfe Frauenheilkd. 1985;45(8):503–10.

7. Serlo W, Kirkinen P, Jouppila P, Herva R. Prognostic signs in fetal hydrocephalus. Childs Nerv Syst. 1986;2(2):93–7.

8. Till SR, Everetts D, Haas DM. Incentives for increasing prenatal care use by women in order to improve maternal and neonatal outcomes. Cochrane Database Syst Rev. 2015;(12):CD009916.

9. Ammar A, Al-Jama F, Rahman S, Anazi AR, Muazen Y, Sibai H. Prolonged intrauterine transabdominal ventricular external drainage. A method to decompress dilated fetal ventricles. Minim Invasive Neurosurg. 1996;39:1–3. George Thieme Verlag Stuttgart, New York.

10. Glick PL, Harrison MR, Halks-Miller M, Adzick NS, Nakayama DK, Anderson JH, Nyland TG, Villa R, Edwards MS. Correction of congenital hydrocephalus in utero II: efficacy of in utero shunting. J Pediatr Surg. 1984;19(6):870–81.

11. Glick PL, Harrison DK, Nakayama MS, Edwards MS, Filly RA, Chinn DH, Callen PW, Wilson SL, Golbus MS. Management of ventriculomegaly in the fetus. J Pediatr. 1984;105(1):97–105.

12. Rosseau GL, McCullough DC, Joseph AL. Current prognosis in fetal ventriculomegaly. J Neurosurg. 1992;77(4):551–5.

13. Clewell WH, Johnson ML, Meier PR, Newkirk JB, Hendee RW Jr, Bowes WA Jr, Zide SL, Hecht F, Henry G, O'Keeffe D. Placement of ventriculo-amniotic shunt for hydrocephalus in a fetus. N Engl J Med. 1981;305(16):955.

14. Chiu TH, Haliza G, Lin YH, Hung TH, Hsu JJ, Hsieh TT, Lo LM. A retrospective study on the course and outcome of fetal ventriculomegaly. Taiwan J Obstet Gynecol. 2014;53(2):170–7.

15. Oi S, Matsumoto S, Katayama K, Mochizuki M. Pathophysiology and postnatal outcome of fetal hydrocephalus. Childs Nerv Syst. 1990;6(6):338–45.

16. Ortega E, Muñoz RI, Luza N, Guerra F, Guerra M, Vio K, Henzi R, Jaque J, Rodriguez S, McAllister JP, Rodriguez E. The value of early and comprehensive diagnoses in a human fetus with hydrocephalus and progressive obliteration of the aqueduct of Sylvius: Case Report. BMC Neurol. 2016;16:45.

17. Pagani G, Thilaganathan B, Prefumo F. Neurodevelopmental outcome in isolated mild fetal ventriculomegaly: systematic review and meta-analysis. Ultrasound Obstet Gynecol. 2014;44(3):254–60.

18. Bruner JP. Intrauterine surgery in myelomeningocele. Semin Fetal Neonatal Med. 2007 Dec;12(6):471–6.

19. Gupta JK, Bryce FC, Lilford RJ. Management of apparently isolated fetal ventriculomegaly. Obstet Gynecol Surv. 1994;49(10):716–21.

20. Hill LM, Breckle R, Gehrking WC. The prenatal detection of congenital malformations by ultrasonography. Mayo Clin Proc. 1983;58(12):805–26.

21. Vintzileos AM, Campbell WA, Weinbaum PJ, Nochimson DJ. Perinatal management and outcome of fetal ventriculomegaly. Obstet Gynecol. 1987;69(1):5–11.

22. Holzgreve W, Feil R, Louwen F, Miny P. Prenatal diagnosis and management of fetal hydrocephaly and lissencephaly. Childs Nerv Syst. 1993;9(7):408–12.

23. Oi SZ, Yamada H, Kimura M, Ehara K, Matsumoto S, Katayama K, Mochizuki M, Uetani Y, Nakamura H. Factors affecting prognosis of intrauterine hydrocephalus diagnosed in the third trimester—computerized data analysis on controversies in fetal surgery. Neurol Med Chir (Tokyo). 1990;30(7):456–61.

24. Ammar A. Ammar Shunt: an option to improve the outcome of hydrocephalus detected in utero. Child Nerv Syst. 1995;11(7):421–3.

25. Ammar A, Nasser M. Long term complications of buried valves. Neurosurg Rev. 1955;18:65–7.

26. Birnholz JC, Frigoletto FD. Antenatal treatment of hydrocephalus. N Engl J Med. 1981;304(17):1021–3.

27. Garne E, Loane M, Wellesley D, Barisic I. Eurocat Working Group. Congenital hydronephrosis: prenatal diagnosis and epidemiology in Europe. J Pediatr Urol. 2009;5(1):47–52.

28. Ville Y. Recent developments in fetal surgery. Technical, organizational and ethical considerations. Bull Acad Natl Med. 2008;192(8):1611–21; discussion 1621–4.

29. Cavalcanti DP, Salomao MA. Incidence of congenital hydrocephalus and the role of prenatal diagnosis. J Pediatr (Rio J). 2003;79(2):135–40.

30. Clewell WH, Johnson ML, Meier PR, Newkirk JB, Zide SL, Hendee RW, Bowes WA Jr, Hecht F, O'Keeffe D, Henry GP, Shikes RH. A surgical approach to the treatment of hydrocephalus. N Engl J Med. 1982;306(22):1320–5.

31. Dai L, Zhou GX, Miao L, Zhu J, Wang YP, Liang J. Prevalence analysis on congenital hydrocephalus in Chinese perinatal from 1996 to 2004. Zhonghua Yu Fang Yi Xue Za Zhi. 2006;40(3):180–3.
32. Diemert A, Diehl W, Glosemeyer P, Deprest J, Hecher K. Intrauterine surgery—choices and limitations. Dtsch Arztebl Int. 2012;109(38):603–38.
33. Dukanac Stamenkovic J, Steric M, Srbinovic L, Janjic T, Vrzic Petronijevic S, Petronijevic M, Cetkovic A. Fetal ventriculomegalies during pregnancy course, outcome, and psychomotor development of born children. Clin Exp Obstet Gynecol. 2016;43(1):63–9.
34. Frigoletto FD Jr, Birnholz JC, Greene MF. Antenatal treatment of hydrocephalus by ventriculoamniotic shunting. JAMA. 1982;248(19):2496–7.
35. von Koch CS, Gupta N, Sutton LN, Sun PP. In utero surgery for hydrocephalus. Childs Nerv Syst. 2003;19(7–8):574–86.
36. Araujo J, Eggink AJ, van den Dobbelsteen J, Martins WP, Oepkes D. Procedure-related complications of open vs endoscopic fetal surgery for treatment of spina bifida in an era of intrauterine myelomeningocele repair: systematic review and meta-analysis. Ultrasound Obstet Gynecol. 2016;48(2):151–60.
37. Awary B, El Lardi A, El Najashi S, El Umran K, Ammar A. Prevalence of hydrocephalus, myelomeningocele, Dandy Walker Syndrome and anencephaly in Saudi Arabia. Pan Arab Neurosurg J. 1997;1:31–5.
38. Meiniel A. The secretory ependymal cells of the subcommissural organ: which role in hydrocephalus? Int J Biochem Cell Biol. 2007;39(3):463–8. Epub 2 Nov 2006.
39. Al-Anazi A, Al-Mejhim F, Al-Qahtani N. In uteroventriculo-amniotic shunt for hydrocephalus. Childs Nerv Syst. 2008;24(2):193–5.
40. Bruner JP. Intrauterine surgery in myelomeningocele. Semin Fetal Neonatal Med. 2007;12(6):471–6.
41. Chervenak FA, Berkowitz RL, Tortora M, Hobbins JC. The management of fetal hydrocephalus. Am J Obstet Gynecol. 1985;151(7):933–42.
42. Chervenak FA, Duncan C, Ment LR, Hobbins JC, McClure M, Scott D, Berkowitz RL. Outcome of fetal ventriculomegaly. Lancet. 1984;2(8396):179–81.
43. Valat AS, Dehouck MB, Dufour P, Dubos JP, Djebara AE, Dewismes L, Robert Y, Puech F. Fetal cerebral ventriculomegaly. Etiology and outcome, report of 141 cases. J Gynecol Obstet Biol Reprod (Paris). 1998;27(8):782–9.
44. Cavalheiro S, Moron AE, Zymberg ST, Dastoli P. Fetal hydrocephalus—prenatal treatment. Childs Nerv Syst. 2003;19(7–8):561–73.
45. Xie D, Yang T, Liu Z, Wang H. Epidemiology of birth defects based on a birth defect surveillance system from 2005 to 2014 in Hunan Province, China. PLoS One. 2016;11(1):e0147280.
46. Davis GH. Fetal hydrocephalus. Clin Perinatol. 2003;30(3):531–9.
47. Scala C, Familiari A, Pinas A, Papageorghiou AT, Bhide A, Thilaganathan B, Khalil A. Perinatal and long-term outcomes in fetuses diagnosed with isolated unilateral ventriculomegaly: systematic review and meta-analysis. Ultrasound Obstet Gynecol. 2017;49(4):450–9.

Surgical Management of Fetal Hydrocephalus

25

Abdulrahman Al Anazi

25.1 Introduction

Hydrocephalus is frequently defined as an abnormal accumulation of cerebrospinal fluid (CSF) resulting to enlargement of the ventricular system in which the intracranial pressure is known or suspected to be elevated [1]. Neonatal hydrocephalus is one of the most common central nervous system (CNS) congenital anomalies, with an incidence range of 0.3–4.2 per 1000 live births [2]. In Saudi Arabia, the prevalence rate of congenital hydrocephalus is 1.6–1.8 per 1000 live births [3–5].

In general, the mortality of fetuses with hydrocephalus varies directly with the presence and severity of extra-CNS anomalies, whereas the neurologic outcome is determined by the underlying CNS malformation. In a report of Cherwenak FA et al. [6] of 50 fetuses with hydrocephalus, 72% died in immediate neonatal period, and 84% had one or more major CNS anomalies and/or extra-CNS anomalies (49%). The overall uncorrected mortality rate was 67%, with death occurring in all fetuses with multiple extra-CNS anomalies, 57% of fetuses with an isolated extra-CNS anomaly [7]. Oi S et al. [8] classified 61 cases of fetal hydrocephalus as primary (communicating hydrocephalus, aqueductal stenosis, foramen atresia, and other forms of obstructive hydrocephalus), dysgenetic (associated with spina bifida, bifid cranium, Dandy-Walker malformation, holoprosencephaly, hydranencephaly, lissencephaly, congenital cysts, and others), or secondary (because of brain tumor, hemorrhagic or other vascular disease, infection, trauma, subdural fluid collection, and others). Nineteen cases (31%) were primary, 34 (56%) were dysgenetic, and 8 (13%) were secondary. The mean intelligence quotient was 74.2 (range, 20–132) in patients with primary hydrocephalus, 52.4 (range 20–120) in patients with dysgenetic hydrocephalus, and 26 (range 5–70) in patients with secondary hydrocephalus [8].

A. Al Anazi
Department of Neurosurgery, King Fahd Hospital of the University, Imam Abdulrahman Bin Faisal University, Al Khobar, Saudi Arabia
e-mail: prof.anazi@gmail.com

© Springer International Publishing AG 2017
A. Ammar (ed.), *Hydrocephalus*, DOI 10.1007/978-3-319-61304-8_25

Thus, only those fetuses with isolated obstructive hydrocephalus are candidates for prenatal treatment.

The concept of shunting hydrocephalus is based on the assumption that the ventriculomegaly is due to the intraventricular hypertension. This elevated pressure is presumed to be because of obstruction of the normal circulation of cerebrospinal fluid through the ventricular system, the cerebral aqueduct, or the subarachnoid space. Ultrasonography (US) criteria for the diagnosis of congenital obstructive hydrocephalus include isolated marked lateral and third ventricular enlargement with macrocephaly or progressive enlargement on serial examinations. In the presence of marked ventriculomegaly, however, the posterior fossa and cerebellum are usually examined only with great difficulty. MRI has a different biophysical basis than US, with superior soft tissue contrast. MRI also allows sagittal and coronal scans of fetal organs. MRI provided better visualization of the cerebellum and corpus callosum and is able to image the pons, the base of the skull, and the cervical spine [9].

The use of MRI with T2-weighted sequence is most satisfactory, as it is too fast to be affected by the fetal movements and allows a perfectly accurate observation mainly of the posterior fossa structures such as the brainstem and the cerebellum, thus increasing diagnostic possibilities. Small hemorrhages formerly not visualized by ultrasonography can now be more easily detected by fetal MRI [8, 10–14]. Fetal hydrocephalus is a serious malformation, which, behind a ventriculomegaly, hides a large number of different defects. Each one of them with different evolution depends on the type of the disease given to cause hydrocephaly [15]. The fetal moment, characterized as the period during which the fetus suffers the insult culminating to hydrocephalus, is also one of the most important factors to consider when establishing a prenatal prognosis. Fetal hydrocephaluses occurring on the last trimester of gestation have much more satisfactory evolution than those diagnosed at the beginning of gestation. Experimental studies in animal models have demonstrated that the more precocious is the treatment for fetal hydrocephalus, the more it is effective. This type of result is not verified in the daily clinical practice due to the large variety of diseases a fetus is subject to. In the case of malformative hydrocephalus, many patients present with multiple associated malformations, which compromise a good outcome [16–20]. The real incidence of fetal hydrocephalus is probably underestimated because many cases of fetal death at the beginning of gestation are often discarded without being studied. In fact, it is not even known how often abortions are performed among mothers of hydrocephalic fetuses.

25.2   Pathophysiology of Fetal Hydrocephalus

Intrauterine diagnostic neuroimaging including US and MRI [21–28] has been applied to the prenatal diagnosis of fetal hydrocephalus in recent years. It has become possible to classify, long with morphologic analyses of the CNS structure of fetal hydrocephalus, the pathologic types. However, these conditions have to be discussed with the specific clinical category of hydrocephalus in utero because of the essential difference in the pathophysiology including the dynamics of ICP and CSF circulation, changes chronologically depending on the fetal period being considered [29]. Oi et al. have analyzed fetal ICP dynamics under hydrocephalic condition [30]. The fetuses underwent transabdominal or transvaginal cephalocentesis in the prenatal period, and ICP was measured during drainage of CSF. The results suggested that, in such cases, the fetal brain is subjected to extremely high ICPs, resulting from a mixture of hydrocephalic pressure and intermittent uterine constriction. Immediately after birth, the biparietal diameter was found to be increased by an average of 7.7 mm, and the hydrocephalic state was converted to that seen in the neonatal type, characterized by macrocephaly and a relatively low ICP [30]. Based on their findings, it is assumed that the hydrocephalic fetal brain is subjected to extremely high ICP, being the net product of both hydrocephalic pressure and sporadic uterine contractions.

Michejda et al. [31] initiated a series of studies on the antenatal treatment of hydrocephalus in the rhesus monkey model and reported that the ICP range in the normal fetal cranium was 55–66 mmH_2O, while in hydrocephalic fetuses it was 100–250 mmH_2O. This is a very different situation from that found in human neonatal hydrocephalus, which is usually characterized by macrocephaly and relatively low ICP [32] at normal atmospheric pressures. It is suggested that the pressure dynamics in cases of fetal hydrocephalus should be evaluated, bearing in mind that the fetal cranium is a compartment within the intrauterine structure. In particular, the draining of CSF into the amniotic fluid cavity by ventriculo-amniotic shunt [33] may not be the best decompressive procedure for healing with fetal hydrocephalus if the ICP remains high. The results of experimental in utero shunt procedures in primate models are encouraging, in that the computer tomographic findings [31] and histologic demonstration of cortical mantle reconstitution [18] suggested that there is postoperative improvement. In humans, however, hydrocephalus has a variety of anomalies and underlying conditions associated with it. Morphologic analysis of the intracranial structure in fetal hydrocephalus has depended largely on ultrasonography. Thickman et al. also emphasize that those extensive morphologic investigations using MRI, [34] in addition to ultrasonography, should provide more precise data on the anomalies linked with hydrocephalus.

25.3 Clinicoembryologic Staging

The above-described specific pathophysiology of fetal hydrocephalus characterizes the clinical presentation and postnatal prognosis. Although there have been various classifications for congenital hydrocephalus applied in clinical practice, none of them can be used for a complete analysis of factors affecting the postnatal prognosis at the present time, as none of the presently available classifications include the chronologic change of hydrocephalic state from fetal to neonatal and then to infantile hydrocephalus nor does any reflect the underlying developmental or embryologic stages of the brain, especially its neuronal maturation process. From these standpoints, Oi et al. proposed a new classification of congenital hydrocephalus: perspective classifications of congenital hydrocephalus (PCCH) [29, 35, 36]. This multifactorial classification includes factors possibly affecting congenital hydrocephalus. The major factors are as follows:

1. Clinicoembryologic stage (stages I–V): I, 8–21 weeks; II, 22–31 weeks, III, 32–40 weeks in the gestational period; IV, 0–4 weeks; and V, 5–50 weeks in the postnatal period
2. Clinicoembryologic type (type P, primary or simple hydrocephalus without the associated lesion, and type D, dysgenetic hydrocephalus associated with non-malformative lesion)
3. Clinical category (category F, fetal hydrocephalus; category N, neonatal hydrocephalus; category I, infantile hydrocephalus)

   Regarding the clinicoembroyologic stage, each stage reflects both clinical management considerations and embryological development in the neuronal maturation process in the hydrocephalic fetus or infant, as follows:

   Stage I: Between 8 and 21 weeks of gestational age, which is the legal period within which termination is permitted in Japan, cell proliferation is the main process in the neuronal maturation.

   Stage II: Between 22 and 31 weeks of gestational age, which is the period of intrauterine preservation of the fetus while waiting for the pulmonary maturation to be completed, and cell differentiation and migration are the main process in the neuronal maturation.

   Stage III: Between 32 and 40 weeks of gestation age, which is the period of possible premature preterm neonatal hydrocephalus, axonal maturation is the main process in the neuronal maturation.

   Stage IV: Between 0 and 4 weeks of postnatal age, which is the period of neonatal hydrocephalus, dendritic maturation is the main process in the neuronal maturation.

   Stage V: Between 5 and 50 weeks of postnatal age, which is the period of infantile hydrocephalus, myelination is the main process in neuronal maturation.

   In the individual stages, cases with different aspects of hydrocephalus can be classified along with the embryologic or developmental background of the affected brain and CSF circulation and in each pathologic type with subtypes. These conditions are discussed in the standard clinical categories of fetal, neonatal, and infantile hydrocephalus, mainly because of the essential differences in pathophysiology, including dynamics of the ICP and the CSF circulation.

25.4   History of Fetal Hydrocephalus

Hydrocephalus has amazed and challenged clinicians throughout the history of medicine. To trace the history of the treatment of hydrocephalus, in many respects, is to document the parallel development of medicine as a whole; when one reviews the treatment of hydrocephalus, the integral relationship between basic science and therapy is reaffirmed. As we progress further in this new millennium, it is appropriate to reflect on the past understanding and treatment of this disorder, review strategies to curb this disease process, and consider therapies and possibly cures that will be available in the future.

Prior to the late nineteenth century, treatment for "water on the brain" involved more observation than intervention. Hippocrates [37] (fifth century B.C.), the father of medicine, is thought to be the first physician to attempt and document the treatment of hydrocephalus [38, 39].

The advances of diagnostic tools for fetal anomalies, US, and MRI have made it possible to detect early enlargement of the ventricles and diagnose hydrocephalus as early as 15 weeks of gestation. Bors [40] performed the first successful fetal procedure in animals in 1925, in which he amputated the limbs of guinea pig fetuses. Several CNS procedures were performed in utero including spinal cord transection in fetal rats, cortical resection in fetal labs and monkey, and ventriculopleural and ventriculojugular shunts in sheeps [41–44]. In the early 1980s, much of the focus regarding the potential for fetal therapy centered on obstructive ventriculomegaly. The interest in the disorder emerged from the relative ease of diagnosis by US and was amplified by the success rates of shunting procedures performed in neonates. The concept, as developed in animal models, was that early shunting of hydrocephalus in utero might prevent the irreversible brain damage caused by prolonged increased ICP. This potential for recovery is because of elasticity of normal cells and the fact that a reduction in ICP at this stage may prevent cortical ischemia, as well as restore synaptic neurotransmitter formation and function. The first clinical trial of intrauterine shunting in a human hydrocephalus fetus was attempted by Clewell et al. [45] at the University of Colorado in 1982. The shunt often called the Denver shunt is introduced by a sonographically guided needling technique through the maternal abdominal and uterine walls. There were several trials of intrauterine shunting, but most of them ended with complications such as shunt migration (intracranially or extracranially), obstruction, infection, and malposition [46–48].

To overcome the previously mentioned complications, the author has invented a special ventriculo-uterine shunt called Al-Anazi ventriculo-uterine shunt (KACTS Patent No. 2289).

The International Fetal Surgery Registry set a guideline for patient selection for the intrauterine treatment of hydrocephalus [49]:

1. A singleton pregnancy
2. Absence of any other significant anomalies
3. Progressive ventricular dilatation
4. Presence of multispecialty team
5. A normal karyotype
6. Viral culture
7. Gestational age less than 32 weeks or lung immaturity
8. Adequate follow-up
9. A consensus by the team to proceed

Cavalheiro S. et al. [50] presented their experience in 57 fetuses with evolutive hydrocephaly, whose gestational ages were between 24 and 32 weeks and that were analyzed and subjected to fetal neurosurgical procedures: 26 underwent repeated cephalocenteses; three underwent endoscopic third ventriculostomy (which was possible in only one of them—for the other two, the procedure had to be changed into ventriculo-amniotic shunting). Thirty fetuses received ventriculo-amniotic shunting.

The repeated cephalocenteses were guided by ultrasound. The mother was sedated with opiates. The volume of liquor removed varied from 20 to 120 mL. The fetal heartbeat was monitored throughout the procedure, and removal of liquor would be discontinued as soon as any deceleration should occur in the fetal heart rate. Ventriculo-amniotic shunting was performed percutaneously, under ultrasound guidance, and a pigtail catheter (KCH-Rocket Medical PLC, New England) was inserted. One tip of the catheter was left in the fetal lateral ventricle and the other in the amniotic cavity. Third ventriculostomy was performed under fetal anesthesia, the umbilical cord was punctured, the umbilical vein was catheterized, and a total dose of 5 µg/kg of fentanyl citrate and 0.1 mg/kg pancuronium bromide was injected. Five minutes after fetal anesthesia, a small incision was made in the mother's abdominal skin with an 11-blade scalpel, with a 2.5-mm-diameter needle, always under ultrasound guidance, the fetal skull was punctured on the brim of the bregmatic fontanelle, and the lateral ventricle could be accessed. As soon as the mandrel was withdrawn, liquor came out with increased pressure. A 2.3-mm-diameter neuroendoscope (NeuroView, flexible scope, 25C, Traatek, USA) was inserted through the needle, as well as a 1 mm working channel connected to a 300-W xenon lighting system. The Monro's foramen could be identified, and the endoscope was inserted into the third ventricle—its floor was opened, and the fetal basilar artery could be visualized. The opening was sufficiently enlarged with a 2-Fr Fogarty catheter, and the endoscope was withdrawn along with the needle. A small occlusive dressing was applied to the mother's abdomen.

Out of the 57 fetuses treated in utero, 26 underwent repeated cephalocenteses. Thirty fetuses were subjected to ventriculo-amniotic shunting. Five fetuses underwent two procedures due to migration or obstruction of the shunting system. In ten cases, the shunt migrated to the uterine cavity, and in six cases, it migrated to the ventricular cavity. In the cases where the catheter had migrated to the ventricular cavity, it was removed after birth by means of a neuroendoscopic procedure followed by third ventriculostomy. After birth, the ventriculo-amniotic shunts were removed, and the newborn infants underwent either ventriculoperitoneal shunting or endoscopic third ventriculostomy. No porencephalic cysts were observed in the patients subjected to repeated cephalocenteses. The number of punctures varied from two to five. The deliveries took place after lung maturity had been evidenced. Preterm labor occurred in four cases after cephalocentesis. Endoscopic third ventriculostomy was attempted in three fetuses, but due to technical and anatomical problems, it could only be achieved in one of them. The ventriculoperitoneal system inserted after birth in all the cases was a low-pressure-type Pudens (Codman's Accuo-Fluo) without the reservoir and the 4–5 cm right-angle ventricular catheter.

Eighteen patients were subjected to endoscopic third ventriculostomy after birth. In 11 of these cases, no shunting was required. Thirty-nine patients were followed up for more than 3 years and had their intelligence coefficients assessed, with the following results: 26 were considered normal (IQ above 70); 6 had mild or moderate handicaps (IQ between 35 and 70); and 7 were severely handicapped (IQ below 35). The best results were obtained in those cases where hydrocephaly had been diagnosed in a later period (third gestational trimester).

25.5   Al-Anazi Ventriculo-Uterine Shunt

The Al-Anazi ventriculo-uterine (VU) shunt (Figs. 25.1 and 25.2) consists of a short 25–30 mm catheter with a longitudinal central opening of an internal diameter (at least 1.25 mm). The paired wings are adjacent to a proximal end securing the shunt to the skull. Four openings, two on each side, admit the cerebrospinal fluid (CSF) from the ventricular cavity to the central opening of the catheter up to a one-way valve at the proximal end of the shunt outside of the skull. The valve lets fluid out into the uterine cavity of the mother and prevents backflow of amniotic fluid in to the ventricular cavity of the fetuses.

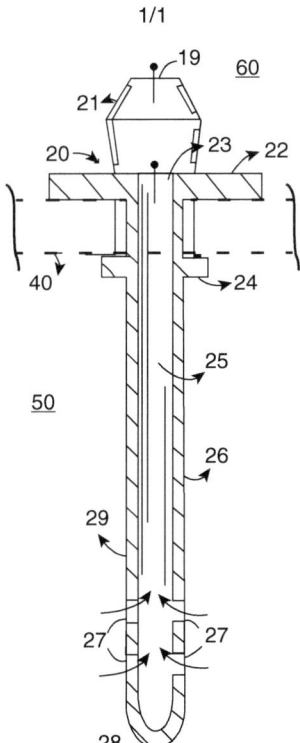

Fig. 25.1 Al-Anazi VU shunt valve outline

Fig. 25.2 Al-Anazi VU shunt

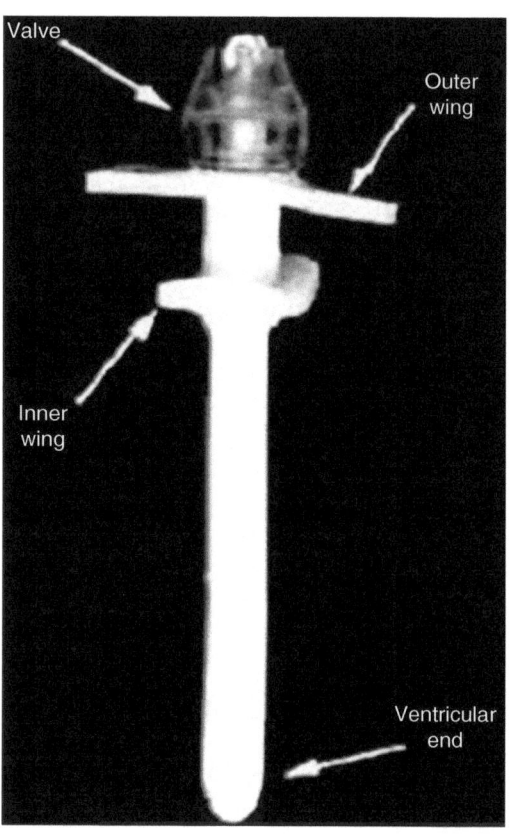

Between January 2005 and December 2007, Al-Anazi VU shunts were placed in five fetuses with isolated obstructive hydrocephalus. Diagnostic transabdominal amniocentesis was performed, and samples of the amniotic fluid were taken for fetal karyotype, cytomegalovirus, and toxoplasma by polymerase chain reaction and culture. Fetal magnetic resonance imaging (MRI) was performed (Fig. 25.3). Permission from the ethics committee of the hospital and parental consent were obtained. Patients were between 27 and 32 weeks of gestation. Under general anesthesia and ultrasonic (US) guidance, the uterus was exposed through the usual bikini incision; then, after confirmation of an appropriate site for the shunt insertion by US, a 1 cm hysterotomy was performed. The brain cannula touched by the monopolar was used to create a small opening in the scalp, skull, and the dura of the fetus' head; then, the Al-Anazi VU shunt (KACTS Patent No. 2289) was inserted [51–53]. Patients were kept in the hospital for 1 week for close monitoring of both the mother and the fetus. After discharge, the patients were followed on weekly visits for evaluations of both the mother and the fetus. Cesarean delivery through the earlier incision was anticipated, with the timing determined by the usual obstetrical factors. Immediately after delivery, the shunt device was pulled out and the hole in the scalp

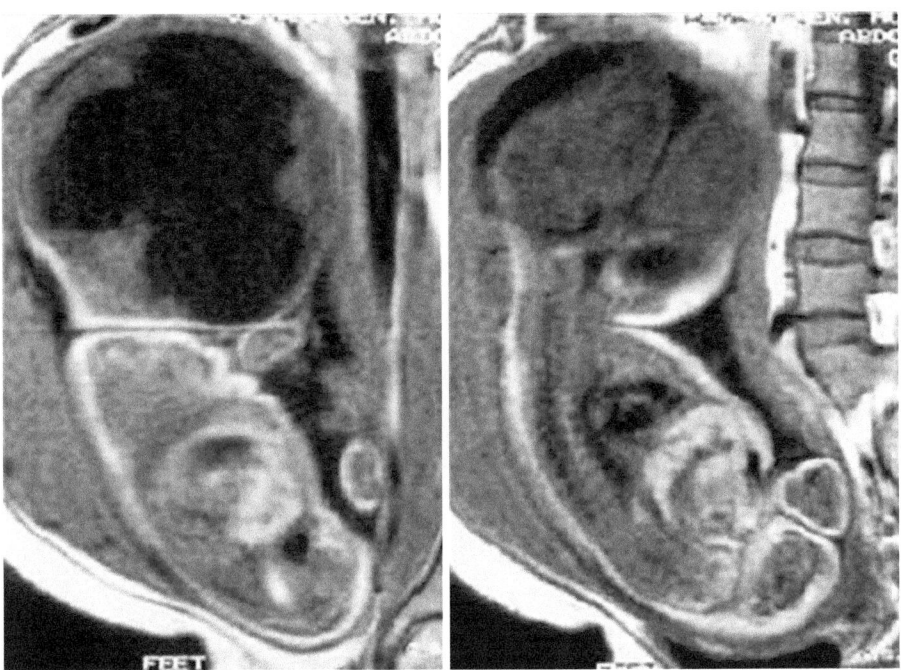

Fig. 25.3 Preoperative fetal MRI showed clear evidence of hydrocephalus

Table 25.1 Patient's demographic data and hospital course

Case	Maternal age	Gravity/ parity	GA at the time of procedure	GA at the time of delivery	Length of surgery (min)	Intrao-perative compli-cation	Immediate posto-perative compli-cation	Length of stay (d)
1	32	G6 P4 + 1	31	31	45	None	Abruptio placentae	5
2	34	G4 P3	32	37	35	None	None	10
3	25	G3 P0 + 2	28	35	30	None	None	7
4	23	G1 P0	32	37	25	None	None	4
5	20	G1 P0	27	36	25	None	None	4

was closed with one stitch. The mean gestation age at the time of intrauterine shunt placement was 30 weeks. In the first case, the mother developed abruptio placentae in which the obstetrician decided to deliver the baby 18 h postshunt insertion, whereas the remaining cases completed pregnancy without complications. Demographic features and the hospital course of the patients are listed in Table 25.1. After discharge, all patients had their regular outpatient follow-up. The shunt functioned well in all fetuses, with a gradual decrease of ventricular size and biparietal diameter (Fig. 25.4). All patients, however, experienced obstetrical and neonatal complications, which are noted in Table 25.2. The mean gestational age at the time of delivery was 35.2 weeks. The shunts were all secured at the time of delivery (Fig. 25.5). Despite our technical success with the shunts, all outcomes were unfavorable.

Fig. 25.4 A cranial CT scan performed on the day of delivery demonstrating a reduced ventricular size and an overriding of skull sutures

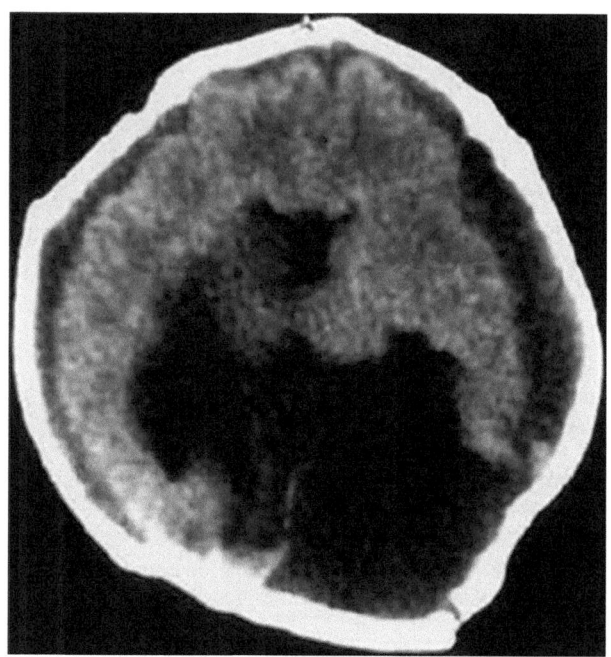

Table 25.2 Postoperative complications experienced by all five cases

Case	GA at the time of delivery	APGAR score	Obstetrical complications	Birth weight (g)	Neonatal complications	Neurologic outcome
1	31	5.6	Abruptio placentae	1600	Cyanotic heart disease	Died 8 months later as a cardiac complication
2	37	8.9	None	2300	Shunt infection	Developmental delay poor vision
3	35	5.8	Preterm premature rupture of membranes	1900	Premature shunt infection, glaucoma	Developmental delay
4	37	8.9	Oligohydramnios premature labor	2800	Staph sepsis	Developmental delay
5	36	6.8	Premature labor	2400	None	Developmental delay

Fig. 25.5 Fetal photograph showing that the shunt is in good position

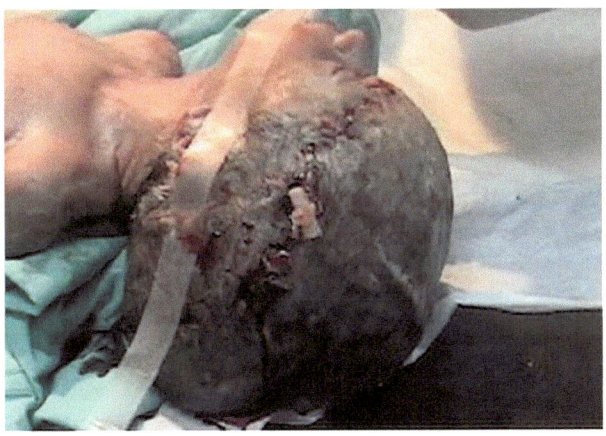

Conclusion

More than 30 years after the first in utero procedure for treatment of fetal hydrocephalus, little progress has been made as concerns neurosurgical techniques for management of that disease during the gestational period. Diagnostic techniques have evolved a lot, and, today, we have better conditions to evaluate more precisely the cases of fetal hydrocephalus and associated malformations of the central nervous system. We believe that in utero fetal procedures should be performed in cases of acute installation of evolutive but not destructive hydrocephaly without any other associated malformation.

References

1. Sutten L, Sun P, Adsick S. Fetal neurosurgery. Neurosurgery. 2001;48:124–44.
2. Bruner J, Davis G, Tulipan N. Intrauterine shunt for obstructive hydrocephalus—still not ready. Fetal Diagn Ther. 2006;21:532–9.
3. Al-Anazi AR, Nasser MJ. Hydrocephalus in the Eastern province of Saudi Arabia. Qatar Med J. 2003;12:133–5.
4. El-Awad ME. Infantile hydrocephalus in south-western region of Saudi Arabia. Ann Trop Paediatr. 1992;12:119–23.
5. Murshid W, Imma Dad M, Jarallah J. Epidemiology of infantile hydrocephalus in Saudi Arabia: birth prevalence and associated factors. Pediatr Neurosurg. 2003;32:119–23.
6. Cherwenak FA, Duncan C, Ment LR, et al. Outcome of fetal ventriculomegaly. Lancet. 1984;2:179–81.
7. Nyberg DN, Mac LA, Hirch J, et al. Fetal hydrocephalus. Radiology. 1987;163:187–91.
8. Oi L, Honda Y, Hidaka M, et al. Intrauterine high-resolution magnetic resonance imaging in fetal hydrocephalus and prenatal estimation of postnatal outcomes with "perspective classification". J Neurosurg. 1998;88:685–94.
9. Kirkinen P, Partanen K, Vainio P, et al. MRI in obstetrics: a supplementary method for ultrasonography. Ann Med. 1996;28:131–6.
10. Benacerraf BR, Shipp TD, Bromley BR, Levine D. What does magnetic resonance imaging add to the prenatal sonographic diagnosis of ventriculomegaly? J Ultrasound Med. 2007;26:1513–22.

11. Dill P, Poretti A, Boltshauser E, Huisman TA. Fetal magnetic resonance imaging in midline malformations of the central nervous system and review of the literature. J Neuroradiol. 2009;36:138–46.
12. Jeanty P, Dramaix-Wilmet M, Delbeke D, Rodesch F, Struyven J. Ultrasonic evaluation of fetal ventricular growth. Neuroradiology. 1981;21:127–31.
13. Leidig E, Dannecker G, Pefeifer KH, Salinas R, Pfeifer G. Intrauterine development of posthaemorrhagic hydrocephalus. Eur J Pediatr. 1988;147:26–9.
14. Levine D, Trop I, Mehta TS, Barnes PD. MR imaging appearance of fetal cerebral ventricular morphology. Radiology. 2002;223:652–60.
15. Manning FA, Harrison MR, Rodeck C, Members of the International Fetal Medicine and Surgery Society. Catheter shunts for fetal hydronephrosis and hydrocephalus: report of the international fetal surgery registry. N Engl J Med. 1986;315:336–40.
16. Chervenak FA, Ment LR, McClure M, Berkowitz RL, Duncan C, Hobbins C, Scott D. Outcome of fetal ventriculomegaly. Lancet. 1984;2:179–81.
17. Clark RG, Milhorat TH. Experimental hydrocephalus—light microscopic findings in acute and subacute obstructive hydrocephalus in the monkey. J Neurosurg. 1970;32:400–13.
18. Edwards MSD, Harrison MR, Halks-Miller M, Nakayama DK, Berger MS, Glick PL, Chinn DH. Kaolin induced congenital hydrocephalus in utero in fetal lambs and rhesus monkeys. J Neurosurg. 1984;60:115–22.
19. Stein SC, Feldman JG, Apfel S, Kolh SG, Casey G. The epidemiology of congenital hydrocephalus. A study in Brooklyn, N.Y. 1968–1976. Childs Brain. 1981;8:253–62.
20. Cavalheiro S, Moron AF, Zymberg ST, Dastoli P. Fetal hydrocephalus—prenatal treatment. Childs Nerv Syst. 2003;19:561–73.
21. Hoffman-Tretin JC, Horoupian DS, Koenigsberg M, Schnur MJ, Llena JR. Lobar holoprosencephaly with hydrocephalus: antenatal demonstration and differential diagnosis. J Ultrasound Med. 1986;5:691–7.
22. Bronshtein M, Zimmer E, Greshonu-Baruch R, Yoffe N, Meyer H, Blumenfield Z. First and second trimester diagnosis of fetal ocular defects and associated anomalies: report of eight cases. Obstet Gynecol. 1991;77:443–9.
23. Patten RM, Mack LA, Finberg HJ. Unilateral hydrocephalus: prenatal sonographic diagnosis. AJR Am J Roentgenol. 1991;156:359–63.
24. McGahan JP, Haesslein HC, Meyers M, Ford KB. Sonographic recognition of in utero intraventricular hemorrhage. AJR Am J Reontgenol. 1984;142:171–3.
25. Hanigan WC, Gibson J, Kelopoulos NJ, Cusack T, Zwicky G, Wright RM. Medical imaging of fetal ventriculomegaly. J Neurosurg. 1986;64:575–80.
26. Montegudo A, Reuss L, Tiomr-Tritsch IE. Imaging the fetal brain in the second and third trimester using transvaginal sonography. Annu Rev Hydroceph. 1992;10:53s55.
27. Clark SL, deVore GR, Sabey P. Prenatal diagnosis of cysts of the choroid plexus. Obstet Gynecol. 1988;72:585–7.
28. Oi S, Matsumoto S, Katayama K, Mochizuki M. Prenatal neuroimaging in fetal dysraphism. Neurosonology. 1990;3:90–6.
29. Oi S, Sato O, Matsumoto S. A new classification for congenital hydrocephalus and postnatal prognosis (Part I). A proposal of a new classification of fetal/neonatal/infantile hydrocephalus based on neuronal maturation process and chronological changes. Jpn J Neurosurg (Tokyo). 1994;3:122–7.
30. Oi S, Matsumoto S, Katayama K, Mochizuki M. Pathophysiology and postnatal outcome of fetal hydrocephalus. Childs Nerv Syst. 1990;6:338–45.
31. Michejda M, Queenan JT, McCullough D. Present status of intrauterine treatment of hydrocephalus and its future. Am J Obstet Gynecol. 1986;155:873–82.
32. Shapiro K, Fred F, Marmarou A. Biomechanical and hydrodynamic characterization of hydrocephalic infant. J Neurosurg. 1985;63:69–75.
33. Clewell WH, Johnson ML, Meier RP, et al. A surgical approach to the treatment of fetal hydrocephalus. N Engl J Med. 1982;306:1320–5.

34. Thickman D, Mints M, Mennuti M, Kressel HY. MR imaging of cerebral abnormalities in utero. J Comput Assist Tomogr. 1984;8:1058–61.
35. Oi S. Classification of hydrocephalus: critical analysis of classification categories and advantages of "Multi-categorical hydrocephalus classification" (Mc HC). Childs Nerv Syst. 2011;27:1523–33.
36. Oi S. Hydrocephalus research update: controversies in definition and classification of hydrocephalus. Neurol Med Chir (Tokyo). 2010;50:859–69.
37. Hippocrates: De Morbis. Cited by Whytt R: observations on the dropsy in the brain. Edinburgh: Balfour; 1768. p. 4.
38. Davidoff LE. Treatment of hydrocephalus. Arch Surg. 1929;18:1737–62.
39. Drake JM, Sainte-Rose C. The shunt book. Cambridge: Blackwell Science; 1995. p. 3–12.
40. Bors E. Bie methodic der intrauterine operation am uberlebenden saugerties foetus. Arch EntwckIngsmechn. Organ. 1925;105:655–66.
41. Barron D. An experimental analysis of some factors involved in the development of the fissure pattern of the cerebral cortex. J Exp Zool. 1950;113:553–81.
42. Hooker D, Nicholas J. Spinal cord section in rat fetuses. J Comp Neurol. 1930;50:413–67.
43. Nakayama DK, Harrison MR, Berger MS, et al. Correction of congenital hydrocephalus in utero: part I- the model: intracisternal kaolin produces hydrocephalus in fetal lambs and rhesus monkeys. J Pediatr Surg. 1983;18:331338.
44. Rakic P, Goldman-Rakic P. Use of fetal neurosurgery for experimental studies of structural and functional brain development in non-human primates. In: Perinatal neurology and neurosurgery. New York: SP Medical and Scientific Books; 1985. p. 1–15.
45. Clewell WH, Johnson ML, Meier PR, et al. A surgical approach to the treatment of fetal hydrocephalus. N Engl J Med. 1982;306:1820–5.
46. Micheida M. Intrauterine treatment of hydrocephalus. Fetal Ther. 1986;1:75–9.
47. Reynolds J, Pernoll M, Gill W, et al. A case of ventricular-amniotic shunting. South Med J. 1985;78:203–5.
48. Ammar A, Rahman S, Anazi AR, Muazen Y, Sibai H. Prolonged intrauterine transabdominal ventricular external drainage. A method to decompress dilated fetal ventricles. Minim Invasive Neurosurg. 1996;39:1–3.
49. Harrison MR, Filly RA, Golbus MS, et al. Fetal treatment. N Engl J Med. 1982;307:651–2.
50. Cavalheiro S, Fernandes AM, Almodin CG, et al. Fetal hydrocephalus. Childs Nerv Syst. 2011;27:1575–83.
51. Al-Anazi A. Novel shunt device for intrauterine treatment of hydrocephalus. Pan Arab J Neurosurg. 2007;11:37–40.
52. Al-Anazi AR, Al-Mejhim F, Al-Qahtani N. In-utero ventriculoamniotic shunt for hydrocephalus. Childs Nerv Syst. 2008;24:193–5.
53. Al-Anazi AR. In-utero ventriculouterine shunt treatment of fetal hydrocephalus: preliminary study of Al-Anazi ventriculouterine shunt. Neurosurg Q. 2010;20:1–4.